Microbial Surfactants

A sustainable class of versatile molecules

Online at: https://doi.org/10.1088/978-0-7503-5989-4

Microbial Surfactants

A sustainable class of versatile molecules

Edited by

Divya Tripathy

Chemistry Division, School of Basic Sciences, Galgotias University (India), Greater Noida, India

Anjali Gupta

Division of Chemistry, School of Basic Sciences, Galgotias University, Greater Noida, India

IOP Publishing, Bristol, UK

ISBN 978-0-7503-5989-4 (ebook)
ISBN 978-0-7503-5987-0 (print)
ISBN 978-0-7503-5990-0 (myPrint)
ISBN 978-0-7503-5988-7 (mobi)

DOI 10.1088/978-0-7503-5989-4

Version: 20240601

IOP ebooks

British Library Cataloguing-in-Publication Data: A catalogue record for this book is available from the British Library.

Published by IOP Publishing, wholly owned by The Institute of Physics, London

IOP Publishing, No.2 The Distillery, Glassfields, Avon Street, Bristol, BS2 0GR, UK

US Office: IOP Publishing, Inc., 190 North Independence Mall West, Suite 601, Philadelphia, PA 19106, USA

Contents

Preface

Microbial surfactants or biosurfactants describe surface-acting substances that enhance the surface-surface intersection of two fluid phases by generating micelles in their natural habitat, like plants, animals and microorganisms. Biosurfactants are produced by different classes of microorganisms on their surface like bacteria (*Acinetobacter*, *Pseudomonas*, and *Bacillus*) yeast (saccharomyces, *pseudonym Rhodotorula)*, and fungi (*Trichoderma*, *Fusarium*, *Penicillium fusarium*) which is the reason they are known as Biosurface-active compounds. Biosurfactants have amphiphilic nature which include hydrophilic as well as hydrophobic moieties. In some applications, biosurfactants have been utilized to lower the interfacial surface tension at solution–surface, oil–water, and air–water interfaces. In other situations, the addition of surfactants results in a decrease in surface tension up to the level at which the surfactants begin to form a structure like micelles, vesicles, and bilayers; often this crucial level is referred to as the critical micelle concentration. Biosurfactants have gained considerable interest in recent years, unlike their synthetic counterparts (chemical surfactants), due to their biodegradable and non-toxic properties of by-products. Due to this compositional diversity and notable properties, biosurfactants occupy a significant position in almost all industrial sectors ranging from biomedicine, cosmetics, environmental remediation, agriculture, textile industry, and enhanced oil recovery making them a multifunctional material of the 21st century.

Chapters 1 and 2 focus on an overview of biosurfactants, their types, properties, and the potential applications of biosurfactants. It also discusses various characteristics of biosurfactants, including their low toxicity, biodegradability, availability, surface and interfacial tension, emulsification properties, and micelle formation. It explores the different types of biosurfactants based on their chemical composition and molecular mass, their sources (produced by various microorganisms like yeast, fungi, or bacteria), and their production and applications in diverse fields.

Chapter 3 enters the ground-breaking field of metabolic technology, explaining its principles and strategies for optimizing the production process. The accumulated knowledge highlights the biotechnological potential of microbial cell plants and opens new pathways for sustainable and efficient production.

In chapter 4, kinetics of biosurfactant production explore the timing and quantity of biosurfactant synthesis by microorganisms. Additionally, the chapter provides an overview of methods for measuring microbes and biosurfactants, including optical density (OD) measurement, colony count techniques, dry weight measurement, and surface tension measurement. Furthermore, it delves into the various agricultural, industrial, and environmental applications of biosurfactants. It emphasizes the importance of these techniques and their potential for a wide range of scientific and industrial uses.

Chapter 5 highlights the significance of effective downstream processing techniques to guarantee the economic viability and purity of biosurfactant products, as well as the importance of optimizing bioreactor design and operational parameters for increased biosurfactant yields. With implications for a range of commercial applications, this thorough analysis advances knowledge of the bioproduction of biosurfactants from fermentation to extraction.

Chapter 6 emphasizes the extensive applications of biosurfactants across various industries, including nanotechnology product formulation, petroleum, personal healthcare product manufacturing, pharmaceuticals, agriculture, agrochemicals, and food processing.

Chapter 7 provides a comprehensive overview of their role as bioremediation agents for heavy metals in contaminated soil, emphasizing their biodegradability, low toxicity, and efficiency in sequestering heavy metals.

Chapter 8 expounds the different properties of an ideal biosurfactant that could be used for bioremediation, its mechanism of action, factors affecting biosurfactant production and its various applicability in the biodegradation of water pollutants. The authors also recapitulate some eminent studies on screening of efficient microbial biosurfactant producers and its production for wider usage.

Chapter 9 explores how microbial surfactants can contribute to the development of new formulas that improve texture, shelf life and overall food quality along with pharmaceutical industry where microbial surfactants have shown great potential in drug delivery and technology development.

Chapters 10 and 11 discuss the potential and commercial applications of microbial surfactants like medicinal, agriculture, cosmetics, personal care products, etc.

Anjali Gupta
Divya Tripathy

Editor biographies

Divya Tripathy

Professor (Dr) Divya Tripathy has research and teaching experience of more than 13 years. She has more than 80 research publications in reputed publications with Sci/Scopus indexed journal article, book chapters including seven edited books with reputed publishers and nine published Indian patents. She has guided 15 master's research students. Currently, she is supervising seven research scholars in the field of science. She has been awarded a women scientist fellowship from the Department of Science and Technology, Government of India for the duration of 2015–18. She has guest editorship in the journal of current material sciences published by Bentham Science publishers, *Der Chemica Sinicea*. She has reviewed more than 25 publications of reputed journals. She has a research collaboration with Professor M A Quraishi, King Fahd University, Saudi Arab, Professor Seifedine Kadry, Noroff School of Technology and Media, Norway, Dr Anjali Gupta, Galgotias University, Greater Noida, Professor Anuradha Mishra, Gautam Buddha University, Greater Noida. She has been the organizer, jury member and participant in many national/international conferences, workshops and academic meetings.

Anjali Gupta

Prof. Anjali Gupta is currently working as Professor in the Department of Chemistry, School of Basic & Applied Sciences, Galgotias University, Greater Noida, India. She has research and teaching experience of around 14 years. She is the recipient of the Young Scientist Award by the Department of Science & Technology and Dr. D.S. Kothari postdoctoral fellowship by University Grants Commission, Senior Research Fellowship from CSIR, India and JRF from DBT. Her research area is In-silico screening and synthesis of naturally occurring bio-active analogs. She has 14 published patents with around 50 research publications in reputed Journals/Book chapters/Conference proceedings. She has 7 Ph.D. and 10 Masters research students under her guidance. She has been principal investigator in DST sponsored research projects. She did her graduation, postgraduate and doctorate from University of Delhi, Delhi.

List of contributors

Dr Ashish Kumar Agrahari
Translational Health Science and Technology Institute, Faridabad, Haryana 121001, India

Dr Garima Bartariya
School of Life Science and Technology IMT University, Meerut, Uttar Pradesh, India

Dr Senthilkumar Dharmaraj
School of Biological and Life Sciences, Galgotias University, Greater Noida, Uttar Pradesh 203201, India

Dr Payal Gulati
Gene Regulation Laboratory, National Institute of Immunology, Aruna Asaf Ali Marg, New Delhi 110067, India

Dr Sam Joy
Indian Council of Medical Research, New Delhi, India

Ms Ashlesha Kawale
Department of Chemistry, Guru Ghasidas Central University, Bilaspur, Chhattisgarh, India

Mr Royal Kumar
School of Biological and Life Sciences, Galgotias University, Greater Noida, Uttar Pradesh 203201, India

Ms Pramilaa Kumar
School of Bio-Sciences and Technology, Vellore Institute of Technology, Vellore, Tamil Nadu 632 014, India

Dr Shruti Mishra
School of Biological and Life Sciences, Galgotias University, Greater Noida, Uttar Pradesh 203201, India

Dr Lekshmi Narendrakumar
Translational Health Science and Technology Institute, Faridabad, Haryana 121001, India

Ms Soghra Nashath Omer
School of Bio-Sciences and Technology, Vellore Institute of Technology, Vellore, Tamil Nadu 632 014, India

Ms Pallawi
School of Biomedical Sciences, Galgotias University, Greater Noida, Uttar Pradesh, India

Dr Achyut Pandey
School of BioMedical, Galgotias University, Greater Noida, Uttar Pradesh 203201, India

Ms Riya Pandey
School of Biological and Life Sciences, Galgotias University, Greater Noida, Uttar Pradesh 203201, India

Dr Deepjyoti Paul
Translational Health Science and Technology Institute, Faridabad, Haryana 121001, India

Dr Anupam Prakash
School of Biological and Life Sciences, Galgotias University, Greater Noida, Uttar Pradesh, India

Ms Shruti Roy
School of Biological and Life Sciences, Galgotias University, Greater Noida, Uttar Pradesh, India

Dr Pragati Sahai
AP-II, Amity University, sector-125, Gautam Budh Nagar, Uttar Pradesh, India

Dr Pragati Saini
School of Biological and Life Sciences, Galgotias University, Greater Noida, Uttar Pradesh, India

Dr Venkat Kumar Shanmugam
School of Bio-Sciences and Technology, Vellore Institute of Technology, Vellore, Tamil Nadu 632 014, India

Dr Vishal Sharma
School of Biological and Life Sciences, Galgotias University, Greater Noida, Uttar Pradesh 203201, India

Dr Neha Sharma
Amity Institute of Virology and Immunology, Amity University, Uttar Pradesh, Noida India

Dr Cheshta Sharma
University of Texas Health Science Center, San Antonio, Texas, USA

Mr Nishant Shekhar
Department of Chemistry, Guru Ghasidas Central University, Bilaspur, Chhattisgarh, India

Dr Anuradha Singh
School of Biological and Life Sciences Galgotias University, Greater Noida, Uttar Pradesh, India

Dr Pradeep Kumar Singh
Department of Zoology Maharana Pratap Government Post Graduate College, Gadarwada, Madhya Pradesh, India

Dr Pratichi Singh
School of Biological and Life Sciences, Galgotias University, Greater Noida, Uttar Pradesh 203201, India

Dr Arti Srivastava
Department of Chemistry, Guru Ghasidas Central University, Bilaspur, Chhattisgarh, India

Dr Pravat Swain
Department of Chemistry, Dr J N College, Rasalpur, Balasore, Odhisha, India

Dr Savitri Tiwari
School of Biological and Life Sciences, Galgotias University, Greater Noida, Uttar Pradesh 203201, India

Dr Vivek Kumar Yadav
School of Biological and Life Sciences Galgotias University, Greater Noida, Uttar Pradesh, India

Dr Mohd Yusuf
School of Life and Natural Sciences, Glocal University, Mirzapur Pole, Saharanpur, Uttar Pradesh 247121, India

IOP Publishing

Microbial Surfactants
A sustainable class of versatile molecules
Divya Tripathy and Anjali Gupta

Chapter 1

Microbial surfactants: overview

Royal Kumar, Riya Pandey, Savitri Tiwari and Pratichi Singh

Biosurfactants, also known as surface-active biomolecules, are synthesized by microbes using renewable resources like sugar or plant oil, and they find applications in laundry products, cleaning agents, and cosmetics. Their structural diversity, stemming from various microorganisms, is a key advantage. They can be broadly categorized as either anionic or neutral molecules, with varying molecular weights, and possess antioxidant, antibacterial, and anti-adhesive properties. These molecules play a vital role in the development of microorganisms, aiding in surface adhesion under challenging conditions. Depending on their chemical composition and the organisms involved, biosurfactants fall into different categories, with fungi, yeasts, and bacteria being among the many microbial producers. The production process is influenced by factors like microbial strains, fermentation methods, and substrate selection. Biosurfactants provide a number of advantages over chemical surfactants, such as better biodegradability, lower toxicity, decreased surface tension, increased selectivity, increased foaming capacity, and efficacy in challenging environments with fluctuating salinity, temperature, and pH. This chapter focuses on various characteristics of biosurfactants, including their low toxicity, biodegradability, availability, surface and interfacial tension, emulsification properties, and micelle formation. It also explores the different types of biosurfactants based on their chemical composition and molecular mass, their sources (produced by various microorganisms like yeast, fungi, or bacteria), and their production and applications in diverse fields such as the pharmaceutical industry, environmental applications, cosmetics industry, agriculture, and the food industry.

1.1 Introduction to microbial surfactants

Microbial surfactants are also termed biosurfactants. The word ‘biosurfactants’ has been used to describe surface-acting substances that can enhance the surface–surface intersection of two fluid phases by generating micelles in their natural habitat, such as plants, microorganisms, and animals [1]. Biosurfactants are produced by different

doi:10.1088/978-0-7503-5989-4ch1

classes of microorganisms on their surface which is why they are also known as biosurface-active compounds. The different class of microorganisms that produces biosurfactants are bacteria (*Acinetobacter Pseudomonas, and Bacillus*), yeast (saccharomyces, *pseudonym Rhodotorula*), and fungi (*Trichoderma, Fusarium, penicillium, fusarium*) [2]. Biosurfactants are amphiphilic in nature including both hydrophilic and hydrophobic moieties. In aqueous solution and hydrocarbon mixture, this composition gives surface-active features like the ability to lower surface tension and interfacial tension. In some applications, biosurfactants have been utilized to lower the interfacial surface tension at solution–surface, oil–water, and air–water interfaces. In other situations, the addition of surfactants results in a decrease in surface tension up to the level at which the surfactants begin to form structures like micelles, vesicles, and bilayers; often this crucial level is referred to as the critical micelle concentration [3, 4]. Based on their molecular mass, biosurfactants are classified into two groups: high molecular weight and low molecular weight. Based on their chemical makeup, they are divided into the following categories: glycolipid (phospholipid, lipopeptides, rhamnolipid, sophorolipid, trehalolipid, and mannosyl, erythritol lipid) (surfactin, lichenin, iturin, fencing) biosurfactants polymeric and neutral lipids (spiculisporic acid and phosphatidylethanolamine) This classification is made possible by spectroscopic and chromatographic analyses [5]. Microbial surfactants, which are regarded as secondary metabolites, are important for the survival of microorganisms that produce biosurfactants by enabling food transportation and promoting interaction between the microbes and hosts. Microorganisms that can produce biosurfactants are found in water, soil, and extremely harsh conditions [6].

1.2 Historical developments of microbial surfactants

Arima *et al* described and isolated the first biosurfactant 'surfactin' [7]. The usage of biosurfactants is growing daily in industry because of their various environmental benefits, which make them extremely important in a variety of industries. The global market is expected to grow to $52.4 billion by 2025, with surfactant demand expected to rise at a pace of 35% per year [8–11].

1.3 Molecular structure of biosurfactants

1.3.1 Types of biosurfactants

There are two types of biosurfactants: those with a high molecular weight and those with a low molecular weight. The types of biosurfactants with their microbial source and application are shown in table 1.1.

1.3.1.1 High molecular weight biosurfactants

High molecular weight substances called bioemulsifiers are better at creating and preserving water–oil or oil–water emulsions without changing the interfacial or surface tensions [13, 14]. High molecular weight biosurfactants consist of polymeric and particulate surfactants [6].

Table 1.1. Types of biosurfactants with their microbial source and applications.

Microorganism	Biosurfactants	Applications	References
1. Biosurfactants producing bacteria			
Pseudomonas aeruginosa	Rhamnolipid	Antimicrobial, anti-aging, foaming	[18]
Micrococcus luteus BN56	Trehalose tetra ester	Bioremediation of oil-contaminated environment	[19]
Bacillus licheniformis	Lipopeptides	Used in cosmetics	[8]
Rhodococcus erythropolis	Trehalose lipid	Determination of molecular weight of glycolipids present	[19]
Bacillus subtillis	Surfactin	Antibacterial, antiviral, antitumor, and Haemolytic effect	[18]
2. Biosurfactant-producing yeasts			
Candida bombicola	Sophorolipid	Environmental application	[20]
Candida lipolytica	Sophorolipid	Oil recovery	[19]
Candida glabrata UCP1002	Protein carbohydrates lipid complex	Oil recovery from sand	[19]
3. Biosurfactants producing fungi			
Candida utilise	Emulsifier	Emulsifier	[8]
Pseudozymaflocculosa	Flocculosin	Antifungal agents and antimicrobial agents	[17]
Penicillium	Corynomycolic acid	—	[17]
Aspergillus ustus	Glycoprotein	—	[8]

1.3.1.2 Low molecular weight biosurfactants

Because of their low critical micelle concentration (CMC) and ability to reduce surface and interfacial tension between two phases, low molecular weight biosurfactants are excellent surface tensioners that can maintain emulsions with molecular weights between 500 and 1500 Da. This group of biosurfactants includes neutral lipids, fatty acids, glycolipids, phospholipids, and lipoproteins [13].

1.3.1.2.1 Glycolipid

The most prevalent biosurfactants, known as glycolipids, are carbohydrates and combine carbohydrates with long-chain aliphatic acid or hydroxyl aliphatic acids. Either the ester group or ether is used for the connection. The rhamnolipids, trehalolipids, and sophorolipids are the well-known members of this taxon [15].

1.3.1.2.2 Lipopeptides

This family of biosurfactants typically consists of cyclic peptides joined to a fatty acid. *Bacillus subtilis* produces surfactin, a lipopeptide. Surfactin is made up of amino acids joined by lactone bonds to the hydroxyl and carboxyl groups of 14 carbon acids. Surfactin can break down human red blood cells, resulting in the creation of spheroplasts [13].

1.3.1.2.3 Neutral lipids, phospholipids, and fatty acids

When bacteria and yeast grow on n-alkane they produce significant amounts of phospholipid and fatty acid surfactants. *Rhodococcus erythropolis* produces phosphatidylethanolamines when it is cultivated on n-alkane, which lowers the CMC to 30 mg l^{-1} and interfacial tension between hexadecane and water to less than 1 mN m^{-1} [16].

1.3.1.2.4 Polymeric biosurfactants

The most well-known polymeric biosurfactants include emulsans, alasan, liposan, and lipomanan. To emulsify a hydrocarbon–water mixture at low concentration, the emulsion is recognized as an effective bio-emulsifier [6]. *Candida lipolytica* produces the emulsifier known as liposan, which is mostly made up of proteins (17%) and carbohydrates (83%). *Yarrowia lipolytica* also produces such a glycoprotein complex [13].

1.3.1.2.5 Particulate biosurfactants

These biosurfactants are external membrane vesicles that separate hydrocarbons that combine to create microemulsion and are crucial for the microbial cell to absorb or uptake alkane, such as in the case of *Acinetobacter* spp. [17].

1.3.1.3 Chemical composition and structure of biosurfactants

Based on their polarity, biosurfactants can be classified as either anionic or neutral molecules. comprising a hydrophobic tail and a hydrophilic head. The hydrophilic head includes substances like amino acids, phosphate groups, carbohydrates, and other substances. On the other hand, a hydrophobic tail typically contains fatty acid derivatives or a long-chain fatty acid [12]. The structure of a few biosurfactants is shown in figure 1.1.

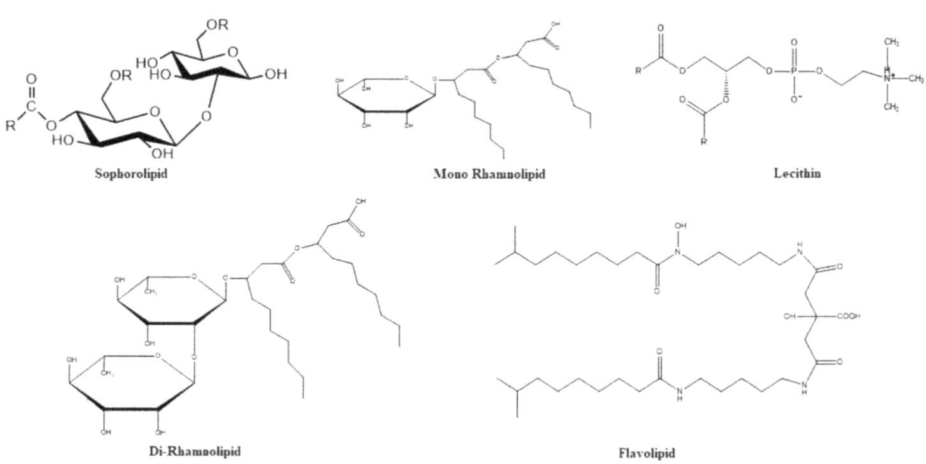

Figure 1.1. Structure of biosurfactants.

1.3.2 Surface activity of biosurfactants

1.3.2.1 Surface and interfacial activity

Biosurfactants decrease both interfacial pressure and surface tension. *B. subtilis* produces surfactin that can decrease the water's surface tension to 25 mN m^{-1} and the interfacial tension of water/hexadecane to under 1 mN m^{-1} [21, 22].

1.3.2.2 Biodegradability

Since bacteria and other microorganisms can quickly break down biosurfactants in soil or water, they are suitable for waste treatment and bioremediation applications [23]. On the biodegradation of biosurfactants, limited research has been done.

According to the OECD guidelines for testing of chemicals, an experiment on the non-pathogenic yeast *Candida bombicola* that produces sophorolipid revealed that after cultivation, biodegradation of biosurfactants begins instantly. Additionally, after 8 days of cultivation, the biodegradability of sophorolipids was determined in the form of TOD (total oxygen demand) and BOD (biochemical oxygen demand) [18]. Surfactin and arthrofactin, two more biosurfactants, were similarly easily biodegradable as sophorolipids. However synthetic biosurfactants exhibited no biodegradability after 8 days.

1.3.2.2.1 Tolerance to temperature, pH, and ionic strength

Environmental factors like pH and temperature have an impact on many biosurfactants and their activity. For example, the lipopeptide from *Bacillus licheniformis* can be resistant to a pH range of 5–12 and temperatures up to 140 h at 75 °C. Additionally, biosurfactants are tolerant of high salt up to five times (2%), which is enough to inactivate synthetic surfactants [24].

1.3.2.2.2 Availability

Biosurfactants can be produced by using abundantly available, extremely cheap raw materials. The hydrocarbons and carbohydrates or lipids are sources of carbon supply. Which can be utilized separately or in combination with others [18].

1.3.2.2.3 low toxicity

Compared to the chemical-derived surfactants, biosurfactants have lower toxicity. Biosurfactants can be used in cosmetics, pharmaceuticals, and food due to their low toxicity. For environmental applications, low toxicity is also of utmost significance. Industrial waste and available raw materials can both be used to produce biosurfactants [25].

1.3.2.2.4 Emulsification

Biosurfactants can operate as both an emulsifier and a de-emulsifier. There are two types of emulsions: oil-in-water emulsions and water-in-oil emulsions. Emulsions made with two different phase solutions are typically unstable. When biosurfactants are added they enable the dispersion of one liquid into another and facilitate the mixing of two immiscible liquids [26].

1.3.2.2.5 De-emulsification

During this process, emulsions are broken by destroying stable surfaces between the internal and bulk phases. This method aids in addressing issues caused by natural emulsifying substances in the production process such as the petroleum industry equipment and oil recovery [26]. According to Cooper and Goldenberg (1987), emulsification is calculated by calculating the emulsification index. This process involves mixing an equal proportion of hydrocarbon-based compound and culture supernatant for 120 s before cooling at room temperature for 24 h. Emulsions and liposans are used as bioemulsifiers produced by strains of *Acinetobacter calcoaceticus* and candida lipoma to increase emulsification activity.

1.3.3 Micelle formation and critical micelle concentration (CMC)

Amphiphilic in nature, biosurfactants have a hydrophilic head and a hydrophobic tail. The term 'biosurfactants' refers to surface-acting substances that can enhance surface–surface interaction by generating micelles in their natural habitats, such as plants, microorganisms, and animals. When surfactants are added to a mixture, they position themselves, so that the hydrophilic portion is in the water and the hydrophobic portion is in the air at the air–water interface. When surfactants are added to an oil–water or air–water system, the surface tension is reduced to a certain point at which the surfactants start to form micelle, bilayer, and vesicles. This point is known as the CMC [3, 12].

1.3.4 Thermodynamic properties of biosurfactants

Lipopolysaccharides, phospholipids, glycolipids, lipoproteins–lipopeptides, cross-linked and hydroxylated fatty acids are the six main categories of biosurfactants whose physical properties have been extensively researched, but little is known about their thermodynamic properties. Because of the necessity to protect the environment, biosurfactants are produced in bulk form by fermentation process using yeast, fungi, and bacteria. These substances can generate a micelle and exhibit high surface and interface activity. Rhamnolipid and surfactin are the most recognizable biosurfactants for their superior biological and interfacial properties. Both biosurfactants are particularly used in the protection of surgical devices and biofilm formation inhibitory capabilities. The characteristics of the surface layer at the air–water interface are important in several of their applications. They consist of surfactant molecule density and layering. The area that the surfactant's molecule occupies at the air–water interface is directly related to the density, which is a measurement of the effectiveness of surfactants. However, the effectiveness of absorption is related to the surfactant concentration in the aqueous solution's bulk phase at which the lowest area consumed by a molecule is achieved. On the other hand, the absorption's effectiveness and efficiency are strongly correlated with its thermodynamic properties. The CMC, at which surfactants start to assemble in the bulk phase, is where this tendency first appears. The biosurfactant's solution surface tension and its relationship to concentration are frequently used to calculate CMC. The density, viscosity, and conductivity of a substance may all be used to measure its physicochemical qualities. These characteristics alter as a result of variations in biosurfactant (BS) concentration. As a result, the changes in density, viscosity, and conductivity that occur in lipopeptides BS concentration may be used to calculate its CMC. The approaches that are more common for estimating the standard Gibbs free energy of absorption were taken into consideration, and it was suggested based on CMC, the surfactant's surface tension of hydrophobic part, and the maximum amount of surfactants that might be present on a given surface were recommended. Multiple non-ionic and ionic surfactants were also tested by using this method as well [27, 28].

1.3.4.1 Gibbs free energy of absorption

In this chapter, a potent lipopeptide biosurfactant known as surfactin's adsorptions at the air–liquid interface is studied. The adsorption occurred from a buffered solution with high quantities of surfactin co- and counterions.

1.3.4.2 Experimental mechanism

1.3.4.2.1 Material

Surfactin and rhamnolipid (supplied by Sigma-Aldrich) were utilized without additional purification. Both deionized water and distilled water, which had an internal resistance level of 18.2, were used to make the aqueous solution of these biosurfactants. Surface tension readings taken before solution preparation also served to control the water purity. Rhamnolipid and surfactin concentrations were altered from 2×10^{-4} to 40 mg dm^{-3}.

1.3.4.2.2 Measurements

Using the platinum ring detachment method (du Nouy's method) the equilibrium surface tension of the surfactin and rhamnolipid solution was measured by the Krüss K9 tensiometer. The ring was washed with distilled water before the surface tension measurement, and a Bunsen burner was used to heat it to a red color. In each case, more than ten consecutive measurements were taken.

1.3.4.2.3 Calculation

At constant temperature and pressure, the absorption and aggregation of surfactant produce changes in the system's Gibbs free energy, satisfying equation (1.1).

$$\Delta \mathbf{G} = \Delta \mathbf{H} - \mathbf{T}\Delta \mathbf{S} \tag{1.1}$$

Thus, at the specific temperature (T), ΔH and ΔS, represent the change in entropy and enthalpy [28, 29].

1.3.4.3 Rheological properties of biosurfactants

Surfactant molecules' tendency to aggregate can be considerably influenced by a variety of external stimuli, including pressure, temperature, flow, etc. The majority of now ionic and ionic surfactants at low deformation rates are regarded as Newtonian fluids. However, surfactant solutions frequently display complex rheological properties as the deformation rate increases. For instance, a surfactant solution with a micelle structure resembling a worm exhibits a large amount of viscoelastic behaviour [30].

1.3.4.3.1 Viscosity and shear thinning

It is crucial to understand surfactant solutions' rheology to employ them more effectively under diverse flow conditions for the effective creation of industrial goods such as pharmaceuticals, plastics, pain, food, cosmetics, etc. First, we examine how the shear viscosity varies with the shear rate at various surfactant concentrations at 25 °C in order to achieve this. Regardless of the surfactant type and concentration, a shear-thinning behaviour between the low and intermediate shear rate regimes is always observed. This results from the fact that the inner entangled network framework of the micelle is predicted to be disrupted at a rate that is higher than the rate of structure reformation when the shear rate steadily increases. The shear viscosity also exhibits a plateau zone at low shear rates, which is followed by an order of magnitude decrease in value at intermediate shear rate [30, 31].

1.4 Biodegradation and environmental properties of biosurfactants

Reducing environmental toxins through the use of biological methods is known as bioremediation. It includes the modification or transformation of organic pollutants, producing less dangerous chemicals [32]. Pollutants are naturally broken down into carbon dioxide and water by microorganisms during this process. Following creation, the products are integrated into several biogeochemical cycles [33]. Microorganisms cannot easily access the components of hydrocarbon pollutants, despite the fact that they are frequently linked to soil elements. As a result, they become less accessible, and the biodegradation process is inefficient [34]. Biosurfactant synthesis can occur as cell-bound compounds or extracellularly. Biosurfactants generation outside of cells results in emulsification of substrate (hydrocarbon). Microorganisms manufacture biosurfactants as a component of their cell membranes, which serves to make it easier for substrates to move through the membrane [35]. Environmental contaminants can be effectively treated microbiologically by using organisms that produce biosurfactants to degrade hydrocarbons and biopreparation of contaminated areas [36].

1.4.1 Sustainable production of biosurfactants

Production of biosurfactants can be carried out on a large scale (fermentation) in the laboratory [37]. From the sample site, production of biosurfactants begins. Diesel-polluted soil, extreme environments, an oil reservoir, and a sea harbor, from unpolluted soil are a few places where microbes have been found that produce biosurfactants [5]. Most mineral medium (MSM) that incorporate organic substrate as nutrient broth and carbon source are employed for the isolation of microorganisms [38]. Microorganisms are introduced in their appropriate broth medium that contains appropriate amounts of carbon and nitrogen sources [38] and then incubated in a controlled environment under optimum conditions. This then undergoes a large-scale process (fermentation) [39].

1.4.1.1 Factors affecting the biosurfactants production

1.4.1.1.1 Nutrient factor
The carbon source affects the quantity and quality of biosurfactant production. For the synthesis of biosurfactants, glycerol, sucrose, and glucose are good sources of carbon [40]. Nitrogen is a limiting ingredient that is crucial for microbial growth, protein synthesis, and enzyme synthesis. Good sources of nitrogen for the production of biosurfactants are urea, peptone, nitrate, ammonium sulfate, yeast, meat, and malt extracts. For the growth of microorganisms, phosphate also plays an important role and it is typically offered in triphosphate form [41].

1.4.1.1.2 Environmental factors
Environmental elements that impact the generation and characteristics of biosurfactants include temperature, pH, agitation speed, and aeration. The growth of biosurfactant-producing organisms is influenced by temperatures between 25 °C and 37 °C. It has been observed that biosurfactant growth is optimum when the pH of the medium is around 8. There are rare exceptions but generally, yeast and fungi prefer an acidic environment and bacteria prefer an alkaline environment for their proper growth. Incubation time is also important for the production of biosurfactants because different microorganisms possess different incubation times for their proper growth [42].

1.5 Application of biosurfactants

Biosurfactants are utilized in a variety of industries such as agriculture, cosmetics, the environment, food, and pharmaceuticals. The use of biosurfactants in various industries is shown in figure 1.2.

1.5.1 In cosmetic industry

For practical cosmetic use, many chemicals are synthesized. One example is the enzymatic modification of hydrophobic compounds by whole cells and different lipases. The biosurfactants that are approved for cosmetic use, should be usable for at least three years. Surfactants used in cosmetics are used to make glycerol from tallow (1:5:2) using lipase from pseudomonas fluorescence (use percentage = 90%). They offer a dermal and moisturizing effect due to their lower toxicity [43].

1.5.2 In the pharmaceutical industry

Pharmaceutical and biomedical applications:

a. Antimicrobial properties: biosurfactants possess inherent antimicrobial properties, making them suitable for use in various therapeutic applications [44].
b. Anti-adhesive and biofilm disruption: they can act as anti-adhesive surfaces and disrupt biofilm structures, which are crucial in preventing infections and treating diseases [45].

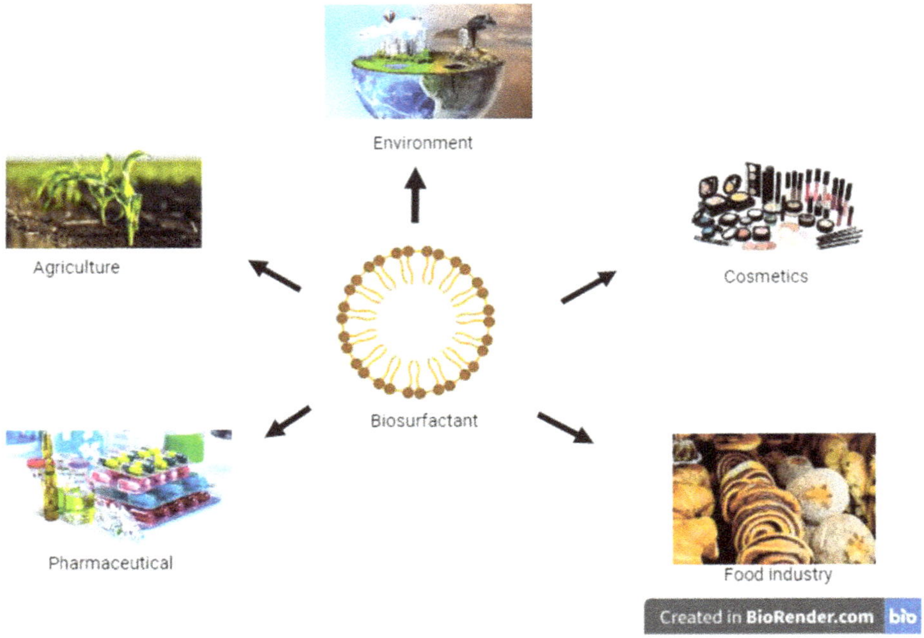

Figure 1.2. Potential use of biosurfactants.

c. Respiratory failure: biosurfactants may find use in treating respiratory failure conditions.
d. Immunological adjuvants: they can serve as immunological adjuvants, potentially enhancing the effectiveness of vaccines and therapies [46].
e. Inhibition of pathogens: biosurfactants can inhibit pathogens, aiding in the prevention and treatment of infections.
f. Stimulation of skin fibroblast metabolism: useful in skincare and wound healing products [47].
g. Potential in cancer: they may have applications in cancer treatment and research.

1.5.3 In agriculture

In agriculture, biosurfactants are used for a variety of things including soil quality improvement, elimination of common water-soluble pollutants assisting in the eradication of plant diseases, promoting beneficial plant–microbes interaction, pesticide preparation, etc. Pentachlorophenol (PCP) was eliminated by rhamnolipid from sandy-silty soil (61%) and sand soil (60%). *Rhizoctonia solani* in maize crop was treated with surfactin, which stimulates the synthesis of a defense enzyme. Due to such quality, biosurfactants are useful in the control of phytopathogenic organisms [12].

1.5.4 In the food industry

In the food industry surfactants can serve a variety of functions. A few examples are to prevent fat globules from aggregating, enhance the consistency and shelf life of products containing starch, enhance the consistency and texture of products containing fat, stabilize the aerated system, and alter the rheological characteristics of wheat dough. In the food industry, rhamnolipids are used to enhance the qualities of cream, frozen, confectionery, and butter items [26].

1.5.5 Environmental application

Biosurfactants have proved to be effective at washing out crude oil from contaminated ground or sand columns. The biodegradation of oils is also greatly accelerated by biosurfactants. Conversely, the use of synthetic surfactants with a span of Tween 80 and 40 did not accelerate the breakdown of petroleum hydrocarbons; rather, the application of biosurfactants produced by *candida Antarctica* accelerated the rate of petroleum hydrocarbon degradation. In terms of increasing the rate of phenanthrene biodegradation, biosurfactants produced by *Pseudomonas aeruginosa* ATCC 9027 and PCG3 strain PCG3 strain were more efficient than synthetic triton *x*-100 and less efficient than Tween 8o [24].

1.5.6 Application of biosurfactants in bio-nanotechnology

The application of biosurfactants in bio-nanotechnology is an emerging and promising field with potential benefits for various applications. Here are some key points regarding the use of biosurfactants in bio-nanotechnology:

a. Green synthesis of nanoparticles: biosurfactants can act as eco-friendly stabilizers and reducing agents in the synthesis of nanoparticles. This approach is in line with green chemistry principles as it reduces the need for hazardous chemicals and energy-intensive processes [48].
b. Environmental remediation: biosurfactant-mediated nanoparticles have great potential in remediating contaminated environments. These nanoparticles can be designed to target and remove specific pollutants, including heavy metals, organic compounds, and even oil spills, from soil and water.
c. Microbial biosynthesis: microbes, such as bacteria and fungi, can produce biosurfactants that are capable of stabilizing and reducing nanoparticles. This biologically mediated nanoparticle production offers a sustainable and environmentally friendly alternative to conventional chemical methods [49].
d. Stability and longevity: biosurfactants help in stabilizing nanoparticles and preventing their aggregation. This enhanced stability allows the nanoparticles to remain effective for extended periods, making them suitable for various applications [50].
e. Eco-friendly materials: biosurfactant-mediated nanoparticles can serve as eco-friendly materials for a wide range of products and services. These nanoparticles can be incorporated into products such as cosmetics, pharmaceuticals, and even food products.

f. Cost-effectiveness: to be practical and widely adopted, biosurfactant-mediated nanoparticle synthesis must be cost-effective. Researchers are working on optimizing the production processes to ensure that the use of biosurfactants remains economically viable.
g. Toxicant removal efficiency: One important consideration is how well biosurfactant-mediated nanoparticles remove environmental toxins. The goal of research is to increase their efficacy in eliminating pollutants while reducing any possible damage to the environment.
h. Environmental compatibility: biosurfactant-mediated nanoparticles should be designed to have minimal environmental impact. This includes considerations for biodegradability and eco-toxicity to ensure they do not introduce new environmental problems [50].

1.5.7 Application of biosurfactants in the petroleum industry for oil recovery

Biosurfactants have several valuable applications in the petroleum industry for oil recovery and environmental remediation. Here are some key applications and benefits of biosurfactants in the petroleum industry:

a. Enhanced oil recovery (EOR): biosurfactants are used to enhance the recovery of oil from reservoirs that have become less productive over time. They help by reducing the interfacial tension between oil and water, thereby improving the mobility of oil within the reservoir. This makes it easier to extract trapped oil, increasing overall oil production [51].
b. Oil spill cleanup: biosurfactants can be employed to clean up oil spills in marine environments. They break down the oil into smaller droplets, making it more accessible for microbial degradation. Biosurfactants are advantageous in this context because they are biodegradable and less toxic than synthetic chemical dispersants.
c. Oil residue removal: in storage tanks and equipment used in the petroleum industry, there is often a buildup of oil residues. Biosurfactants can be used to efficiently clean and remove these residues, ensuring the equipment operates efficiently and safely.
d. Microbial enhanced oil recovery (MEOR): biosurfactants can play a crucial role in MEOR processes. They can stimulate the growth of oil-degrading microorganisms in reservoirs, leading to improved oil recovery rates. Biosurfactants are more compatible with microbial activity in the reservoir than synthetic surfactants.
e. Bioremediation: biosurfactants can assist in the bioremediation of petroleum pollutants in the environment. They enhance the solubility and bioavailability of hydrophobic pollutants, making it easier for microorganisms to degrade them. This is particularly useful in situations where oil has contaminated soil or groundwater [52].
f. Emulsification/de-emulsification: biosurfactants can act as emulsifiers or de-emulsifiers, depending on the need. In the petroleum industry, they can help

break down oil–water emulsions, facilitating the separation of oil from water, which is essential in oil processing and storage [53].

g. Environmental compatibility: biosurfactants are often preferred over synthetic surfactants due to their higher emulsifying activities and greater ecological compatibility with the environment. They are less toxic and biodegradable, making them a greener choice for various applications.
h. Low toxicity: biosurfactants are generally less toxic than chemical surfactants, reducing potential harm to the environment and aquatic life when used in oil spill cleanup or other remediation efforts [54].
i. Biodegradability: biosurfactants are biodegradable, which means they can be broken down naturally by microorganisms in the environment, minimizing their long-term impact on ecosystems [55].

1.5.8 General industrial applications

a. Detergent: biosurfactants enhance the performance of detergents [56].
b. Ornamentals: they can improve pesticide and nutrient dispersion in horticulture.
c. Wide range of products: biosurfactants are used in various products such as antifungal, antibacterial, germicidal, anti-cancer products, wound dressings, and skin and personal care products [57].
d. Emulsification and de-emulsification: they play roles in emulsification, foaming, phase dispersion, and emulsion polymerization in industrial processes.
e. Stability: biosurfactants remain stable in extreme conditions of temperature, pH, and salt concentration [58].

1.5.8.1 Biotechnological potential

Biosurfactants have significant biotechnological potential, but further research and development are needed to optimize their production on a large scale.

Biosurfactants offer a wide array of applications across multiple industries, owing to their versatile properties, stability, and environmentally friendly nature. Continued research and advancements are likely to unlock their full potential for large-scale commercial use in various products and processes. Table 1.2 summarizes the applications of biosurfactants.

1.6 Challenges and limitations

Biosurfactants, also known as microbial surfactants, offer numerous advantages in everyday life and various industries, including pharmaceuticals, detergents, cosmetics, and food. However, alongside their benefits, they do come with certain limitations, primarily related to purification and production costs. The purification of surfactants faces specific constraints, and the biotechnological methods employed for surfactant production can be relatively expensive [60]. To discover new biosurfactants, an extensive range of bacterial species must undergo initial screening using basic qualitative assays to identify biosurfactant producers. Subsequently, a

Table 1.2. Applications of biosurfactants [59].

Industry	Biosurfactant	Field	Mechanism/functioning
Agriculture and food industry	Glycolipid, rhamnolipids, lipopeptides	Soil quality, plant protection	By changing the cell surface structures of the pathogen, bioremediation of toxic metals and other chemicals associated with soil impacts the target cell.
	Lipopeptides	Pest control	The feature of detergency shows toxicity towards nematodes and insects.
	Rhamnolipids	Food stabilizer	Food that has been altered to have the proper texture and consistency
Bioremediation/ cleaning oils-spills	Glycolipid and trehalose	Spill remediation	Oil spills that are soluble are accessible to hydrocarbon degraders, facilitating their quicker biodegradation.
	Rhamnolipids, lipopeptides	Treatment of soil and wastewater	By serving as emulsifiers and de-emulsifiers, bioavailability enhancers, tension reducers, and mobilizers, remove oil and pollutants from the soil.
	Rhamnolipids, sophorolipids	Hydrocarbon remediation	Increase the pollutants' bioavailability for biodegradation by solubilizing them into the aqueous phase.
	Rhamnolipids	Heavy metal remediation	Heavy metals can be extracted from soils using a variety of techniques, including binding, desorption, interactions, trapping of metals, and ion exchange.
Medicine/ pharmaceuticals/ bioprocessing	Rhamnolipids, sophorolipids	Antimicrobial agent,	Function as an antimicrobial agent, demonstrating actions akin to those of a detergent
		anti-cancer activity	Check cell replication as an antiviral agent in favour of cell differentiation.
	Sophorolipid	Antiviral activity	Inactivation of viral capsids and lipid envelopes
	Rhamnolipids	Antibiotics recovery	Using their surfactant properties to remove medications and proteins
Mining precious metals	Biodispersan	Precious metal recovery	Using an enzyme like nitrate reductase, biosurfactant-producing microorganisms transform (Ag–Au) NO_3 into silver/gold particles.
Petroleum/enhance oil recovery	Glycolipids and lipopeptide	Crude oil extraction from reservoirs	Destroy the rock oil layer, enhance stable emulsion formation, and reduce tensional and capillary forces that impede oil flow through rock pores.
	Emulsan, alasan, biodispersan	Crude oil pipelines/ Transport	To facilitate the oil's movement, lower its viscosity, and prevent drop coalescence, maintain a constant water-in-oil emulsion.

smaller selection of candidates is chosen for pilot-scale testing, where the commercial potential of novel biosurfactants can be assessed. This evaluation involves increasingly precise, time-consuming, and costly quantitative assays. Surfactin, a cyclic lipopeptide expressed by *Bacillus subtilis*, was the first biosurfactant isolated from bacteria (surface tension from 72 mN m^{-1} to 27 mN m^{-1}). Despite the wealth of research in this field, it remains uncertain whether significantly more potent biosurfactants have been uncovered in recent investigations. This uncertainty arises from the challenge of comparing work conducted over the past 30–40 years [61].

1.7 Conclusions

Biosurfactants are natural surface-active agents produced by microorganisms on their surfaces. Notable microorganisms known for biosurfactant production include *Pseudomonas, Bacillus, Acinetobacter, Aspergillus niger, Lactobacillus, Rhodococcus, Fusarium, Candida, Gordonia*, and more. These compounds can effectively reduce both surface tension and interfacial tension within mixtures or solutions. Biosurfactants are highly sought after in the global market because they can serve as eco-friendly alternatives to synthetic surfactants. Their exceptional qualities have led to significant demand across various industries, including cosmetics, healthcare, detergent production, petroleum exploration, agriculture, food processing, pharmaceuticals, and environmental applications. This review primarily delves into the different types of biosurfactants, their characteristics, structural components, production methods, and a wide array of applications across diverse sectors.

Abbreviations

BOD	Biochemical oxygen demand
CMC	Critical micelle concentration
EOR	Enhanced oil recovery
HMW	High molecular weight
LMW	Low molecular weigh
MSM	Most mineral medium
MEOR	Microbial enhanced oil recovery
OCED	Organisation for Economic Co-operation and Development
PCP	Pentachlorophenol
TOD	Total oxygen demand

References

[1] Jahan R, Bordratti A M, Tsianou M and Alexandridis P 2019 Biosurfactants, natural alternatives to synthetic surfactants: physiochemical properties and application *Adv. Colloid Interface Sci.* **275** 102061

[2] Fenibo E O, Ijoma G N, Selgarajan R and Chikere C B 2019 Microbial surfactants: the next generation multifunctional biomolecules for application in the petroleum industry and its associated environmental remediation *Microorganisms* **7** 581

[3] Kumar A *et al* 2021 Microbial biosurfactant: a new frontier for sustainable agriculture and pharmaceutical industries *Antioxidants* **10** 1472

[4] Sáenz-Marta C I, de Lourdes Ballinas-Casarrubias M, Rivera-Chavira B E and Nevárez-Moorillón G V 2015 Biosurfactants as useful tools in bioremediation *Advances in Bioremediation of Wastewater and Polluted Soil* (London: InTechopen) vol 5
[5] Vasanthabharthi V and Mwajar M 2019 *Ngeiywa, Maduka, production, classification, properties, characterization* 51198 (JAMB)
[6] Prasad B, Kaur HP and Kaur S 2015 Potential biomedical and pharmaceutical applications of microbial surfactants *World J Pharm Pharm Sci* **4** 1557–75
[7] Sekhon Randhawa K K and Rahman P K S M 2014 Rhamnolipids biosurfactants—past, present, future scenario of global market *Front. Microbiol.* **5** 454
[8] Bjerk T R, Severino P, Jain S, Marques C, Silva A M, Pashirova T and Souto E B 2021 Properties and applications in drug delivery, biotechnology and ecotoxicology *Bioeng. (Basel)* **8** 115
[9] Burilova E A, Pashirova T N, Lukashenko S S, Sapunova A S, Voloshina A D, Zhiltsova E P, Campos J R, Souto E B and Zakharova L Y 2018 Synthesis, biological evaluation and structure-activity relationships of self-assembled and solubilization properties of amphiphilic quaternary ammonium derivatives of quinuclidine *J. Mol. Liq.* **272** 772–30
[10] Pashirova T N, Sapunova A S, Lukashenko S S, Burilova E A, Lubina A P, Shaihutdinova Z M, Gerasimova T P, Kovalenko V, Voloshina A D, Souto E B *et al* 2020 Synthesis, structure–activity relationship and biological evaluation of tetra cationic gemini dabco-surfactants for transdermal liposomal formulations *Int. J. Pharm.* **575** 118953
[11] Zakharova L Y, Pashirova T N, Doktorovova S, Fernandes A R, Sanchez-Lopez E, Silva A M, Souto S B and Souto E B 2019 Cationic surfactants: self-assembly, structure-activity correlation, and their biological applications *Int. J. Mol. Sci.* **20** 5534
[12] Pardhi D S, Panchal R R, Raval V H, Joshi R G, Poczai P, Almalki W H and Rajput K N 2022 A journey from fundamentals to recent advances *Front. Microbiol.* **13** 982603
[13] Adetunji A I and Olaniran A O 2021 Production and potential biotechnological application of microbial surfactants: an overview *Saudi J. Biol. Sci.* **28** 669–79
[14] Uzoigwe C, Burgess J G, Ennis C J and Rahman P K S M 2015 Bioemulsifiers are not biosurfactants and require different screening approaches *Front. Microbiol.* **6** 245
[15] Dhiman R, Raj Meena K, Sharma A and Kanwar S S 2016 Biosurfactants and their screening methods *Res. J. Recent Sci.* **5** 1–6
[16] Shoeb E, Akhlaq F, Badar U, Akhter J and Imtiaz S 2013 Classification and industrial applications of biosurfactants *Academic Research International* **4** 243
[17] Bhattacharya B, Ghosh T K and Das N 2017 Application of biosurfactants in cosmetics and pharmaceutical industry *Scholars Acad. J. Pharm. (SAJP)* **6** 320–9
[18] Sansarode D S and Shahasrabudhe S 2018 Biosurfactants: classification, properties and recent applications in cosmetics *J. Emerg. Technol. Innov. Res. (JETIR)* **5** 160–7
[19] Shah N, Nikam R, Gaikwad S, Sapre V and Kaur J 2016 Biosurfactants: types, detection methods, importance and applications *Ind. J. Microbiol. Res.* **3** 5–10
[20] Roy A 2017 A review on the biosurfactants: properties, type and its applications *J. Fundam. Renew. Energy Appl.* **8** 1000248
[21] Cooper D G, Macdonald C R, Duff S J B and Kosaric N 1981 Enhanced production of surfactin from *Bacillus subtilis* by continuous product removal and metal cation additions *App. Environ. Microbiol.* **42** 408–12
[22] Tripathy D B and Mishra A 2016 Sustainable biosurfactants *Sustain. Inorg. Chem.* **1** 175–92

[23] Cavalero D A and Cooper D G 2003 The effect of medium composition on the structure and physical state of sophorolipids produced by *Candida bombicola* ATCC 22214 *J. Biotechnol.* **103** 31–41

[24] Klosowska-Chomiczewska I E, Medrzycka K and Karpenko E 2013 Biosurfactants—biodegradability, toxicity, efficiency in comparison with synthetic surfactants *Research and Application of New Technologies in Wastewater Treatment and Municipal Solid Waste Disposal in Ukraine, Sweden and Poland, 17* Article 17

[25] Mary Swarnalatha S and Chinna Rani J 2019 Biosurfactants: unique properties and their versatile applications *Pharm. Innov. Int. J.* **8** 684–7

[26] Vijayakumar S and Saravanan V 2015 Biosurfactant—types, sources and applications *Res. J. Microbiol.* **10** 181–92

[27] Vaidya S and Ganguli A K 2019 *Microemulsion Methods for Synthesis of Nanostructured Materials* (Amsterdam: Elsevier)

[28] Singh P and singh cameotra S 2004 Potential applications of microbial surfactants in biomedical sciences *Trends Biotechnol.* **22** P142–6

[29] Manko D, Zdziennicka A and Janczuk B 2016 Absorption and aggregation activity of sodium dodecyl sulfate and rhamnolipid mixture *J. Surfact. Deterg.* **20** 411–23

[30] Adamson A W and Gast A P 1997 *Physical Chemistry of Surfaces* (New York: Wiley-Interscience)

[31] Khan M B and Sasmal C 2022 A detailed and systematic study on rheological and physiochemical properties of Rhamnolipid biosurfactants solution *JCIS Open* **8** 100067

[32] Khan M B and Sameer Khan M 2018 Experimental investigation on effect of alkali, salt on the rheological properties of polyacrylamide and mixed assembly of polymer-surfactant *Chem. Data Collec.* **13–14** 60–72

[33] Olasanmi I O and Thring R W 2018 The role of biosurfactants in the continued drive for environmental sustainability *Sustainability* **10** 4817

[34] Cameotra S S and Bollag J 2003 Biosurfactant-enhanced bioremediation of polycyclic aromatic hydrocarbons *Crit. Rev. Environ. Sci. Technol.* **33** 111–26

[35] Sobrinho H B S, Luna J M, Rufino R D, Porto A L F and Sarubbo L A 2013 Biosurfactants: classification, properties and environmental applications *Recent Developments in Biotechnology* ed J N Govil (Studium Press) 1st edn

[36] Lai C-C, Huang Y-C, Wei Y-H and Chang J-S 2009 Biosurfactant-enhanced removal of total petroleum hydrocarbons from contaminated soil *J. Hazard. Mater.* **167** 609–14

[37] Santos D K F, Rufino R D, Luna J M, Santos V A and Sarubbo L A 2016 Biosurfactants: multifunctional biomolecules of the 21st century *Int. J. Mol. Sci.* **17** 401

[38] Dhail S 2012 Isolation of potent biosurfactant-producing bacteria from oil-spilt marine water and marine sediments *Afr. J. Biotechnol.* **11** 16751–7

[39] Elazzazy A M, Abdelmoneim T S and Almaghrabi O A 2015 Isolation and characterization of biosurfactant production under extreme environmental conditions by alkali-halo-thermophilic bacteria from Saudi Arabia *Saudi J. Biol. Sci.* **22** 466–75

[40] Nwachi A C, Onochie C C, Iroha I R, Agah V M, Agumah B N, Moses I B and Ogbeja D O 2016 Extraction of biosurfactants produced from bacteria isolated from waste-oil contaminated soil in Abakaliki Metropolis, Ebonyi State *J. Biotechnol. Res.* **2** 24–30

[41] Md F 2012 Biosurfactant: production and application *Petrol. Environ. Biotechnol.* **3** 1000124

[42] Rahman P K and Gakpe E 2008 Production, characterisation and applications of biosurfactants-review *Biotechnology* **7** 360–70

[43] Fontes G C, Amaral F, Filomena P, Nele M, Coelho Z and Alice M 2010 Factorial design to optimize biosurfactant production by *Yarrowia lipolytica Bio. Med. Res. Int.* **2010** 821306
[44] Shakeri F, Babavalian H, Ali Amoozegar M, Ahmadzadeh Z, Zuhuriyanizadi S and Pahlevan Afshrian M 2021 Production and application of biosurfactants in biotechnology *Biointerface Res. Appl. Chem.* **11** 10446–60
[45] Chen J, Wu Q, Hua Y, Chen J, Zhang H and Wang H 2017 Potential applications of biosurfactant rhamnolipids in agriculture and biomedicine *Appl. Microbiol. Biotechnol.* **101** 8309–19
[46] Chakraborty J and Das S 2017 Application of spectroscopic techniques for monitoring microbial diversity and bioremediation *Appl. Spectrosc. Rev.* **52** 1–38
[47] Vecino X, Cruz J M, Moldes A B and Rodrigues L R 2017 Biosurfactants in cosmetic formulations: trends and challenges *Crit. Rev. Biotechnol.* **37** 911–23
[48] Christopher F C, Ponnusamy S K, Ganesan J J and Ramamurthy R 2018 Investigating the prospects of bacterial biosurfactants for metal nanoparticle synthesis–a comprehensive review *IET Nanobiotechnol.* **13** 243–9
[49] Joanna C, Marcin L, Ewa K and Grażyna P 2018 A nonspecific synergistic effect of biogenic silver nanoparticles and biosurfactant towards environmental bacteria and fungi *Ecotoxicology* **27** 352–9
[50] Gómez-Graña S, Perez-Ameneiro M, Vecino X, Pastoriza-Santos I, Perez-Juste J, Cruz J M and Moldes A B 2017 Biogenic synthesis of metal nanoparticles using a biosurfactant extracted from corn and their antimicrobial properties *Nanomaterials* **7** 139
[51] Karlapudi A P, Venkateswarulu T C, Tammineedi J, Kanumuri L, Ravuru B K, ramu Dirisala V and Kodali V P 2018 Role of biosurfactants in bioremediation of oil pollution—a review *Petroleum* **4** 241–9
[52] Liu G, Zhong H, Yang X, Liu Y, Shao B and Liu Z 2018 Advances in applications of rhamnolipids biosurfactant in environmental remediation: a review *Biotechnol. Bioeng.* **115** 796–814
[53] McClements D J and Gumus C E 2016 Natural emulsifiers—biosurfactants, phospholipids, biopolymers, and colloidal particles: molecular and physicochemical basis of functional performance *Adv. Colloid Interface Sci.* **234** 3–26
Meenakshisundaram M and Pramila M 2017 Detoxification of heavy metals using microbial biosurfactant *Int. J. Curr. Microbiol. Appl. Sci.* **6** 402–11
[54] Arora N K 2018 Biodiversity conservation for sustainable future *Environ. Sustain.* **1** 109–11
[55] Arora N K 2018 Environmental sustainability—necessary for survival *Environ. Sustain.* **1** 1–2
[56] Bratovcic A, Nazdrajic S, Odobasic A and Sestan I 2018 The influence of type of surfactant on physicochemical properties of liquid soap *Int. J. Mat. Chem.* **8** 31–7
[57] Dave N and Joshi T 2017 A concise review on surfactants and its significance *J. Appl. Chem.* **13** 663672
[58] Dhundale V R, Hemke V M, Salve S, Sharyu G, Budhwant J, Aglave T and Desai D 2018 Production and stability studies of the biosurfactant isolated from alkaliphilic bacterium SJS1 *Bio. Sci. Res. Bull.* **34** 7
[59] Ambaye T G, Vaccari M, Prasad S and Rtimi S 2021 Preparation, characterization and application of biosurfactant in various industries: a critical review on progress, challenges and perspectives *Environ. Technol. Innov.* **24** 102090

[60] Bajpai Tripathi D and Mishra A 2016 Sustainable biosurfactants *Encyclopedia of Inorganic and Bioinorganic Chemistry* (Wiley)
[61] Fechtner J, Cameron S, Deeni Y Y, Hapca S M, Kabir K, Mohammad I U and Andrew J 2008 Limitation of biosurfactants strength produced by bacteria *Biosurfactants: Occurrences, Applications and Research* ed R Upton (Nova Science)

Chapter 2

Biosurfactants and bioemulsifiers: introduction, types, synthesis, properties and potential applications

Sam Joy, Pallawi and Neha Sharma

Biosurfactants are a diverse range of biomolecules that include polymers such as lipoproteins, polysaccharide-protein fatty acid complexes having high molecular weight, and lipopolysaccharide-protein complexes, as well as glycolipids, lipopeptides, flavolipids, and phospholipids having low molecular weight with a broad range of structural and functional variability. Biosurfactants have gained considerable interest in recent years, unlike their synthetic counterparts (chemical surfactants), due to their biodegradable and nontoxic properties of by-products. Due to this compositional diversity and notable properties, biosurfactants occupy a significant position in almost all industrial sectors ranging from biomedicine, cosmetics, environmental remediation, agriculture, the textile industry, and enhanced oil recovery, making them a multifunctional material of the 21st century. Over the decades the biosurfactant industry has undergone a constant transition with changes in the use of different raw materials for production based on the need, applications, cost, availability, and specific demands of the domestic and industrial end-users. In this chapter, we discuss an overview of biosurfactants, their types, properties, and the potential applications of biosurfactants.

2.1 Introduction to microbial surfactants

Surface active agents which are generally referred to as 'surfactants' are organic amphiphilic compounds made up of hydrophilic and hydrophobic entities. These amphiphilic compounds separate specifically at the interface between fluid phases such as oil/water or air/water interfaces with different degrees of polarity and hydrogen bonding altering the free energy of these surfaces or interfaces [1]. The partitioning ability imparts to them the key unique properties which include the lowering of surface or interfacial tensions, micellization, emulsification, foaming,

doi:10.1088/978-0-7503-5989-4ch2

wetting, and dispersing abilities, thereby making them versatile compounds significant for different industrial sectors and applications [2, 3].

2.1.1 Brief history

The oldest known evidence of production of surface-active compounds dates to as early as 2800 B.C. in ancient Babylon in the form of soap-like materials for cleaning purposes. An ancient Egyptian medical document, the Ebers papyrus, dating to early 1500 B.C. describes the use of soap-like material for washing and treatment of skin diseases using a combination of vegetable oils and animal fats with alkaline salts. In the 18th century the discovery of a soda ash-making process from common salt by Nicholas Leblanc in 1791, increased the manufacturing of detergents/soaps by combining either animal fat or vegetable oil, which continued for centuries [4, 5]. In 1916, the first synthetic surface-active-agents were developed from coal tar in Germany in response to the shortage of natural fats and oils during World War I. Later, from the 1930s to the late 1940s, synthetic surfactants such as alkylnaphthalene sulfonates, alcohol sulfates, and branched alkylbenzene sulfonates (BAS) dominated the surfactant markets. In the 1950s, the products having surface activity such as foaming agents, emulsifiers, wetting agents, detergents, and dispersants were all represented for the first time by the word 'surfactant' which was coined by Antara products [6]. Towards the end of World War II, from 1950–1965 the development of propylene tetramer-based alkylbenzene sulfonates (ABS), a type of anionic surfactant dominated more than half of the global surfactant market due to its versatility, cost, and ease in manufacturing [7].

Towards the late 1960s, the issues regarding the increased accumulation of alkylbenzene sulfonates in the environment and their harmful effects were raised, which forced their replacement by modified linear alkylbenzene sulfonates (LABS) which still occupy more than 75% of the market distributions [8, 9].

Over the decades, the surfactant industry has undergone a constant transition with changes in the basic chemical composition using different raw materials based on the need, applications, cost, availability, and specific demands of the domestic and industrial end-users [10, 11].

2.2 Synthetic surfactants

Synthetic surfactants are an essential group of chemicals that are extensively used in numerous industrial processes and formulations due to their surface-active and emulsifying properties. Due to their amphiphilic structures consisting of hydrophilic hydrophobic moieties, they accumulate at liquid surfaces or the interface between two immiscible liquids and influence the surface-active properties resulting in the formation of micelles and stable emulsions. A wide variety of synthetic surfactants exist in the current market and are generally synthesized by alkylation, ethoxylation, or sulphation of petrochemicals or petroleum feedstock like benzene, chloroalkane, paraffin, alkene, alkyl sulfates, ethene, epoxyethane, long-chain alkenes, alkyl alcohol and esters of linear alcohols [12]. They are generally categorized into three broad classes based on the charge of the hydrophilic head group: ionic (anionic and cationic), nonionic, and amphoteric/zwitterionic surfactants. The hydrophobic tail is generally made of alkyl chains of varying lengths [13].

(i) **Anionic surfactants**

Anionic surfactants display better foaming and detergency properties and are widely used for industrial/household cleaning and pesticide formulations. These surfactants with negatively charged hydrophilic head groups bind to positively charged particles like clay, oil, and dirt which results in their removal on washing [14]. As per the market analysis, anionic surfactants constitute about 75% of total surfactants in the market owing to their efficiency in different environments and relatively lower cost of manufacturing [15]. The most prominent representative examples of commercially used anionic surfactants include ammonium lauryl sulfate (ALS), sulfonic acid salts, alkylbenzene sulfonates (ABS), sodium laureth sulfate (SLES), sodium lauryl sulfate (SLS), alcohol ether sulfates (AES), alcohol sulfates (AS), secondary alkane sulfonates, and phosphoric acid esters [16–18].

(ii) **Cationic surfactants**

Cationic surfactants are positively charged and are widely applied for surface modification in fabric softeners, shampoos, and hair conditioners as they tend to rapidly adsorb and not desorb from the surface (negatively charged) it binds such as fabric and hair. Besides these, other applications include anticorrosive agents, dispersants, antistatic agents for plastics/fabric, anticaking agents for fertilizers, and bactericides in hygiene and cosmetic products. The most important cationic surfactants are the quaternary ammonium complexes such as; dodecyl trimethyl ammonium chloride, ditallow dimethyl ammonium chloride, and dialkyl dimethyl ammonium chloride [19, 20].

(iii) **Nonionic surfactants**

Nonionic surfactants carry no charge and represent the second most widely used group of surfactants in cleaning products. The most commonly used nonionic surfactants are alcohol ethoxylates, polysorbates (Tween), alcohol alkoxylates, poly-glycerol alkyl ethers, alkyl polyglucosides, alkylphenol ethoxylates, sorbitan esters [21, 22].

(iv) **Amphoteric surfactants**

Amphoteric biosurfactants carry both cationic and anionic groups in their hydrophilic head, giving them a net charge of zero. The action of these surfactants depends on the pH of the working systems and finds applications as mild surfactants in baby shampoos, facial cleansers, eye disinfectants, and baby washes. They are also used as ‘coupling agents,’ along with anionic and cationic surfactants which hold the surfactants, solvents, and inorganic salt components together. The commonly used amphoteric surfactants are lauro ampho acetate, cocamido propyl-betaine, sodium acyl ampho-propionate, etc [23, 24].

2.3 Need for an alternative and safer substitute

The massive use of synthetic surfactants in domestic and industrial sectors have resulted in their accumulation in different aquatic and terrestrial ecosystems, posing serious environmental and ecological threats due to their toxic, non-degradable, and recalcitrant properties [25, 26].

In the terrestrial environment, significant amounts of surfactants penetrate the soil from domestic discharges, industrial effluents, and through continuous use of various agrochemicals such as chemical fertilizers and pesticides. These persist in soil for years due to their recalcitrant structures and alter the soil properties even at low concentration affecting nutrient mineralization and cycling, soil buffering, nutrient holding capacity, increasing soil cation exchange, pH, and altering soil enzyme activities [27, 28].

In aquatic environments, the presence of surfactant results in several negative abiotic and biotic effects. Abiotic effects lead to the deterioration of water quality, creation of hypoxic conditions due to foaming, turbidity, change in the physio-chemical properties (pH, hardness, dissolved oxygen), and eutrophication while the biotic effects include the reduction of plant growth and its photosynthetic ability, abnormal spore formation, impaired reproductive activities with stunted growth in fishes and acute toxicity in embryos and adult aquatic organisms [29–31].

In humans, the exposure causes eye and skin irritations, and the intake can induce severe damage to vital organs, growth retardation, and even hormonal and enzyme disturbances [31–33]. For example, nonionic surfactants such as alkylphenolpolyethoxylates (APEs) and nonylphenolethoxylates (NPEs) used in different industrial applications degrade to form alkylphenols (APs) and nonylphenol (NP) which have been proved to be endocrine-disrupting chemicals [30].

Thus, the cumulative factors of increasing concerns about surfactant toxicity, awareness of environmental safeguards along with rising price of petrochemical-based raw materials have driven the synthetic surfactant industry to search for alternative and safer substitutes as biosurfactants, derived from plants and microbes [34].

2.4 Types of microbial surfactant

Microbial biosurfactants are produced by a diverse group of microbes ranging from bacteria, fungi, yeast, and actinomycetes with varying chemical compositions [35, 36]. Unlike the synthetic composition of the hydrophilic (carboxylate, sulfate, sulfonate, phosphates quaternary ammonium, betaines, and ethoxylates groups) and hydrophobic (polyoxymethylene, alkyl chain, siloxane chains, fluorocarbon chains) moieties of chemical surfactants, the hydrophilic moieties of biosurfactants consist of either carbohydrate or protein molecules and the hydrophobic entity consists of saturated and unsaturated fatty acids with varying chain lengths [37].

2.5 Biosurfactant

Biosurfactants are a broad class of biologically derived amphipathic compounds derived from a wide variety of plants and microorganisms. Saponins, lecithin, and soy proteins are plant biosurfactants derived from different plant parts such as roots, stems, bark, leaves, and seeds. These are reported to have good emulsification properties but have issues of low productivity and solubility [38]. Microbial biosurfactants are superior in comparison to biosurfactants obtained from plants, on account of rapid production, scale-up capacity, and multifunctional properties [39].

Biosurfactants exhibit wide structural diversity and superior properties such as low toxicity, biodegradability, better dispersion and emulsification ability, higher efficiency and selectivity, lower CMCs, stability at broad temperatures, pH and salinity ranges, and production feasibility from cheap or low-cost feedstocks [40, 41]. These significant properties and advantages have projected biosurfactants as a potential substitute for chemical surfactants in various industrial sectors [42, 43].

2.6 Properties of biosurfactants

2.6.1 Non-toxic

Biosurfactants due to their microbial origin are considered to be nontoxic. The nontoxic property makes them suitable for pharmaceutical, food, and cosmetic formulations and also for environmental applications. Patowary *et al* [44] reported that rhamnolipid biosurfactant derived from *Pseudomonas aeruginosa* showed no cytotoxic effect on the L292 cell line (mouse fibroblast), implying its applicability for different biological applications. Araujo *et al* [1] studied the application of anionic biosurfactant produced from *Serratia marcescens*, which did not show any repressive effects on seed germination and root elongation of Brassica *oleracea*, rather it resulted in enhancing the germination process and growth of seedlings by producing secondary roots and leaves on the seeds. Similar reports of non-toxicity of lipopeptide biosurfactant on the seeds of lettuce and cabbage produced from *Streptomyces* sp. were reported by Santos *et al* [45]. Acute toxicity tests carried out using surfactin biosurfactant produced from *Bacillus subtilis* with *Daphnia magna*, *Vibrio fischeri* and *Selenastrum capricornutum* (green alga) and the outcome showed that the biosurfactant was less toxic than its equivalents which are derived chemically such as alkyl polyglucosides and sodium dodecyl sulfate (SDS) [46].

2.6.2 Bio-degradable

The toxicity of any biomolecule is not only defined by its non-hazardous properties but also by its biodegradability and insistence in terrestrial and aquatic environments [47]. Unlike chemical surfactants, biosurfactants are biodegradable due to their simpler natural composition (carbohydrates, proteins, fatty acids, and lipids) and are particularly appropriate for various environmental applications such as enhanced oil recovery (EOR), dispersion and cleaning up of oil spillages and bioremediation of contaminated sites. Olivera *et al* [46] tested the biodegradability of crude surfactin extract from *B. subtilis* and found it to be degraded by both *Pseudomonas putida* and a consortium of microbes from a sewage treatment plant. Degradation analysis of rhamnolipid biosurfactants carried out by Hogan *et al* [48], a maximum degradation of 92% was achieved.

2.6.3 Temperature, pH, and salinity stability

An extensive array of thermal, pH, and salinity stability of any bioproduct is an important criterion to ensure its wide industrial applicability. In most biosurfactants, the surface-active properties are not affected by changes in physical conditions such as temperature, pH, and salinity. The wide uses of biosurfactants in several sectors such as EOR, bioremediation, and food industries where extreme temperatures, pH, and salinity are encountered, indicates the broad-spectrum of stability of these biomolecules under extensive variations of temperatures, pH, and saline conditions [49, 50]. Astuti *et al* [51] reported that biosurfactants feasible for applications involving extreme conditions such as EOR must be stable at temperature ranges of 50 °C–80 °C, a broad range of pH, and higher salinity concentrations. Culture filtrate containing lipopeptide biosurfactant produced from *Streptomyces* sp. exhibited broad stability with respect to temperature (4 °C–80 °C), NaCl (salinity) concentrations of 1%–12% (*w/v*), and pH range of 4–10 (152). Astuti *et al* [51] studied the crude biosurfactant produced from *Pseudoxanthomonas* sp. G3 was found to be stable even at higher temperatures up to 100 °C and could maintain a stable emulsion up to 50% in a broad pH range of 2.0–12.0 and NaCl concentrations up to 10% (*w/v*). Khademolhosseini *et al* [52] reported the significant application of biosurfactant in microbial enhanced oil recovery (MOER) using rhamnolipid produced from *P. aeruginosa*, which showed excellent stability at temperatures from 40 °C–121 °C, pH of 3.0–10 and salt concentration up to 10% (*w/v*). Zhang *et al* [53] reported stable lipopeptide biosurfactants from *Bacillus atrophaeus* showing a broad array of temperature (20 °C–121 °C), pH [6–13], and salinity (up to15% NaCl) stability.

2.6.4 Surface and interface activity

Biosurfactants due to their amphiphilic structures show improved surface-active properties in comparison to chemical surfactants. Biosurfactant production is reported to be an adaptation of the producing organism in which the biomolecule results in the accumulation at interfaces between fluids with varying polarities present in the surrounding environments leading to the lowering in the interfacial and surface tensions. The reduction of the interfacial (liquid–liquid) and surface (air–liquid) tension results in the reduction of the repulsive forces concerning two different phases thereby allowing these two different segments to interact and mix proficiently [54, 55]. The effectiveness and efficiency are essential characteristics of a good surface-active agent. The effectiveness is measured by the lowering of interfacial and surface tensions, whereas efficiency is related to critical micelle concentrations [56]. Biosurfactants with efficient lowering of interfacial and surface tension are essential for various applications since surface tension plays a vital role in different industrial processes, such as wetting, dye fixing, emulsification, softening, lubrication, and foaming [57]. The lowering of surface tension is generally used as a key measure for screening and selecting potent biosurfactant-producing microorganisms in which higher surface activity of the biosurfactant is indicated by lower surface tension values [1]. The ability of the biosurfactants to lower the surface tension is attributable to the specific concentration of the biosurfactants. Various biosurfactants produced from a diverse range of microbes are reported to reduce surface tension of distilled water from 72 to 35–28 mN m^{-1} and the interfacial tension of hexadecane from 40 to 1.0 mN m^{-1} [56, 58, 59]. The lipopeptide biosurfactant, surfactin (cyclic lipopeptide) produced by different species of *Bacillus* genera is primarily known as one of the most efficient biosurfactants and can reduce surface tension of water from 72 to 28–27 mN m^{-1} at a concentration even as low as 50 mg l^{-1} [60].

2.6.5 Emulsification

Emulsification constitutes an important property of biosurfactants, in which the emulsion formed consists of a heterogeneous system of a non-miscible liquid dispersed in another liquid in the form of droplets. The emulsifying property and its stability are often used as a screening method for isolating potent biosurfactant producers and as an indicator of a good surfactant [61]. Biosurfactants, due to their surface-active properties aid in the breakdown of a large hydrophobic/oil particle into smaller particles followed by the formation of a stable suspension. The stability of the emulsion is affected by pH, type biosurfactants concentration, temperature, and ionic strengths [56]. There are two kinds of emulsions: oil-in-water (O/W) in which the oil is solubilized into smaller micelles within an aqueous (water) continuous phase, whereas water-in-oil (W/O) emulsions are characterized by the formation of reverse micelles with water surrounded by the continuous oil phase [62]. Rhamnolipid type of biosurfactant shows excellent emulsifying properties under high temperature and pH conditions and thus finds applications in the agriculture and food industries, biomedicine and bioremediation for producing stable emulsions and also in oil reservoirs for the process of EOR [63, 64]. Li *et al* [61] reported rhamnolipid biosurfactant produced by *P. aeruginosa* which showed excellent emulsifying properties with hydrocarbons liquid paraffin (E_{24} 58%), kerosene (E_{24} 62%) and n-hexane (E_{24} 57%) in comparison to synthetic surfactants sodium dodecyl sulfate (SDS) and tetradecyl trimethyl ammonium bromide (TTAB) with emulsion stability (E_{24}) lower than 10%. Zhang *et al* [53] reported stable emulsion using lipopeptide biosurfactant produced from *B. atrophaeus* in which the emulsification index (E_{24}%) remained unchanged on exposing the culture filtrate containing biosurfactant to extreme temperatures (100 °C–121 °C), pH (6.0–13) and salinity (1.0%–15% NaCl) conditions. Janek *et al* [168] reported a pseudofactin, a cyclic lipopeptide produced from *Pseudomonas fluorescens* BD5 with better emulsification of 20%–50% than synthetic surfactants—Triton X-100 and Tween 20.

2.6.6 Lower critical micelle concentrations (CMC)

The critical micelle concentration (CMC) is the minimum concentration of a surface-active agent at which self-assembled organized structures called micelles are formed and also correlates to the concentration that is required to achieve a maximum surface tension reduction of water. The CMC is also widely used as one of the variables to determine the efficiency or quality of a surface-active agent [65, 66].

Biosurfactants have lower critical micelle concentration in comparison to chemically produced surfactants, which is among the factors that distinguish biosurfactants from them in terms of value [13, 58]. In biosurfactant-producing organisms, the micelle formation increases the bioavailability of insoluble hydrophobic organic compounds in the surrounding environments allowing it to thrive in hydrophobic environments [55]. The CMC various biosurfactants may vary depending on the composition, especially the chain lengths of the hydrophobic moiety. The CMCs of rhamnolipid (glycolipid) and surfactin (lipopeptide), mannosyl-erythritol lipids (MELs) biosurfactant ranges between 10–400, 10–63, and 20–250 mg l^{-1}, respectively, in comparison to 1.8–2.9 g l^{-1} for anionic synthetic surfactant sodium dodecyl sulfate [52, 67–69].

2.6.7 Market potential of biosurfactants

The unique features of biodegradability, stability, low toxicity, and sustainability of production from cheap renewable feedstocks accompanied with superior surface-active properties such as foaming, emulsification, wetting, dispersion, stabilization, and detergency make biosurfactants a potential substitute suitable for various industrial applications making them a significant biomaterial in the current scenario [56, 70]. Among the various categories of currently known biosurfactants, the glycolipid biosurfactant 'rhamnolipids' has been the most studied because of their excellent and superior properties and standalone nature to witness the highest gain of over 8.0% in the current global biosurfactants market [71]. Thus, the advantages, wide range of applications, environmental sustainability, and promising market demands have led to increased interest and acceptance with studies centered around the isolation of novel rhamnolipid-producing strains, exploring the use of cheap/renewable by-products as substrates, and process development for cheaper and sustainable production of rhamnolipid biosurfactants [72, 73].

2.6.8 Classification of biosurfactants

The diverse range of structures produced by the microorganisms makes up the microbial biosurfactants. Based on the molecular weight, they can be broadly categorized into low- and high-molecular-weight biosurfactants. Biosurfactants with low molecular weight are efficient in lowering the interfacial and surface tension and comprise glycolipids, lipopeptides, and phospholipids with molecular weights ranging from 500–1500 Da. However, the molecular weight biosurfactant comprises polymeric biosurfactants, particulate biosurfactants, and emulsifiers which are more efficient in foaming and stabilizing emulsions (water-in-oil/oil-in-water) but do not necessarily lower surface or interfacial tensions [35, 74].

Depending on their chemical composition, these biomolecules are also classified into five major types: (1) glycolipids (2) lipopeptides or lipoproteins (3) phospholipids, natural lipids or fatty acids (4) polymeric, and (5) particulate biosurfactants [56]. The hydrophilic moiety of the biosurfactant comprises either carbohydrate

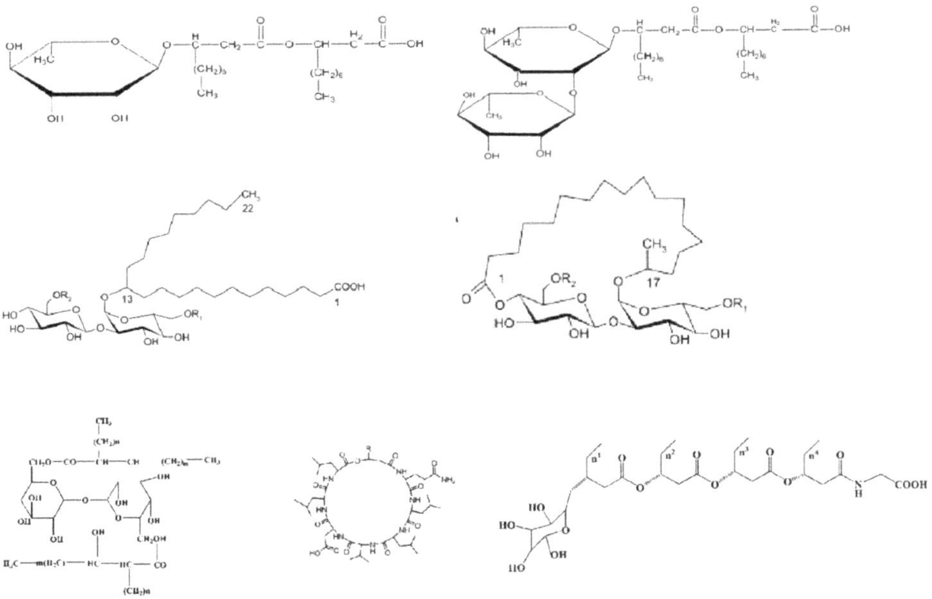

Figure 2.1. Structures of biosurfactant produced by different organisms. (a) Monorhamnolipid (Rha-C_{10}-C_{10}). (b) Dirhamnolipid (Rha-Rha-C_{10}-C_{10}). (c) Sophorolipid in acid form. (d) Sophorolipid in lactone form. (e) Trehalose lipid. (f) Cyclic lipopeptide (surfactin). (g) Polymeric biosurfactants (Emulsan).

(mono-/di-/polysaccharides) or protein (amino acids/peptides) while the hydrophobic region is comprised of saturated or unsaturated fatty acids (figure 2.1).

2.6.8.1 Glycolipids

Glycolipids possess significant commercial importance and are further classified as (a) rhamnolipids (b) sophorolipids (c) trehalose lipids (d) mannosyl-erythritol lipids (MELs) based on the carbohydrate moiety in the hydrophilic head [75]. The hydrophilic moiety consists of a carbohydrate in which the composition may vary from mono-, di-, tri-, and tetra-saccharides as glucose, mannose, sophorose galactose, trehalose, glucuronic acid, rhamnose, and galactose-sulfate. The hydrophobic moiety consists of long-chain aliphatic fatty acids and/or hydroxyl-aliphatic fatty acids [76]. The hydrophobic and hydrophilic are linked together either by an ether or ester group [77]. The glycolipids are further classified as follows.

(a) Rhamnolipids

Rhamnolipids are glycolipid biosurfactants in which the hydrophilic moiety is made up of rhamnose sugar and the hydrophobic region is composed of either one or two saturated or unsaturated β-hydroxy-fatty acids. The hydroxyl group of β-hydroxy-fatty-acid facilitates the glycosidic bonds with the reducing ends of the rhamnose sugar, whereas the hydroxyl group of the second β-hydroxy-fatty acid is accountable for esterification [35]. The production of rhamnolipids is characteristic of the genus *Pseudomonas* and was first described in *P. aeruginosa* species [78].

However, in recent years many other bacterial species have also been reported [63]. The fatty acid chain length in the β-hydroxy-fatty acids ranges between 6 and 24 carbon atoms with 8 to 16 carbons the most reported [63, 79]. A wide structural diversity exists in rhamnolipid biosurfactants and the structural distinctions are observed due to variations in the sugar and the hydrophobic region. The most commonly reported rhamnolipids consists of one/two rhamnose residues and β-hydroxy-fatty acids, with varying number of carbon atoms between 6 and 24, however, in recent years more than 50–60 rhamnolipid congeners have been identified [80].

(b) Sophorolipids

Sophorolipids are glycolipid biosurfactants in which the hydrophilic region comprises the disaccharide sophorose sugar and the hydrophobic moiety of the hydroxylated fatty acid tail of 16 or 18 carbon atoms. The sophorose sugar is a disaccharide sugar with a β-1,2-glycosidic bond that undergoes further acetylation at the 6′ and/or 6″ positions after being integrated into the sophorolipid. Gorin *et al* [81] first described the sophorolipid biosurfactant as a combination of both open (acidic) and closed (lactonized) forms. The lipid portion of the hydroxylated fatty acid tail is connected to the reducing end of the sophorose through a glycosidic linkage and the carboxylic end is either a free acid (acidic form) or intra-moleculary esterified between the carboxylic end and the sophorose sugar at the 4′-position into the closed lactonized form. The lactonic sophorolipids have better surface tension reducing property, whereas the acidic form has better foaming and solubility properties [75]. Sophorolipid production is reported from a wide variety of non-pathogenic yeasts, however, *Candida* and *Torulopsis* are the most commonly reported species [75, 82].

(c) Mannosyl-erythritol lipids (MELs)

Mannosyl-erythritol lipids contain mannose attached to sugar alcohol (erythritol) as a hydrophilic moiety (permanent) attached to a variety of fatty acids and acetyl groups as a hydrophobic chain (variable). The different degrees of acetylation result in the diversity of mannosyl-erythritol lipids namely mannosyl-erythritol lipids A, B, C, and D [67, 83]. A vast variety of MELs has been reported from the genus *Pseudozyma* and several smut fungi of the *Ustilaginaceae* family [68].

(d) Trehalolipids

Trehalose lipids consist of disaccharide trehalose made up of two units of glucose linked at the sixth carbon to two ®-hydroxy-branched fatty acids [79]. The hydroxy acyl chains vary in number, length as well as in position of their glycosidic bond at the glucose sugar rings. Trehalolipids are broadly categorized into anionic trehalose tetra esters, succinoyl trehalose lipids, and nonionic trehalose mono-, di- and trimycolates [84]. The production of trehalose lipids is reported from several Gram-positive bacteria, predominantly from the oleaginous bacterium, *Rhodococcus* sp. The production, however, has also been reported from other genus of *Gordonia*, *Propionibacterium, Mycobacterium*, *Escherichia, Nocardia, Arthrobacter*, and *Corynebacterium* [85, 86].

2.7 Bioemulsifiers

2.7.1 Lipopeptides

Lipopeptide type of biosurfactants is made up of amino acid chains (peptides) which are linked to fatty acids with variable chain lengths. They are produced by a wide variety of organisms ranging from bacteria, yeast, and fungi. They are distinguished by high structural variations and can be broadly divided into cyclic and linear lipopeptides. The hydrophilic moiety of amino acid chain is either cyclic or linear and the length of the hydrophobic fatty acids vary from 13–16 carbon atoms and may be branched. Lipopeptide biosurfactants exhibit remarkable surface activity and are effective in lowering surface and interfacial tensions. Surfactin, fengycin, and iturin are the most commonly reported cyclic lipopeptide biosurfactants and are mainly produced by *B. subtilis* [87, 88]. The cyclic lipopeptides are generally made up of peptides consisting of 7–10 amino acids associated with a long fatty acid chain. One of the most efficient cyclic lipopeptides is surfactin which consists of cyclic heptapeptide with seven amino acids (Glu-Leu-DLeu- Val-Asp-D-Leu-Leu) attached to a fatty acid tail of 13–15 carbon molecules by a lactone linkage [89]. The other types of lipopeptide biosurfactants reported are viscosin, arthrofactin, tensin, pseudofactin, syringomycin, amphomycins, and friulimicin produced from *Pseudomonas* and *Actinoplanes friuliensis* (actinomycetes) [60, 90].

2.7.1.1 Surfactin

Surfactin, a *Bacillus* lipopeptide, is composed of a cyclic heptapeptide head group (Glu-Leu-D-Leu-Val-Asp-D-Leu-Leu) attached to a C13-C15 beta-hydroxy-fatty acid through a lactone linkage. Surfactin is tolerant to degradation by pepsin and trypsin due to its cyclic peptide structure [91]. Surfactin has recently gained interest for possible biological and pharmacological uses. It may attach to proteins (for example, hemoglobin) to generate a fractal structure that looks like a 'necklace model' of micelle-like assemblies inconsistently dispersed throughout the polypeptide chain [92]. Surfactin can also penetrate through cell membranes. The hydrophobic communications between the fatty acid chain of surfactin and the phospholipid chain of the cell membrane primarily regulate the penetration process [93]. With the engineering strategy for surfactin production, 53 genes in *B. subtilis* have been changed, and the production of surfactin reached roughly 42% of the potential yield. As one of the utmost surfactin yields ever documented, this could eventually lead to more surfactin being produced and used commercially as a biosurfactant for industrial applications. [93]

2.7.1.2 Iturin

Iturin is a cyclic lipopeptide (CLP) isolated from *B. subtilis* and other associated bacteria, which has a wide range of inhibitory effects against plant pathogenic microorganisms. *B. subtilis* is a spore-forming gram-positive bacteria. The ability of *B. subtilis* to synthesize basic biological surfactants, especially surfactant-containing CLPs, is of commercial importance [94]. Iturin (Type A, C, D, E), Mycobacterium subtilisin, and types of bacteriomycin (D, F, L) are the key members of the iturin family. These compounds are composed of β-amino acids (b-AA), fatty acids, and acylated peptides with a chain length from C14 to C17. These molecules have a cyclic peptide structure and they differ in content and position of the amino acid and fatty acid chain length [95]. Extraction and purification of iturins is because of their presence in a variety of organic compounds (polysaccharides, peptides, lipids, sterols, humic substances, and polymers) formed in the rhizosphere [96]. The hydrophilic and hydrophobic properties of its surfactants can lead to new and difficult separation steps during the extraction of iturin A [97]. Iturin has a wide range of biological functions and is mainly used for biological control in agriculture. As a result, more research is being done on Iturins to replace pesticide production. On the other hand, the biological effects of iturin produced from subtilis sp. are determined by the amino acid content and peptide ring sequence as well as the type of lipid moiety. Furthermore, the antibacterial activity of iturin often increases with the amount of carbon atoms in the fatty acid side chain [97].

2.7.1.3 Fengycin

Fengycins are a form of lipopeptide generated by some bacteria strains, particularly *Bacillus* species. These lipopeptides are well-known for their antibacterial characteristics, and their prospective uses as biopesticides and biocontrol agents have sparked interest in the fields of biotechnology and agriculture. Fengycins are amphiphilic because they are made up of a cyclic peptide ring connected to a lipid tail. Because of their amphiphilicity, they may interact with and break the cell membranes of target microbes, allowing them to be potent antimicrobial agents. The principal bioactive compounds of Serenade (Bayer), a commercial antifungal and antibacterial agent are fengycins, surfactins, and iturins. It is used to cure plant diseases such as clubroot disease (*Plasmodiophora moniliforme*), barley head blight (*Fusarium graminearum*), powdery cucurbit (*Podosphaera fusca*), black Sigatoka (*Mycosphaerella fijiensis*), gray mold (*Botrytis cinerea*), and soft rot (*Rhizopus stolonifer*) [98]. Fengycin appears to be exclusive to fungal cells which imposes a slight impact on vertebrates. This contributes to its 'green' label because it has no negative impact on the agricultural workers and the animals in the surrounding ecosystem. Also, its medical use has been emphasized due to its potential as an antifungal skin ointment. [99]. Fengycins have been studied for their capacity to suppress the growth of plant diseases such as fungi and bacteria. They can also be used to manage infections in agriculture and protect crops from pathogens. Because they are biodegradable and have a lesser environmental effect, these lipopeptides are seen as a possible alternative to chemical pesticides.

2.7.1.4 Cyclic lipopeptides

Cyclic lipopeptides are a type of natural chemical generated by microorganisms such as bacteria and fungi. They have a cyclic peptide ring structure that is connected to a lipid (fatty acid) tail. Cyclic lipopeptides have unusual features due to the combination of the cyclic peptide and the lipophilic tail, including amphiphilicity, which allows them to interact with and disrupt biological membranes. These substances have a diverse set of biological actions and uses such as antibacterial activity, biocontrol agents, immunomodulatory and anti-inflammatory properties [100]. Cyclic lipopeptides are a broad class of molecules with several bioactive characteristics. Researchers continue to investigate their prospective applications and are looking at ways to manufacture and adapt these molecules for specialized purposes in agriculture, pharmacology, and biotechnology.

2.7.2 Phospholipids

Phospholipid and fatty acids biosurfactants can be produced by several bacteria and yeast when grown on hydrocarbon such as alkanes as a result of microbial oxidation. A wide variety of phospholipids constitute a major component of microbial cell membranes. In certain hydrocarbon-degrading bacteria or yeast, the phospholipid composition in the cell membranes increases greatly when grown on hydrocarbon (alkane) substrates [101]. In *Acinetobacter* sp. vesicles of phosphatidylethanolamine phospholipid are formed, which exhibit enhanced emulsification properties producing microemulsions of alkanes in water which are found to be optically clear. The microemulsion produced using phospholipid biosurfactants can be useful in pharmaceutical and therapeutic applications [13, 56]. Fatty acids and phospholipids with different compositions have been reported from the genus *Arthrobacter, Corynebacterium, Talaramyces, and Acinetobacter* [56, 102, 103].

2.7.2.1 Phosphatidylglycerol (PG)

Membranes are essential biological components that segregate metabolic activities from their surroundings. This distinct feature facilitates the evolution of distinct adaptations and mechanisms that allow cell and organelle membranes the flexibility to respond while maintaining the structural and functional integrity of the cell. Phosphatidylglycerol is made up of a glycerol backbone and two fatty acid chains. It also has a phosphate group attached to the glycerol, which is connected to a hydrophilic head group. Glycerol is the head group in the case of PG. The length and saturation of the fatty acid chains can change, impacting the fluidity and characteristics of the membrane. One such adaptation is the ability to maintain an equilibrium between zwitterionic and anionic phospholipids [104, 105]. The proportion of various phospholipid types is determined by the control of their production pathways [106]. In general, the CDP-DAG system synthesizes anionic phospholipids e.g. phosphatidylserine (PS), the phosphatidylinositol (PI) family, and the phosphatidylglycerol (PG) family. They are the anionic phospholipids which are produced by the CDP-DAG driven pathway, whereas zwitterionic phospholipids are produced by the Kennedy pathway [107, 108]. This family's key members are PG and cardiolipin (CL). Phosphatidylglycerol phosphate (PGP), phosphatidylglycerol phosphate methyl ester (PGPM), and phosphatidylglycerol sulfate (PGS) are domain-specific anionic phospholipids for archaea, and carbohydrate- and amino acid derivatives of PG are carbohydrate—and amino acid derivatives of PG in bacteria [109].

2.7.2.2 Phosphatidylethanolamine (PE)

Phosphatidylethanolamine plays important roles, including acting as a structural component of cell membranes, aiding in membrane curvature, and participating in the processes of membrane fusion and fission. It also participates in cell signaling and acts as a precursor for the production of other lipid compounds. Two types of phospholipids abundant in eukaryotic cells are the zwitterionic glycerophospholipids phosphatidylcholine (PC) and phosphatidylethanolamine (PE). Polyethylene exists in abundance and is a component of membrane phospholipids of most bacteria, followed by phosphatidylglycerol and cardiolipin, and is produced only by decarboxylation of phosphatidylserine (PS) [110]. Changes in the composition of phosphatidylethanolamine in cell membranes are associated with a variety of diseases and conditions, including neurological diseases and cancer.

2.7.2.3 Cardiolipin

Cardiolipin is a type of phospholipid found in the inward membrane of the mitochondria that helps to organize and operate the mitochondria, [107] enabling appropriate ATP generation [111]. Cardiolipin is a dimeric anionic molecule having three glycerol groups, two phosphate groups, and four esterified fatty acyl chains all connected to a compacted polar head group. Cardiolipin contributes to 10%–15% of the phospholipids present in the membrane of the mitochondrion (1%–5% of the outer membrane, 10%–15% of the inner membrane) and stimulates the production of mitochondrial cristae. Cardiolipin is also important in mitochondrial bioenergetics because it promotes the construction and operation of respiratory chain supercomplexes [112, 113]. It is usually confined to the mitochondrial internal membrane and is quickly externalized to the mitochondrial surface when the mitochondrial membrane potential collapses [114, 115]. Cardiolipin production imposes a complicated coordinated sequence of numerous processes that are still unknown. The majority of what we know about cardiolipin synthesis derives from investigations on yeast and culture cells; consequently, factors controlling the cardiolipin production process *in vivo* would be more beneficial to investigate further. The synthesis of cardiolipins begins with the formation of phosphatidic acid (PA) in mitochondrial membranes. Glycerol-3-phosphate acyltransferase (GPAT1–4) catalyzes the assembly of acyl-CoA and glycerol-3-phosphate molecules to form lysophosphatidic acid (LPA), which is then catalyzed by acyltransferase (LPAAT1–4) to form PA. Further, PA is transported into the intermembrane space of mitochondria via unknown mechanisms [116, 117]. Cardiolipin can also be found in signaling pathways within cells. Cardiolipin, for example, can be exposed on the mitochondrial outer membrane under cellular stress or apoptosis (programmed cell death), resulting in the activation of apoptosis-related proteins.

Particulate biosurfactants consist of either the whole cell or extracellular membrane vesicles. The extracellular membrane vesicles partition hydrocarbons to form stable microemulsions which aid the producing organism in the hydrocarbon uptake [56, 118, 119]. *Acinetobacter* sp., when grown on a hydrophobic substrate (hexadecane), accumulates extracellular vesicles of 20–50 mm diameter exhibiting surface-active properties and composed of phospholipids, protein, and lipopolysaccharides [120]. A similar type of extracellular membrane vesicles with surface-active properties have been reported in *Pseudomonas marginalis* [121].

2.8 Polymeric biosurfactants

Polymeric type of biosurfactants is high molecular weight complex biosurfactants made up of polysaccharide-protein complexes. These complex biomolecules are generally made up of three to four repeating units of sugars linked to fatty-acid chains of varying lengths by O-ester linkages and are produced by bacterial and yeast species from different genera such as *Acinetobacter calcoaceticus*, *Sphingomonas paucimobilis*, *Bacillus stearothermophilus*, *Candida utilis* and *Candida lipolytica* [122–124]. The most reported polymeric biosurfactants are liposan, emulsan, lipomanan, and alasan [125]. Araujo *et al* [1] reported a polymeric type of biosurfactant produced from *S. marcescens* with a complex polymeric composition of 43% lipids, 32% proteins, and 11% carbohydrates.

2.9 Production and extraction of microbial surfactants

The production of biosurfactants from microbes represents a unique metabolic and physiological response of adaptation under adverse environmental conditions in diverse habitats. In hydrophobic environments, biosurfactants enable microorganisms to grow by increasing the solubility of hydrophobic substrates through the emulsification process, increasing their surface area and making them easily available for uptake and metabolism. Besides their role in substrate utilization, they are also reported to be involved in other physiological roles such as cell signaling, biofilm formation, cell adherence, mobility, antibiotic resistance, swarming motility, and pathogenicity to help the microbes to survive under hostile environmental conditions [126–129]. Biosurfactants help maintain the stability of biofilms and are a survival adaptation that shields entrapped microorganisms by increasing access to limited nutrients and by inhibiting the penetration of antimicrobials/toxins from the external environment. Similarly, swarming motility which requires the presence of biosurfactants for the coordinated translocation of bacterial populations across solid or semi-solid surfaces helps in finding new habitats for survival under nutrient-limited, hostile conditions and in surface colonization [77, 130, 131]. Biosurfactants restrict the invasion of competing bacterial population by exhibiting antagonistic effect resulting in increasing the chances of survival of the organism producing it [132].

2.10 Microorganisms used for surfactant production

Biosurfactant-producing microorganisms are ubiquitous with widespread distribution across the archeal and eubacterial domains [133, 134]. These are found in both terrestrial and aquatic environments and also thrive under extreme environments with varying temperatures, pH, and salinity conditions such as oil reservoirs, hot springs, desert soil, low/high-temperature shale reservoirs, and salt marshes [135]. Isolation of biosurfactant-producing microorganisms from environments/sites contaminated with hydrocarbons, heavy metals, and organic hydrophobic compounds

Table 2.1. Distribution of biosurfactant-producing bacteria.

S. No.	Microorganisms	Sources of isolation	Phylum	References
1	*Marinobacter* sp. *Pseudomonas mendocina*	Marine-sea	*Proteobacteria*	[139]
2	*Pseudomonas aeruginosa*	Coking wastewater	*Proteobacteria*	[140]
3	*Bacillus altitudinis*	Plant root	*Firmicutes*	[141]
4	*Pseudomonas sp.*	Oil-contaminated soils	*Proteobacteria*	[61]
5	*Pediococcus acidilactici* *Lactobacillus plantarum*	Milk cheese	*Firmicutes*	[142]
6	*Staphylococcus saprophyticus*	Petroleum contaminated sediment	*Firmicutes*	[143]
7	*Serratia marcescens*	Hydrocarbon-contaminated site	*Proteobacteria*	[144]
8	*Pseudomonas aeruginosa*	Organic farm soil	*Proteobacteria*	[145]
9	*Lysinibacillus sphaericus*	Pesticides contaminated soil	*Firmicutes*	[146]
10	*Burkholderia* sp.	PAH contaminated soil	*Proteobacteria*	[147]
11	*Bacillus algicola* *Isoptericola chiayiensis*	Beach sediment soil	*Firmicutes* *Actinobacteria*	[148]
12	*Bacillus subtilis*	Marine water	*Firmicutes*	[149]
13	*Bacillus amyloliquefaciens* *Bacillus subtilis*	Oil contaminated sites	*Firmicutes*	[150]
14	*Brevibacillus* sp.	Oil contaminated soil	*Firmicutes*	[151]
15	*Marinobacter hydrocarbonoclasticus*	Hydrocarbon contaminated sediment	*Proteobacteria*	[152]
16	*Pseudomonas aeruginosa*	Cow dung	*Proteobacteria*	[153]
17	*Lysinibacillus.* sp.	Rhizosphere soil	*Firmicutes*	[154]
18	*Acinetobacter junii*	Water/oil sample from petroleum reservoir	*Proteobacteria*	[155]
19	*Pseudomonas aeruginosa*	Petroleum-contaminated soil	*Proteobacteria*	[156]
20	*Pseudomonas aeruginosa*	Oil contaminated mangrove	*Proteobacteria*	[157]
21	*Pseudomonas aeruginosa*	Hydrocarbon contaminated soil	*Proteobacteria*	[158]
22	*Pseudomonas aeruginosa*	Garage-site with petroleum contamination	*Proteobacteria*	[159]
23	*Bacillus amyloliquefaciens*	Detergent industry-soil sample	*Firmicutes*	[160]
24	*Bacillus thuringiensis*	Petroleum contaminated soil	*Firmicutes*	[161]
25	*Serratia rubideae*	Hydrocarbon contaminated soil	*Proteobacteria*	[162]
26	*Bacillus licheniformis*	Marine samples	*Firmicutes*	[163]

27	*Marinobacter hydrocarbonoclasticus*	Hydrocarbon contaminated marine environment	*Proteobacteria*	[164]
28	*Psychrobacter namhaensis* *Brevibacterium epidermidis*	Sediments samples of oil refinery & petrochemical plants	*Proteobacteria*	[165]
29	*Exiguobacterium* sp. *Halomonas* sp. *Rhodococcus* sp	Petroleum hydrocarbon contaminated marine environments in Atlantic, Canada	*Actinobacteria* *Proteobacteria* *Actinobacteria*	[166]
30	*Pseudomonas xanthomarina* *Pseudomonas stutzeri*	Oil-polluted sea water	*Proteobacteria*	[167]
31	*Thalassospira* sp. *Pseudovibrio* sp. *Idiomarina* sp. *Cohaesibacter* sp.	Crude-oil-contaminated sites	*Proteobacteria*	[168]
32	*Pseudomonas sp.*	Dye industry	*Proteobacteria*	[169]
33	*Bacillus methylotrophicus*	Petroleum reservoir samples	*Firmicutes*	[170]
34	*Cronobacter sakazakii*	Oil contaminated waste waters	*Proteobacteria*	[171]
35	*Pseudomonas* sp. *Rhodococcus* sp.	Crude-oil contaminated site	*Proteobacteria*	[172]
36	*Staphylococcus* sp.	Crude-oil contaminated soil	*Firmicutes*	[173]
37	*Bacillus amyloliquefaciens*	Oily sludge petroleum-contaminated soil	*Firmicutes*	[174]
38	*Ochrobactrum* sp. *Brevibacterium* sp.	Crude-oil contaminated soil	*Proteobacteria* *Actinobacteria*	[175]
39	*Stenotrophomonas acidaminiphila* *Bacillus megaterium* *Bacillus cereus*	Petrochemical residues	*Proteobacteria* *Firmicutes*	[176]
40	*Escherichia fergusonii*	Oil contaminated soil	*Eubacteria*	[177]
41	*Paenibacillus alvei*	Oil well sample	*Firmicutes*	[178]
42	*Pseudomonas aeruginosa*	Petroleum-contaminated soil samples	*Proteobacteria*	[179]
43	*Bacillus mycoides*	Oil field	*Firmicutes*	[180]
44	*Rhodococcus fascians*	Antarctic soil	*Actinobacteria*	[181]

is the most prominent process [136]. Bodour and Maier [137] reported that the distribution of biosurfactant producers was higher in soils contaminated with hydrocarbon in comparison to uncontaminated soils. The isolation of Gram-negative biosurfactant-producing bacteria from hydrocarbon-contaminated or co-contaminated soils was reported by Bodour *et al* [138]. However, isolates of bacteria that produced biosurfactants and were positive for heavy metals were found in both contaminated and uncontaminated soils. Table 2.1 summarizes the biosurfactant-producing bacterial isolates belonging to different genera from different sources.

2.11 Fermentation conditions for surfactant production

The major attributes of fermentation for biosurfactant production are microorganism selection, nutrient medium, pH, substrate, oxygenation, aeration time, fermentation time, and monitoring and control. The microbe utilized, the kind of surfactant being generated, and the intended product properties can all affect the fermentation conditions during the surfactant synthesis process [182, 183]. Surfactants are substances that reduce the surface tension between two substances, such as water and oil.

Yeasts, fungi, and bacteria (such as *Pseudomonas*, *Bacillus*, and *Lactobacillus* species) are common sources of surfactants [184, 185]. An important step in the process is choosing the appropriate microorganism for the synthesis of biosurfactants. The production of rhamnolipids, a class of glycolipid biosurfactants, is a well-known characteristic of *P. aeruginosa* [186]. Different kinds of biosurfactants, such as glycolipids and lipopeptides, can be produced by *B. subtilis* [187]. Acinetobacter can produce lipopeptides and glycolipids. Candida yeasts and rhodotorula species can produce sophorolipids, and glycolipids, especially rhamnolipids, respectively. *Aspergillus niger*, a type of filamentous fungi can produce glycolipids, lipopeptides, and other types of biosurfactants [188], whereas some Fusarium strains can produce biosurfactants, including lipopeptides. Also, biosurfactants, such as polysaccharide-based surfactants, can be produced by cyanobacteria and microalgae.

The ability and inclination of different bacteria to produce biosurfactants varies and depends on multiple factors such as biosurfactant types, environmental conditions, growth and production rate, availability, genetic modifications, and economic feasibility.

2.12 Downstream processing for surfactant extraction

Downstream processing of biosurfactant includes recovery and purification of biosurfactant through various processes such as extraction, precipitation, and chromatography, to obtain the final product. The particular procedures and methods depend on the kind of surfactant and the features of the fermentation process [189]. To minimize the negative effects on the environment and maximize yield, purity, and cost-effectiveness, downstream processing techniques should be well-designed. Regulatory considerations and adherence to pertinent laws and safety requirements are crucial in the production of biosurfactants for commercial use. Downstream processing is a cascade of the following events to attain maximum efficiency and yield [189–191]:

2.12.1 Centrifugation or filtration

Separation of biomass (cells) out of the fermentation broth is the first step of downstream processing and can be accomplished by filtration or centrifugation. While the liquid fraction contains the surfactants, the solid biomass is usually discarded, and the liquid fraction is utilized for further processing.

2.12.2 Extraction

The aqueous phase of the fermentation broth frequently contains surfactants. To extract surfactants from the aqueous phase, a variety of extraction techniques are available. Adsorption onto solid supports, liquid–liquid extraction, and solvent extraction are common methods. The suitable extraction method depends on the properties of the surfactant and the availability of the resources.

2.12.3 Precipitation

Precipitation techniques can be used to concentrate and separate the surfactants from the solvent or aqueous phase, depending on the type of surfactant and its concentration. Acids, salts, and temperature changes are common precipitating agents.

2.12.4 Solvent recovery

Removal of the organic solvents utilized for extraction can be removed by distillation or evaporation.

2.12.5 Filtration and purification

Further filtration and purification procedures might be necessary to get rid of contaminants, leftover biomass, and other impurities from the surfactant solution after extraction and precipitation. To increase the surfactant content, the surfactant solution can be further concentrated using evaporation or other concentration techniques.

2.12.6 Chromatography

Techniques like column chromatography can be used to separate specific surfactants with higher purity. These techniques can distinguish between various impurities or surfactant components according to their chemical characteristics.

2.12.7 Ultrafiltration or dialysis

Dialysis or ultrafiltration can be used to remove salts and low molecular weight contaminants. The purity of surfactant solution can be increased and refined with the aid of these procedures.

2.12.8 Drying

Leftover solvent or water can be removed by drying techniques which include spray drying or freeze-drying depending on the type of surfactant.

2.12.9 Characterization and quality control

It is important to characterize the purified surfactant product to make sure it satisfies the necessary quality requirements, which include composition, purity, and functional qualities.

2.12.10 Packaging and storage

When the finished surfactant product is used in different applications, it is packaged and kept in the right environment to preserve its stability and quality.

2.13 Applications of microbial surfactants

The key characteristics of microbial surfactants include biodegradability, self-assembly, structural diversity, low toxicity, and efficiency in lowering surface tension and increasing hydrophobic chemical solubility. There follow some key points to consider regarding biosurfactant applications.

2.14 Enhanced oil recovery

Amphiphilic compounds with hydrophilic and hydrophobic groups are biosurfactants [19]. The hydrophilic and hydrophobic groups of the biosurfactant are dissolved in the aqueous and oil phases of the oil layer, respectively. This unique property lowers oil–water repulsion, changes water resistance, and emulsifies crude oil. Biosurfactants are superior to chemical surfactants in terms of biodegradability, low toxicity, and high stability [49]. For example, lipopeptides and glycolipids are extensively used to enhance oil recycling [154]. Microbial enhanced oil recovery (MEOR) uses *ex situ/in situ* microbes and their primary and secondary metabolites (e.g. biosurfactants, biopolymers, bio enzymes, biogas, solvents, biogenic acids, etc) to be a tertiary oil recovery technology. To improve oil recovery, the properties of the remaining oil in the tank are analyzed [159]. *In situ* production of biosurfactants usually begins with the injection of biosurfactant-producing bacteria and then the administration of nutrients into the reservoir. *Ex situ* biosurfactants can also be created in industrial bioreactors for injection into CO_2 reservoirs. Microorganisms create emulsifiers and surfactants which facilitate the reduction in surface tension which further leads to the removal of trapped oil. Biosurfactants change the water-holding capacity of injected CO_2 and the behavior of CO_2–salt water–rock at the interface and increase oil recovery by increasing the filtering effect of permeated fluid and CO_2 [192]. *P. aeruginosa* rhamnolipid extract helped to recover the average oil up to 50.45% and the presence of bacteria improved by 11.91%. This study proposes a higher oil recovery rate by using microbial surfactants as compared to synthetic surfactants [89].

2.15 Bioremediation

Biosurfactants are stable in the presence of extremes of temperature, salinity, and environmental friction. Removal of petroleum components and heavy metals contaminated sites using biosurfactant-based remediation methodologies and biosurfactant-producing microorganisms has been well documented [51]. It also helps break down hydrophobic pollutants present in the water bodies. Organic and metal pollutants are inaccessible to bacteria, which increases their persistence in soil for long periods. Furthermore, insufficient interactions between contaminants and bacteria and the environment hinder the bioremediation process. Consequently, biosurfactants produced by bacteria and fungi play an essential function in the solubilization of hydrophobic pollutants, enabling direct elimination [160]. *Stenotrophomonas* sp. S1VKR-26 biosurfactants can be used to bioremediate petroleum-contaminated wastewater [94]. *Pseudomonas* species are major producers of biosurfactants in hydrocarbon-polluted environments [65]. According to a recent study, marine bacteria such as *B. subtilis* AS2, *Bacillus licheniformis* AS3 and *Bacillus valezensis* AS4 are reported in this habitat. These strains are capable of producing biosurfactants with high emulsification rates for low molecular weight hydrocarbon and crude oil degradation efficiency of 88%, 92%, and 97%, respectively. *Bacillus cereus* UCP 1615, a lipopeptide-type biosurfactant, has the ability to scavenge oil spills [104]. A biosurfactant derived from *Rhodococcus erythropolis* HX-2 improves the solubility of hydrophilic molecules while also accelerating petroleum biodegradation [193].

2.15.1 Biostimulation for petroleum bioremediation

To increase the activity of the microorganisms that naturally degrade environmental toxins, biostimulation is a technique used in bioremediation. When it comes to petroleum bioremediation, biostimulation is the process of promoting the development and metabolic activity of natural microorganisms that break down hydrocarbons present in petroleum-based goods. Petroleum pollutants in soil or water may naturally degrade more quickly as a result of this process. The degree of interaction between inorganic and natural toxins is represented by the complex physical-mixture interactions at interfaces. A liquid known as the non-aqueous phase liquid (NAPL) is created when toxins attach to the soil's surface and take up residence there. Natural compounds that are insoluble in water and cause long-term pollution are referred to as NAPL. The quantity of polluted particles that can biodegrade in the aqueous phase is decreased when dissipated concentrations are consistently replaced with them [194]. Bioremediation is the process of accelerating the breakdown of biological materials to decrease the harmful effects of toxins. To facilitate pollutant degradation, the bioaugmentation technique introduces cultivated bacteria. This is further supported by the use of surfactants, which facilitate pollutant desorption and accelerate the breakdown of hydrocarbons [195]. Surfactants are utilized for recovering secondary oil through oil washing and oil pipe cleaning. Different types of synthetic and biosurfactants are used for desorption. In bioremediation, biosurfactants are not commonly used because of their high production costs, one potential solution to this problem is to culture the microorganisms that produce these microbial surfactants using less expensive substrates, such as agro-industrial waste and industrial oil waste. The quantity and quality of carbon and nitrogen in the substrate and soil have a major impact on the sort of biosurfactant that is produced. The production of additional biosurfactants is aided in bioremediation as a result of the reduced availability of nitrogen. Nitrogen is commonly left out of the supply of bioremediation systems in an effort to boost the microbial activity that leads to a greater synthesis of biosurfactants [35, 196]. Much research has been conducted throughout the years to assess the effectiveness of microbial strains that produce biosurfactants for the biodegradation of hydrocarbon molecules [197]. Chang *et al* examined how *P. aeruginosa* ATCC 9027, a strain that produces biosurfactants, affected the breakdown of soil-treated with phenanthrene [193]. Chebbi *et al* used a strain of *Pseudomonas* sp. that has the capacity to create biosurfactant to study the degradation of polyaromatic hydrocarbons found in motor oil-contaminated soil [198]. It is essential to remain cognizant that the effectiveness of biostimulation is dependent on site-specific factors. To create an efficient bioremediation strategy, one must carefully consider the type and extent of contamination, the surrounding environment, and the existing microbial community. Furthermore, in some circumstances, using biostimulation techniques may require regulatory clearances before implementation.

2.15.2 Bioaugmentation for petroleum bioremediation

Bioaugmentation is defined as a process of administration of microorganisms into the soil to promote bioproduction [199]. Studies have shown that the introduction of biosurfactant-producing bacteria to the contaminated environment leads to augmentation of hydrocarbon solubilization resulting in biodegradation efficiencies ranging from 32.67% to 87.54%. It has been shown that bioaugmentation causes competition between indigenous and imported microbes [200]. The findings highlighted that the amount of diesel removed from the soil was equivalent for bioaugmentation treatment (57.92%) and control (about 58%). *Pseudomonas* sp. E39, E311, and E313 were isolated from hydrocarbon-polluted soil and tested for biosurfactant potential and diesel biodegradability [201]. This study showed a 70% mitigation in total petroleum hydrocarbons in bioreactors infected with *R. erythropolis* T902.1. TPH eradication rate (about 55%) was obtained for the isolated strains E311, E39 and E313. The E311 strain produced the least amount of lipopeptide but had the best biodegradation efficiency, suggesting that a low dose of biosurfactant product may be adequate for the bioaugmentation process.

2.16 Medical application

Lipopeptides are recognized as biosurfactants with notable pH stability across a broad range, and subjecting them to elevated temperatures does not diminish their surface-active properties. According to recent studies, they also have antimicrobial properties. *Pseudomonas* rhamnolipids, for example, have been shown to exhibit significant antibacterial properties. While mono-rhamnolipids are bacteriostatic, di-rhamnolipids p. is bactericidal against aeruginosa [202]. Some biosurfactants interact synergistically with antibiotics, increasing their absorption efficiency into the cell [203]. Gene-releasing biosurfactants, drugs, antiviral abilities, and antioncogenic applications are some of the significant areas of study using these biomolecules in the field of biomedicine [93]. Recent studies have focused on the use of biosurfactants as Covid-19 management strategies and as disinfectants, eco-friendly sanitizers, cleansers, antiviral agents, and anti-inflammatory preparations. According to new research [204], sophorolipids can be used as therapeutic agents against SARS-CoV-2 virus. It is a positive sense pleomorphic or circular single-stranded RNA envelope virus. The genome encodes four fundamental structural proteins, namely the nucleocapsid protein (NC), membrane protein (M), encapsulation protein (E), and spike protein (S). Additionally, it includes five to eight non-structural accessory proteins. The spike glycoprotein binds to the viral particle via the ACE2 receptor present in the host's respiratory epithelial cells [205]. The anionic nature of sophorolipids destroys the viral coat, dissolves structural components, and interferes with interactions between viral proteins and host receptor proteins. Furthermore, they are suitable for future advances such as nanobiotechnology and efficient pharmaceutical delivery systems. They have attracted the attention of scientific groups owing to their potent medicinal benefits. These properties render them valuable in treating SARS-CoV infections, as well as for their applications in antiviral, immunomodulatory, anti-cancer, wound healing, and other medical contexts [206].

2.17 Cosmetics and personal care

Biosurfactants are necessary physiochemical components for a healthy skin surface. For example, the fatty acid ends of their molecules help to moisturize the rough and arid regions of the skin. The easily accessible fatty acids may also function as antioxidants, reducing free radical production produced by UV exposure [207]. *Staphylococcus aureus*, *P. aeruginosa*, *Candida acnes*, and *Streptococcus pyogenes* have all been found to be efficiently reduced by the application of a range of biosurfactants. As a result, although their bactericidal action is often restricted, biosurfactants are being recommended as a viable replacement for standard antibiotics [208]. Biosurfactants are gentle on the skin and do not irritate it. The antimicrobial actions of biosurfactants have proven to be of great imminence to the cosmetics industry [209]. Evonik, a German chemical business, is developing technology to synthesize rhamnolipid as a foaming agent in cosmetic goods, assuring biosurfactant use as one of the safe and active components in cosmetic product formulation [210]. Glycolipid and lipopeptide have been known to reduce toxicity, antibacterial activity, and cutaneous hydrating properties, making them significantly superior to chemical surfactants in contemporary cosmetic and personal skincare product needs [211]. In Japan, Kao Co. Ltd is a well-known manufacturer of sphorolipid substances for use as a hydrating ingredient in a range of commercial products such as hair conditioners, skin moisturizers, and lip balms. Sophorolipids have been shown to increase hair development and protect the skin [212]. By stimulating adipocyte production of the hormone leptin, these biosurfactants may also assist in reducing fat deposits on the skin. Several investigations have found that rhamnolipids are biocompatible and a potential ingredient for incorporation into medical cosmetic formulations as well as topical dermatological treatment [211]. In a recent study on biosurfactants in the personal care industry,

Resende *et al* explored the use of biosurfactants and chitosan in toothpaste preparations. As biosurfactants, *P. aeruginosa*, *Bacillus methylotrophicus*, and *C. bombicola* were used, while chitosan was extracted from the biomass of the fungus Mucorales [213]. The pH, foamability, cytotoxicity, antibacterial activity, and inhibitory capability against *Streptococcus mutans* biofilm were all tested in the toothpaste. All formulations decreased *S. mutans* cellular viability in the biofilm, and the results were comparable to commercial toothpaste, indicating that they had market potential. A comparable study combining biosurfactant and chitosan compositions in mouthwash was conducted in the personal care business. The researchers tested the toxicity and antibacterial activity of various formulations including the three surfactants, chitosan, and peppermint essential oil used in the prior study. All of the mouthwash samples had a minimum inhibitory concentration that was optimal for cariogenic bacteria. All formulations assessed in the study demonstrated substantially lower toxicity compared to commercial mouthwash, and their safe and efficacious use as a potential natural substitute for commercial mouthwash [214].

2.18 Agriculture

Biosurfactant's versatile properties have found application in agriculture, particularly in replacing synthetic surfactants in pesticide and agrochemical formulations, promoting the growth of 'environmental chemistry' in the agricultural sector, responding to the requirement to mitigate or eliminate adverse effects on the environment and human health associated with the over-use of chemical compounds [215, 216]. As per the literature, rhamnolipid and lipopeptide biosurfactants contribute to enhancing soil quality, a critical factor in crop production. These natural compounds serve as bioremediation agents for hydrophobic organic compounds, including polycyclic aromatic hydrocarbons, in various environmental factors such as soils, surface water, groundwater, and waste streams [217, 218]. Because of their antimicrobial activity, biosurfactants generated from the lipid classes mannosyl-erythritol, rhamnolipids, and lipopeptides can be used as biopesticides to manage a wide range of pests, illnesses, phytopathogenic fungi, and weeds [219, 220]. *P. putida* and *P. fluorescens* lipopeptides cause *Phytphthora capsici* zoospore lysis, resulting in cucumber 'damping off' [221]. as well as inhibit the growth of phytopathogens such as *Pythium ultimum*, *Fusarium oxysporum*, and *Phytophthora cryptogea*. Lipopeptides and glycolipids, in addition to avoiding microbial infections and pest infestations by suppressing the activity of aphids, mosquitos, and toxic toxins generated by the fungus *Aspergillus parasiticum* in peanut, cotton, and maize crops [222], have also shown encouraging results in this sector.

2.19 Food industry

The utilization of biosurfactants in food has grown in prominence in recent years, owing to customer demand for sustainably produced ingredients as well as vegetarian and vegan food items. As some of these natural compounds have low toxicity and have no negative effects on human health, some of these natural compounds can be employed to enhance formulations by modifying viscosity, altering textural aspects, and inhibiting the growth of certain pathogenic microorganisms. This, in turn, contributes to extending the shelf life, improving quality, and ensuring the safety of food products [223]. Hence, food products incorporating biosurfactants derived from microorganisms with a Generally Regarded as Safe (GRAS) status may exhibit enhanced resistance to oxidation attributed to the antioxidant properties of certain biosurfactants. Additionally, these biosurfactants can contribute to stability in the face of acidity, alkalinity, and temperature variations [224]. The yeasts *Starmerella bombicola*, *Candida sphaerica*, *C. lipolytica*, *C. utilis*, *Saccharomyces cerevisiae*, and *Meyerozyma guilliermondii*, which have the potential to produce chemicals with emulsification and surfactant activities, as well as antimicrobial and antioxidant properties, are among the microorganisms recently reported for biosurfactant production [225]. Customer desire for more natural meals with less artificial and chemically generated components is met by the low toxicity of nanoscale surfactants. Thus, biosurfactants can be used to replace chemicals already found in food, which can be hazardous to long-term health if consumed in large quantities [226]. Biosurfactants have demonstrated significant antibacterial activity against common bacterial illnesses such as *B. subtilis* and *E. coli.* Sophorolipids might be employed as food emulsifiers and antibacterial agents. Despite the numerous food application possibilities, various studies are needed to generate a practical application with suitable function performance in food complex matrices under varying processing conditions. To build an economically viable application, it is necessary to devise methodologies that utilize these biomolecules at the lowest possible concentration for optimal performance [227, 228].

2.20 Regulatory concerns

The diverse applications of synthetic surfactants in various fields—paints, pesticides, dye, textile, paper, cosmetics, pharmaceuticals, plastics, food, detergents, and personal care products—have driven the global surfactant market to grow from 29.93 billion US Dollars in 2014 to 43.65 billion US Dollars in 2017. The surfactant market is further projected to keep growing and is estimated to reach 64.40 billion US dollars by 2025, growing at a CAGR of 5.4% from 2017 to 2025. As per a market survey in 2017, the anionic and nonionic surfactant segments occupied up to two-thirds of the global surfactants market volume [71, 229].

2.21 Conclusion

Biosurfactants, with their structural diversity and eco-friendly properties, have become pivotal players in numerous industries. Despite having many distinct advantages and commercially attractive properties in comparison to synthetic surfactants, biosurfactants have not yet been used extensively on industrial scales due to several factors. They are not yet competitive with their synthetic counterparts from the economic perspective as the high production and downstream processing limit their successful commercialization. To overcome these bottlenecks and rationalize the replacement of chemical surfactants by biosurfactants it is necessary to find more economical and feasible production processes. The successful commercialization of biosurfactants thus depends on the utilization of specifically abundant and cheaper raw materials like industrial or agricultural wastes, optimization of the operational cultivation conditions, novel efficient downstream processing methods, and the improvement of existing strains along with the isolation of novel efficient strains. Hence, through the adoption of feasible and efficient production technology, the biosurfactant industry is poised to offer innovative solutions to address challenges within the competitive surfactant market. Additionally, it can play a pivotal role in addressing the intertangled problem of environmental pollution and greener production strategies.

References

[1] Araújo H W C, Andrade R F S, Montero-Rodríguez D, Rubio-Ribeaux D, Alves da Silva C A and Campos-Takaki G M 2019 Sustainable biosurfactant produced by Serratia marcescens UCP 1549 and its suitability for agricultural and marine bioremediation applications *Microb. Cell Fact* **18** 2

[2] Karthick A, Roy B and Chattopadhyay P 2019 A review on the application of chemical surfactant and surfactant foam for remediation of petroleum oil contaminated soil *J. Environ. Manage.* **243** 187–205

[3] Le Guenic S, Chaveriat L, Lequart V, Joly N and Martin P 2018 Renewable surfactants for biochemical applications and nanotechnology *J. Surfactants Deterg.* **22** 5–21

[4] NBC Universal Media LLC Media 2017 A history of soap and detergents https://nbclearn.com/portal/site/HigherEd/flatview?cuecard=54527

[5] Joshi T P 2017 A short history and preamble of surfactants *Int. J. Appl. Chem.* **13** 283–92

[6] Rahman P K S M and Randhawa K K S 2015 Editorial: microbiotechnology based surfactants and their applications *Front. Microbiol.* **6** 1344

[7] Myers D 2005 *Surfactant Science and Technology* (New York: Wiley)

[8] de Freitas F A, Keils D, Lachter E R, Maia C E B, Pais da Silva M I and Veiga Nascimento R S 2019 Synthesis and evaluation of the potential of nonionic surfactants/mesoporous silica systems as nanocarriers for surfactant controlled release in enhanced oil recovery *Fuel* **241** 1184–94

[9] Pakiet M, Kowalczyk I, Leiva Garcia R, Moorcroft R, Nichol T, Smith T *et al* 2019 Gemini surfactant as multifunctional corrosion and biocorrosion inhibitors for mild steel *Bioelectrochemistry* **128** 252–62

[10] Summerton E, Zimbitas G, Britton M and Bakalis S 2017 Low temperature stability of surfactant systems *Trends Food Sci. Technol.* **60** 23–30

[11] Tadros T F 2014 *An Introduction to Surfactants* (De Gruyter)

[12] Cowan-Ellsberry C, Belanger S, Dorn P, Dyer S, McAvoy D, Sanderson H *et al* 2014 Environmental safety of the use of major surfactant classes in North America *Crit. Rev. Environ. Sci. Technol.* **44** 1893–993

[13] Vijayakumar S and Saravanan V B-T 2015 Sources and applications *Res. J. Microbiol.* **10** 181–92

[14] Gracida J, Ortega-Ortega J, Torres B L G, Romero-Avila M and Abreu A 2017 Synthesis of anionic surfactant and their application in washing of oil-contaminated soil *J. Surfactants Deterg.* **20** 493–502

[15] Martins G and Dias M F R G 2023 Hair cosmeceuticals *Alopecia* (Digital Science) pp 285–93
[16] Kinoshita T, Ishigaki Y and Shibata N 2019 Selective recovery of indium via continuous counter-current foam separation from sulfuric acid solutions I—application of anionic organophosphate surfactant as metal collector *Sep. Purif. Technol.* **212** 563–71
[17] Kang E, Jung G Y, Jung S H and Lee B M 2018 Synthesis and surface active properties of novel anionic surfactants with two short fluoroalkyl groups *J. Ind. Eng. Chem.* **61** 216–26
[18] Zhang P, Liu Y, Li Z, Kan A T and Tomson M B 2018 Sorption and desorption characteristics of anionic surfactants to soil sediments *Chemosphere* **211** 1183–92
[19] Akhter K, Ullah K, Talat R, Haider A, Khalid N, Ullah F *et al* 2019 Synthesis and characterization of cationic surfactants and their interactions with drug and metal complexes *Heliyon* **5** e1885–e5
[20] Garcia M T, Ribosa I, Kowalczyk I, Pakiet M and Brycki B 2019 Biodegradability and aquatic toxicity of new cleavable betainate cationic oligomeric surfactants *J. Hazard. Mater.* **371** 108–14
[21] Huang X-F, Wang Y-H, Shen Y, Peng K-M, Lu L-J and Liu J 2019 Using non-ionic surfactant as an accelerator to increase extracellular lipid production by oleaginous yeast *Cryptococcus curvatus* MUCL 29819 *Bioresour. Technol.* **274** 272–80
[22] Shi Q, Cheng J, Liu Y, Liu C and Qiao W 2018 Effects of non-ionic surfactants on the material exchange between crude oil and scCO2 *J. Mol. Liq.* **269** 23–8
[23] Clendennen S K and Boaz N W 2019 Betaine amphoteric surfactants—synthesis, properties, and applications *Biobased Surfactants* (Amsterdam: Elsevier) pp 447–69
[24] Li Y, Holmberg K and Bordes R 2013 Micellization of true amphoteric surfactants *J. Colloid Interface Sci.* **411** 47–52
[25] Rodríguez-López L, Rincón-Fontán M, Vecino X, Cruz J M and Moldes A B 2019 Preservative and irritant capacity of biosurfactants from different sources: a comparative study *J. Pharm. Sci.* **108** 2296–304
[26] Knoth D, Rincón-Fontán M, Stahr P-L, Pelikh O, Eckert R-W, Dietrich H *et al* 2019 Evaluation of a biosurfactant extract obtained from corn for dermal application *Int. J. Pharm.* **564** 225–36
[27] Mohamed R M, Al-Gheethi A A, Noramira J, Chan C M, Hashim M K A and Sabariah M 2018 Effect of detergents from laundry greywater on soil properties: a preliminary study *Appl. Water Sci.* **8** 16
[28] Ivanković T and Hrenović J 2010 Surfactants in the environment *Arch. Ind. Hyg. Toxicol.* **61** 95–110
[29] Tornero V and Hanke G 2016 Chemical contaminants entering the marine environment from sea-based sources: a review with a focus on European seas *Mar. Pollut. Bull.* **112** 17–38
[30] Gautam G J, Chaube R and Joy K 2015 Toxicity and tissue accumulation of 4-nonylphenol in the catfish *Heteropneustes fossilis* with a note on prevalence of 4-NP in water samples *Endocr. Disruptors* **3** e981442
[31] Rebello S, Asok A K, Mundayoor S and Jisha M S 2014 Surfactants: toxicity, remediation and green surfactants *Environ. Chem. Lett.* **12** 275–87
[32] Jackson M, Eadsforth C, Schowanek D, Delfosse T, Riddle A and Budgen N 2016 Comprehensive review of several surfactants in marine environments: fate and ecotoxicity *Environ. Toxicol. Chem.* **35** 1077–86

[33] Ogundiran M A, Fawole O O, Adewoye S O and Ayandiran T A 2009 Pathologic lesions in the gills of clarias gariepinus exposed to sublethal concentrations of soap and detergent effluents *J. Cell Anim. Biol.* **3** 78–82
[34] Akbari S, Abdurahman N H, Yunus R M, Fayaz F and Alara O R 2018 Biosurfactants—a new frontier for social and environmental safety: a mini review *Biotechnol. Res. Innov* **2** 81–90
[35] Varjani S J and Upasani V N 2016 Carbon spectrum utilization by an indigenous strain of *Pseudomonas aeruginosa* NCIM 5514: production, characterization and surface active properties of biosurfactant *Bioresour. Technol.* **221** 510–6
[36] Sharma D, Saharan B S, Chauhan N, Bansal A and Procha S 2014 Production and structural characterization of *Lactobacillus helveticus* derived biosurfactant *Sci. World J.* **2014** 493548
[37] Nakama Y 2017 Surfactants *Cosmetic Science and Technology* (Amsterdam: Elsevier) pp 231–44
[38] Xu Q, Nakajima M, Liu Z and Shiina T 2011 Biosurfactants for microbubble preparation and application *Int. J. Mol. Sci.* **12** 462–75
[39] Sekhon Randhawa K K and Rahman P K S M 2014 Rhamnolipid biosurfactants-past, present, and future scenario of global market *Front. Microbiol.* **5** 454
[40] Marcelino P R F, Peres G F D, Terán-Hilares R, Pagnocca F C, Rosa C A, Lacerda T M *et al* 2019 Biosurfactants production by yeasts using sugarcane bagasse hemicellulosic hydrolysate as new sustainable alternative for lignocellulosic biorefineries *Ind. Crops Prod.* **129** 212–23
[41] Udoh T and Vinogradov J 2019 Experimental investigations of behaviour of biosurfactants in brine solutions relevant to hydrocarbon reservoirs *Colloids Interfaces* **3** 24
[42] Mouafi F E, Abo Elsoud M M and Moharam M E 2016 Optimization of biosurfactant production by *Bacillus brevis* using response surface methodology *Biotechnol. Rep.* **9** 31–7
[43] Shekhar S, Sundaramanickam A and Balasubramanian T 2014 biosurfactant producing microbes and their potential applications: a review *Crit. Rev. Environ. Sci. Technol.* **45** 1522–54
[44] Patowary K, Patowary R, Kalita M C and Deka S 2017 Characterization of biosurfactant produced during degradation of hydrocarbons using crude oil as sole source of carbon *Front. Microbiol.* **8** 279
[45] Santos A P P, Silva M D S, Costa E V L, Rufino R D, Santos V A, Ramos C S *et al* 2017 Production and characterization of a biosurfactant produced by streptomyces sp. DPUA 1559 isolated from lichens of the Amazon region *Braz. J. Med. Biol. Res* **51** e6657–7
[46] De Oliveira D W F, Cara A B, Lechuga-Villena M, García-Román M, Melo V M M, Gonçalves L R B *et al* 2016 Aquatic toxicity and biodegradability of a surfactant produced by *Bacillus subtilis* ICA56 *J. Environ. Sci. Health Part* A **52** 174–81
[47] Raymond J W, Rogers T N, Shonnard D R and Kline A A 2001 A review of structure-based biodegradation estimation methods *J. Hazard. Mater.* **84** 189–215
[48] Hogan D E, Tian F, Malm S W, Olivares C, Palos Pacheco R, Simonich M T *et al* 2019 Biodegradability and toxicity of monorhamnolipid biosurfactant diastereomers *J. Hazard. Mater.* **364** 600–7
[49] Al-Wahaibi Y, Joshi S, Al-Bahry S, Elshafie A, Al-Bemani A and Shibulal B 2014 Biosurfactant production by *Bacillus subtilis* B30 and its application in enhancing oil recovery *Colloids Surf. B Biointerfaces* **114** 324–33

[50] de Cássia F S, Silva R, Almeida D G, Rufino R D, Luna J M, Santos V A and Sarubbo L A 2014 Applications of biosurfactants in the petroleum industry and the remediation of oil spills *Int. J. Mol. Sci.* **15** 12523–42

[51] Astuti D I, Purwasena I A, Putri R E, Amaniyah M and Sugai Y 2019 Screening and characterization of biosurfactant produced by pseudoxanthomonas sp. G3 and its applicability for enhanced oil recovery *J. Pet. Explor. Prod. Technol.* **9** 2279–89

[52] Khademolhosseini R, Jafari A, Mousavi S M, Hajfarajollah H, Noghabi K A and Manteghian M 2019 Physicochemical characterization and optimization of glycolipid biosurfactant production by a native strain of *Pseudomonas aeruginosa* HAK01 and its performance evaluation for the MEOR process *RSC Adv.* **9** 7932–47

[53] Zhang J, Xue Q, Gao H, Lai H and Wang P 2016 Production of lipopeptide biosurfactants by bacillus atrophaeus 5-2a and their potential use in microbial enhanced oil recovery *Microb. Cell Fact* **15** 168

[54] Peele K A, Ch V R T and Kodali V P 2016 Emulsifying activity of a biosurfactant produced by a marine bacterium *3 Biotech.* **6** 177

[55] Pacwa-Płociniczak M, Płaza G A, Piotrowska-Seget Z and Cameotra S S 2011 Environmental applications of biosurfactants: recent advances *Int. J. Mol. Sci.* **12** 633–54

[56] Santos D K F, Rufino R D, Luna J M, Santos V A and Sarubbo L A 2016 Biosurfactants: multifunctional biomolecules of the 21st century *Int. J. Mol. Sci.* **17** 401

[57] Zdziennicka A, Szymczyk K, Krawczyk J and Jańczuk B 2017 Components and parameters of solid/surfactant layer surface tension *Colloids Surf. A Physicochem. Eng. Asp* **522** 461–9

[58] Ghasemi A, Moosavi-Nasab M, Setoodeh P, Mesbahi G and Yousefi G 2019 Biosurfactant production by lactic acid bacterium pediococcus dextrinicus shu1593 grown on different carbon sources: strain screening followed by product characterization *Sci. Rep.* **9** 5287

[59] Wu L, Lai L, Lu Q, Mei P, Wang Y, Cheng L *et al* 2019 Comparative studies on the surface/interface properties and aggregation behavior of mono-rhamnolipid and di-rhamnolipid *Colloids Surf. B Biointerfaces* **181** 593–601

[60] Hu F, Liu Y and Li S 2019 Rational strain improvement for surfactin production: enhancing the yield and generating novel structures *Microb. Cell Fact* **18** 42

[61] Li L, Shen X, Zhao C, Liu Q, Liu X and Wu Y 2019 Biodegradation of dibenzothiophene by efficient pseudomonas sp. LKY-5 with the production of a biosurfactant *Ecotoxicol. Environ. Saf.* **176** 50–7

[62] Goodarzi F and Zendehboudi S 2019 Effects of salt and surfactant on interfacial characteristics of water/oil systems: molecular dynamic simulations and dissipative particle dynamics *Ind.; Eng. Chem. Res.* **58** 8817–34

[63] Tiso T, Zauter R, Tulke H, Leuchtle B, Li W-J, Behrens B *et al* 2017 Designer rhamnolipids by reduction of congener diversity: production and characterization *Microb. Cell Fact* **16** 225

[64] Haba E, Bouhdid S, Torrego-Solana N, Marqués A M, Espuny M J, García-Celma M J *et al* 2014 Rhamnolipids as emulsifying agents for essential oil formulations: antimicrobial effect against *Candida albicans* and methicillin-resistant *Staphylococcus aureus Int. J. Pharm.* **476** 134–41

[65] Câmara J M D A, Sousa M A S B, Barros Neto E L and Oliveira M C A 2019 Application of rhamnolipid biosurfactant produced by *Pseudomonas aeruginosa* in microbial-enhanced oil recovery (MEOR) *J. Pet. Explor. Prod. Technol.* **9** 2333–41

[66] Ma K-Y, Sun M-Y, Dong W, He C-Q, Chen F-L and Ma Y-L 2016 Effects of nutrition optimization strategy on rhamnolipid production in a *Pseudomonas aeruginosa* strain DN1 for bioremediation of crude oil *Biocatal. Agric. Biotechnol.* **6** 144–51
[67] Morita T, Fukuoka T, Imura T and Kitamoto D 2015 Mannosylerythritol lipids: production and applications *J. Oleo Sci.* **64** 133–41
[68] Konishi M and Makino M 2018 Selective production of deacetylated mannosylerythritol lipid, MEL-D, by acetyltransferase disruption mutant of *Pseudozyma hubeiensis J. Biosci. Bioeng.* **125** 105–10
[69] Kłosowska-Chomiczewska I E, Mędrzycka K, Hallmann E, Karpenko E, Pokynbroda T, Macierzanka A *et al* 2017 Rhamnolipid CMC prediction *J. Colloid Interface Sci.* **488** 10–9
[70] Chrzanowski Ł, Ławniczak Ł and Czaczyk K 2012 Why do microorganisms produce rhamnolipids? *World J. Microbiol. Biotechnol.* **28** 401–19
[71] Grand Market Insights Research 2018 *Biosurfactants Market Worth Over $2.7 bn by 2024* (Global Market Insights, Inc.) https://gminsights.com/pressrelease/biosurfactants-market-size
[72] Tan Y N and Li Q 2018 Microbial production of rhamnolipids using sugars as carbon sources *Microb. Cell Fact* **17** 89
[73] Zhao F, Shi R, Ma F, Han S and Zhang Y 2018 Oxygen effects on rhamnolipids production by *Pseudomonas aeruginosa Microb. Cell Fact* **17** 39
[74] Tripathy D B and Mishra A 2016 Sustainable biosurfactants *Sustain. Inorg. Chem.* **1** 175–92
[75] Jezierska S, Claus S and Van Bogaert I 2017 Yeast glycolipid biosurfactants *FEBS Lett.* **592** 1312–29
[76] Mnif I and Ghribi D 2016 Glycolipid biosurfactants: main properties and potential applications in agriculture and food industry *J. Sci. Food Agric.* **96** 4310–20
[77] Karlapudi A P, Venkateswarulu T C, Tammineedi J, Kanumuri L, Ravuru B K, Dirisala V R *et al* 2018 Role of biosurfactants in bioremediation of oil pollution-a review *Petroleum* **4** 241–9
[78] Jarvis F G and Johnson M J 1949 A glycolipid produced by pseudomonas aeruginosa *J. Am. Chem. Soc.* **71** 4124–6
[79] Varjani S J and Upasani V N 2017 Critical review on biosurfactant analysis, purification and characterization using rhamnolipid as a model biosurfactant *Bioresour. Technol.* **232** 389–97
[80] Irorere V U, Tripathi L, Marchant R, McClean S and Banat I M 2017 Microbial rhamnolipid production: a critical re-evaluation of published data and suggested future publication criteria *Appl. Microbiol. Biotechnol.* **101** 3941–51
[81] Gorin P A J, Spencer J F T and Tulloch A P 1961 Hydroxy fatty acid glycosides of sophorose from torulopsis magnoliae *Can. J. Chem.* **39** 846–55
[82] Oliveira M, Magri A, Baldo C, Camilios-Neto D, Minucelli T, Antonia M *et al* 2015 Review: sophorolipids a promising biosurfactant and it's applications *Int. J. Adv. Biotechnol. Res.* **6** 161–74
[83] Nigam K, Gupta S and Gupta N 2016 Biosurfactants: current perspectives in environmental remediation *J. Appl. Life Sci. Int.* **7** 1–19
[84] Kügler J H, Muhle-Goll C, Kühl B, Kraft A, Heinzler R, Kirschhöfer F *et al* 2014 Trehalose lipid biosurfactants produced by the actinomycetes *Tsukamurella spumae* and *T. pseudospumae Appl. Microbiol. Biotechnol.* **98** 8905–15

[85] Franzetti A, Gandolfi I, Bestetti G, Smyth T J P and Banat I M 2010 Production and applications of trehalose lipid biosurfactants *Eur. J. Lipid Sci. Technol.* **112** 617–27

[86] Ruhal R, Kataria R and Choudhury B 2013 Trends in bacterial trehalose metabolism and significant nodes of metabolic pathway in the direction of trehalose accumulation *Microb. Biotechnol.* **6** 493–502

[87] Zhao H, Shao D, Jiang C, Shi J, Li Q, Huang Q *et al* 2017 Biological activity of lipopeptides from Bacillus *Appl. Microbiol. Biotechnol.* **101** 5951–60

[88] Romano A, Vitullo D, Senatore M, Lima G and Lanzotti V 2013 Antifungal cyclic lipopeptides from *Bacillus amyloliquefaciens* strain BO5A *J. Nat. Prod.* **76** 2019–25

[89] Arima K, Kakinuma A and Tamura G 1968 Surfactin, a crystalline peptidelipid surfactant produced by *Bacillus subtilis*: isolation, characterization and its inhibition of fibrin clot formation *Biochem. Biophys. Res. Commun.* **31** 488–94

[90] Mnif I and Ghribi D 2015 Review lipopeptides biosurfactants: mean classes and new insights for industrial, biomedical, and environmental applications *Pept. Sci.* **104** 129–47

[91] Zhang L, Gao Z, Zhao X and Qi G 2016 A natural lipopeptide of surfactin for oral delivery of insulin *Drug Deliv.* **23** 2084–93

[92] Zou A, Liu J, Garamus V M, Zheng K, Willumeit R and Mu B 2010 Interaction between the natural lipopeptide [Glu1, Asp 5] surfactin-C15 and hemoglobin in aqueous solution *Biomacromolecules* **11** 593–9

[93] Drakontis C E and Amin S 2020 Biosurfactants: formulations, properties, and applications *Curr. Opin. Colloid Interface Sci.* **48** 77–90

[94] Bouassida M, Fourati N, Ghazala I, Ellouze-Chaabouni S and Ghribi D 2018 Potential application of *Bacillus subtilis* SPB1 biosurfactants in laundry detergent formulations: compatibility study with detergent ingredients and washing performance *Eng. Life Sci.* **18** 70–7

[95] Ongena M and Jacques P 2008 Bacillus lipopeptides: versatile weapons for plant disease biocontrol *Trends Microbiol.* **16** 115–25

[96] Kinsella K, Schulthess C P, Morris T F and Stuart J D 2009 Rapid quantification of *Bacillus subtilis* antibiotics in the rhizosphere *Soil. Biol. Biochem.* **41** 374–9

[97] Yaraguppi D A, Bagewadi Z K, Patil N R and Mantri N 2023 Iturin: a promising cyclic lipopeptide with diverse applications *Biomolecules* **13** 1515

[98] Zhou X, Lu Z, Lv F, Zhao H, Wang Y and Bie X 2011 Antagonistic action of *Bacillus subtilis* strain fmbj on the postharvest pathogen *Rhizopus stolonifer J. Food Sci.* **76** 254–9

[99] Vanittanakom N, Loeffler W, Koch U and Jung G 1986 Fengycin-A novel antifungal lipopeptide antibiotic produced by *Bacillus subtilis* F-29-3 *J. Antibiot. (Tokyo)* **39** 888–901

[100] Sarkar T, Chetia M and Chatterjee S 2021 Antimicrobial peptides and proteins: from Nature's reservoir to the laboratory and beyond *Front. Chem.* **9** 691532

[101] Rahman P K S M and Gakpe E 2008 Production, characterisation and applications of biosurfactants-review *Biotechnology (Faisalabad)* **7** 360–70

[102] Liu J-F, Mbadinga S M, Yang S-Z, Gu J-D and Mu B-Z 2015 Chemical structure, property and potential applications of biosurfactants produced by *Bacillus subtilis* in petroleum recovery and spill mitigation *Int. J. Mol. Sci.* **16** 4814–37

[103] Okoliegbe I N and Agarry O 2012 Application of microbial surfactant (a review) *Sch J. Biotechnol.* **1** 15–23

[104] Cerbon J and Calderon V 1994 Surface potential regulation of phospholipid composition and in-out translocation in yeast *Eur. J. Biochem.* **219** 195–200

[105] Karlsson O P and Löfås S 2002 Flow-mediated on-surface reconstitution of G-protein coupled receptors for applications in surface plasmon resonance biosensors *Anal. Biochem.* **300** 132–8
[106] Kumar Saha S, Nishijima S, Matsuzaki H, Shibuya I and Matsumoto K 1996 A regulatory mechanism for the balanced synthesis of membrane phospholipid species in *Escherichia coli Biosci. Biotechnol., Biochem.* **60** 111–6
[107] Chang S C, Heacock P N, Clancey C J and Dowhan W 1998 The PEL1 gene (renamed PGS1) encodes the phosphatidylglycerophosphate synthase of *Saccharomyces cerevisiae J. Biol. Chem.* **273** 9829–36
[108] Kennedy E P and Weiss S B 1956 The function of cytidine coenzymes in the biosynthesis of phospholipides *J. Biol. Chem.* **222** 193–214
[109] Kates M 1993 Biology of halophilic bacteria, Part II—membrane lipids of extreme halophiles: biosynthesis, function and evolutionary significance *Experientia* **49** 1027–36
[110] Gibellini F and Smith T K 2010 The Kennedy pathway-de novo synthesis of phosphatidylethanolamine and phosphatidylcholine *IUBMB Life* **62** 414–28
[111] Prola A and Pilot-Storck F 2022 Cardiolipin alterations during obesity: exploring therapeutic opportunities *Biology (Basel)* **11** 1–18
[112] Mejia E M and Hatch G M 2016 Mitochondrial phospholipids: role in mitochondrial function *J. Bioenerg. Biomembr.* **48** 99–112
[113] Ikon N and Ryan R O 2017 Cardiolipin and mitochondrial cristae organization *Biochim. Biophys. Acta—Biomembr* **1859** 1156–63
[114] Kagan V E, Jiang J, Huang Z, Tyurina Y Y, Desbourdes C, Cottet-Rousselle C *et al* 2016 NDPK-D (NM23-H4)-mediated externalization of cardiolipin enables elimination of depolarized mitochondria by mitophagy *Cell Death. Differ.* **23** 1140–51
[115] Schlattner U, Tokarska-Schlattner M, Epand R M, Boissan M, Lacombe M L and Kagan V E 2018 NME4/nucleoside diphosphate kinase D in cardiolipin signaling and mitophagy *Lab. Invest.* **98** 228–32
[116] Tamura Y, Harada Y, Nishikawa S I, Yamano K, Kamiya M, Shiota T *et al* 2013 Tam41 is a CDP-diacylglycerol synthase required for cardiolipin biosynthesis in mitochondria *Cell Metab.* **17** 709–18
[117] Horvath S E and Daum G 2013 Lipids of mitochondria *Prog. Lipid Res.* **52** 590–614
[118] Saravanan V and Vijayakuma S 2015 Biosurfactants-types, sources and applications *Res J. Microbiol.* **10** 181–92
[119] Gharaei-Fa E 2010 Biosurfactants in pharmaceutical industry (a mini-review) *Am. J. Drug Discov. Dev.* **1** 58–69
[120] Monteiro S A, Sassaki G L, de Souza L M, Meira J A, de Araújo J M, Mitchell D A *et al* 2007 Molecular and structural characterization of the biosurfactant produced by *Pseudomonas aeruginosa* DAUPE 614 *Chem. Phys. Lipids* **147** 1–13
[121] Gautam K K and Tyagi V K 2006 Microbial surfactants: a review *J. Oleo. Sci.* **55** 155–66
[122] Mahmood N N 2018 Effect of biosurfactants purified from saccharomyces cerevisiae against corynebacterium urealyticum *J. Pharm. Sci. Res.* **10** 481–6
[123] Rufino R D, Sarubbo L A and Campos-Takaki G M 2006 Enhancement of stability of biosurfactant produced by *Candida lipolytica* using industrial residue as substrate *World J. Microbiol. Biotechnol.* **23** 729–34
[124] Rosenberg E and Ron E Z 1999 High- and low-molecular-mass microbial surfactants *Appl. Microbiol. Biotechnol.* **52** 154–62

[125] Fouts D E, Matthias M A, Adhikarla H, Adler B, Amorim-Santos L, Berg D E *et al* 2016 What makes a bacterial species pathogenic?: Comparative genomic analysis of the genus leptospira *PLoS Negl. Trop. Dis.* **10** e0004403
[126] Wood T L, Gong T, Zhu L, Miller J, Miller D S, Yin B *et al* 2018 Rhamnolipids from *Pseudomonas aeruginosa* disperse the biofilms of sulfate-reducing bacteria *NPJ Biofilms Microbiomes* **4** 22
[127] Wang S, Yu S, Zhang Z, Wei Q, Yan L, Ai G *et al* 2014 Coordination of swarming motility, biosurfactant synthesis, and biofilm matrix exopolysaccharide production in *Pseudomonas aeruginosa Appl. Environ. Microbiol.* **80** 6724–32
[128] Banat I M, Satpute S K, Cameotra S S, Patil R and Nyayanit N V 2014 Cost effective technologies and renewable substrates for biosurfactants' production *Front. Microbiol.* **5** 697
[129] Walter V, Syldatk C and Hausmann R 2010 Screening concepts for the isolation of biosurfactant producing microorganisms *Advances in Experimental Medicine and Biology* (New York: Springer) pp 1–13
[130] Satpute S K, Banpurkar A G, Dhakephalkar P K, Banat I M and Chopade B A 2010 Methods for investigating biosurfactants and bioemulsifiers: a review *Crit. Rev. Biotechnol.* **30** 127–44
[131] Nickzad A, Lépine F and Déziel E 2015 Quorum sensing controls swarming motility of burkholderia glumae through regulation of rhamnolipids *PLoS One* **10** e0128509–e09
[132] Floris R, Rizzo C and Giudice A L 2018 *Biosurfactants from Marine Microorganisms* ed W N Hozzein (Rijeka: IntechOpen) ch 1
[133] Lenchi N, Inceoğlu O, Kebbouche-Gana S, Gana M L, Llirós M, Servais P *et al* 2013 Diversity of microbial communities in production and injection waters of algerian oilfields revealed by 16S rRNA gene amplicon 454 pyrosequencing *PLoS One* **8** e66588–8
[134] Fathepure B Z 2014 Recent studies in microbial degradation of petroleum hydrocarbons in hypersaline environments *Front. Microbiol.* **5** 173
[135] Tripathi L, Irorere V U, Marchant R and Banat I M 2018 Marine derived biosurfactants: a vast potential future resource *Biotechnol. Lett.* **40** 1441–57
[136] Ndlovu T, Khan S and Khan W 2016 Distribution and diversity of biosurfactant-producing bacteria in a wastewater treatment plant *Environ. Sci. Pollut. Res.* **23** 9993–10004
[137] Bodour A A and Miller-Maier R M 1998 Application of a modified drop-collapse technique for surfactant quantitation and screening of biosurfactant-producing microorganisms *J. Microbiol. Methods* **32** 273–80
[138] Bodour A A, Drees K P and Maier R M 2003 Distribution of biosurfactant-producing bacteria in undisturbed and contaminated arid Southwestern soils *Appl. Environ. Microbiol.* **69** 3280–7
[139] Twigg M, Tripathi L, Zompra K, Salek K, Irorere V, Gutierrez T *et al* 2019 Surfactants from the sea: rhamnolipid production by marine bacteria *Access Microbiol.* **1**
[140] Sun S, Wang Y, Zang T, Wei J, Wu H, Wei C *et al* 2019 A biosurfactant-producing *Pseudomonas aeruginosa* S5 isolated from coking wastewater and its application for bioremediation of polycyclic aromatic hydrocarbons *Bioresour. Technol.* **281** 421–8
[141] Goswami M and Deka S 2019 Biosurfactant production by a rhizosphere bacteria Bacillus altitudinis MS16 and its promising emulsification and antifungal activity *Colloids Surf. B Biointerfaces* **178** 285–96

[142] Yan X, Gu S, Cui X, Shi Y, Wen S, Chen H *et al* 2019 Antimicrobial, anti-adhesive and anti-biofilm potential of biosurfactants isolated from *Pediococcus acidilactici* and *Lactobacillus plantarum* against *Staphylococcus aureus* CMCC26003 *Microb. Pathog.* **127** 12–20

[143] Balan S S, Mani P, Kumar C G and Jayalakshmi S 2019 Structural characterization and biological evaluation of Staphylosan (dimannooleate), a new glycolipid surfactant produced by a marine *Staphylococcus saprophyticus* SBPS-15 *Enzyme Microb. Technol.* **120** 1–7

[144] Fadhile Almansoory A, Abu Hasan H, Idris M, Sheikh Abdullah S R, Anuar N and Musa Tibin E M 2017 Biosurfactant production by the hydrocarbon-degrading bacteria (HDB) *Serratia marcescens*: optimization using central composite design (CCD) *J. Ind. Eng. Chem.* **47** 272–80

[145] Rekadwad B, Maske V, Khobragade C N and Kasbe P S 2019 Production and evaluation of mono- and di-rhamnolipids produced by *Pseudomonas aeruginosa* VM011 *Data Br.* **24** 103890

[146] Gaur V K, Bajaj A, Regar R K, Kamthan M, Jha R R, Srivastava J K *et al* 2019 Rhamnolipid from a *Lysinibacillus sphaericus* strain IITR51 and its potential application for dissolution of hydrophobic pesticides *Bioresour. Technol.* **272** 19–25

[147] Abeer Mohammed A B, Tayel A A and Elguindy N M 2018 Production of new rhamnolipids Rha C16-C16 by *Burkholderia* sp. through biodegradation of diesel and biodiesel *Beni-Suef. Univ. J. Basic. Appl. Sci.* **7** 492–8

[148] Lee D W, Lee H, Kwon B-O, Khim J S, Yim U H, Kim B S *et al* 2018 Biosurfactant-assisted bioremediation of crude oil by indigenous bacteria isolated from Taean beach sediment *Environ. Pollut.* **241** 254–64

[149] Yuliani H, Perdani M S, Savitri I, Manurung M, Sahlan M, Wijanarko A *et al* 2018 Antimicrobial activity of biosurfactant derived from *Bacillus subtilis* C19 *Energy Procedia* **153** 274–8

[150] Sharma R, Singh J and Verma N 2018 Production, characterization and environmental applications of biosurfactants from *Bacillus amyloliquefaciens* and *Bacillus subtilis Biocatal Agric. Biotechnol.* **16** 132–9

[151] Vigneshwaran C, Sivasubramanian V, Vasantharaj K, Krishnanand N and Jerold M 2018 Potential of brevibacillus sp. AVN 13 isolated from crude oil contaminated soil for biosurfactant production and its optimization studies *J. Environ. Chem. Eng.* **6** 4347–56

[152] Zenati B, Chebbi A, Badis A, Eddouaouda K, Boutoumi H, El Hattab M *et al* 2018 A non-toxic microbial surfactant from Marinobacter hydrocarbonoclasticus SdK644 for crude oil solubilization enhancement *Ecotoxicol. Environ. Saf.* **154** 100–7

[153] Woźniak-Karczewska M, Myszka K, Sznajdrowska A, Szulc A, Zgoła-Grześkowiak A, Ławniczak Ł *et al* 2017 Isolation of rhamnolipids-producing cultures from faeces: influence of interspecies communication on the yield of rhamnolipid congeners *New Biotechnol.* **36** 17–25

[154] Aleksic I, Petkovic M, Jovanovic M, Milivojevic D, Vasiljevic B, Nikodinovic-Runic J *et al* 2017 Anti-biofilm properties of bacterial Di-rhamnolipids and their semi-synthetic amide derivatives *Front. Microbiol.* **8** 02454

[155] Dong H, Xia W, Dong H, She Y, Zhu P, Liang K *et al* 2016 Rhamnolipids produced by indigenous acinetobacter junii from petroleum reservoir and its potential in enhanced oil recovery *Front. Microbiol.* **7** 1710

[156] Ji F, Li L, Ma S, Wang J and Bao Y 2016 Production of rhamnolipids with a high specificity by *Pseudomonas aeruginosa* M408 isolated from petroleum-contaminated soil using olive oil as sole carbon source *Ann. Microbiol.* **66** 1145–56
[157] Deepika K V, Kalam S, Ramu Sridhar P, Podile A R and Bramhachari P V 2016 Optimization of rhamnolipid biosurfactant production by mangrove sediment bacterium *Pseudomonas aeruginosa* KVD-HR42 using response surface methodology *Biocatal. Agric. Biotechnol.* **5** 38–47
[158] Ismail W, Shammary S Al, El-Sayed W S, Obuekwe C, El Nayal A M, Abdul Raheem A S *et al* 2015 Stimulation of rhamnolipid biosurfactants production in *Pseudomonas aeruginosa* AK6U by organosulfur compounds provided as sulfur sources *Biotechnol. rep. (Amsterdam, Netherlands)* **7** 55–63
[159] Amani H 2015 Study of enhanced oil recovery by rhamnolipids in a homogeneous 2D micromodel *J. Pet. Sci. Eng.* **128** 212–9
[160] Ben Ayed H, Jemil N, Maalej H, Bayoudh A, Hmidet N and Nasri M 2015 Enhancement of solubilization and biodegradation of diesel oil by biosurfactant from *Bacillus amyloliquefaciens* An6 *Int. Biodeterior. Biodegrad.* **99** 8–14
[161] Deepak R and Jayapradha R 2015 Lipopeptide biosurfactant from bacillus thuringiensis pak2310: a potential antagonist against *Fusarium oxysporum J. Mycol. Med.* **25** e15–24
[162] Nalini S and Parthasarathi R 2014 Production and characterization of rhamnolipids produced by serratia rubidaea SNAU02 under solid-state fermentation and its application as biocontrol agent *Bioresour. Technol.* **173** 231–8
[163] Kavitha V, Mandal A B and Gnanamani A 2014 Microbial biosurfactant mediated removal and/or solubilization of crude oil contamination from soil and aqueous phase: an approach with *Bacillus licheniformis* MTCC 5514 *Int. Biodeterior. Biodegrad.* **94** 24–30
[164] Striebich R C, Smart C E, Gunasekera T S, Mueller S S, Strobel E M, McNichols B W *et al* 2014 Characterization of the F-76 diesel and Jet-A aviation fuel hydrocarbon degradation profiles of *Pseudomonas aeruginosa* and marinobacter hydrocarbonoclasticus *Int. Biodeterior. Biodegrad.* **93** 33–43
[165] Hassanshahian M 2014 Isolation and characterization of biosurfactant producing bacteria from persian gulf (Bushehr provenance) *Mar. Pollut. Bull.* **86** 361–6
[166] Cai Q, Zhang B, Chen B, Zhu Z, Lin W and Cao T 2014 Screening of biosurfactant producers from petroleum hydrocarbon contaminated sources in cold marine environments *Mar. Pollut. Bull.* **86** 402–10
[167] El-Sheshtawy H S, Khalil N M, Ahmed W and Abdallah R I 2014 Monitoring of oil pollution at gemsa bay and bioremediation capacity of bacterial isolates with biosurfactants and nanoparticles *Mar. Pollut. Bull.* **87** 191–200
[168] Rizzo C, Michaud L, Hörmann B, Gerçe B, Syldatk C, Hausmann R *et al* 2013 Bacteria associated with sabellids (Polychaeta: Annelida) as a novel source of surface active compounds *Mar. Pollut. Bull.* **70** 125–33
[169] Singh S P, Bharali P and Konwar B K 2013 Optimization of nutrient requirements and culture conditions for the production of rhamnolipid from *Pseudomonas aeruginosa* (MTCC 7815) using mesua ferrea seed oil *Indian J. Microbiol.* **53** 467–76
[170] Chandankere R, Yao J, Choi M M F, Masakorala K and Chan Y 2013 An efficient biosurfactant-producing and crude-oil emulsifying bacterium *Bacillus methylotrophicus* USTBa isolated from petroleum reservoir *Biochem. Eng. J.* **74** 46–53

[171] Jain R M, Mody K, Mishra A and Jha B 2012 Isolation and structural characterization of biosurfactant produced by an alkaliphilic bacterium *Cronobacter sakazakii* isolated from oil contaminated wastewater *Carbohydr. Polym* **87** 2320–6
[172] Kumari B, Singh S N and Singh D P 2012 Characterization of two biosurfactant producing strains in crude oil degradation *Process. Biochem.* **47** 2463–71
[173] Eddouaouda K, Mnif S, Badis A, Younes S B, Cherif S, Ferhat S *et al* 2011 Characterization of a novel biosurfactant produced by *Staphylococcus* sp. strain 1E with potential application on hydrocarbon bioremediation *J. Basic Microbiol.* **52** 408–18
[174] Liu W, Wang X, Wu L, Chen M, Tu C, Luo Y *et al* 2012 Isolation, identification and characterization of *Bacillus amyloliquefaciens* BZ-6, a bacterial isolate for enhancing oil recovery from oily sludge *Chemosphere* **87** 1105–10
[175] Ferhat S, Mnif S, Badis A, Eddouaouda K, Alouaoui R, Boucherit A *et al* 2011 Screening and preliminary characterization of biosurfactants produced by *Ochrobactrum* sp. 1C and *Brevibacterium* sp. 7G isolated from hydrocarbon-contaminated soils *Int. Biodeterior. Biodegrad* **65** 1182–8
[176] Cerqueira V S, Hollenbach E B, Maboni F, Camargo F A O, Peralba M do C R and Bento F M 2011 Bioprospection and selection of bacteria isolated from environments contaminated with petrochemical residues for application in bioremediation *World J. Microbiol. Biotechnol.* **28** 1203–22
[177] Sriram M I, Gayathiri S, Gnanaselvi U, Jenifer P S, Mohan Raj S and Gurunathan S 2011 Novel lipopeptide biosurfactant produced by hydrocarbon degrading and heavy metal tolerant bacterium *Escherichia fergusonii* KLU01 as a potential tool for bioremediation *Bioresour. Technol.* **102** 9291–5
[178] Najafi A R, Rahimpour M R, Jahanmiri A H, Roostaazad R, Arabian D, Soleimani M *et al* 2011 Interactive optimization of biosurfactant production by *Paenibacillus alvei* ARN63 isolated from an Iranian oil well *Colloids Surf. B Biointerfaces* **82** 33–9
[179] Nie M, Yin X, Ren C, Wang Y, Xu F and Shen Q 2010 Novel rhamnolipid biosurfactants produced by a polycyclic aromatic hydrocarbon-degrading bacterium *Pseudomonas aeruginosa* strain NY3 *Biotechnol. Adv.* **28** 635–43
[180] Najafi A R, Rahimpour M R, Jahanmiri A H, Roostaazad R, Arabian D and Ghobadi Z 2010 Enhancing biosurfactant production from an indigenous strain of *Bacillus mycoides* by optimizing the growth conditions using a response surface methodology *Chem. Eng. J.* **163** 188–94
[181] Gesheva V, Stackebrandt E and Vasileva-Tonkova E 2010 Biosurfactant production by halotolerant *Rhodococcus fascians* from casey station, wilkes land, antarctica *Curr. Microbiol.* **61** 112–7
[182] Liu J H, Chen Y T, Li H, Jia Y P, Xu R D and Wang J 2015 Optimization of fermentation conditions for biosurfactant production by *Bacillus subtilis* strains CCTCC M201162 from oilfield wastewater *Environ. Prog. Sustain. Energy* **34** 548–54
[183] Fernandes N, de A T, Simões L A and Dias D 2023 Biosurfactants produced by yeasts: fermentation, screening, recovery, purification, characterization, and applications *Fermentation* **9** 207
[184] Adetunji A I and Olaniran A O 2021 Production and potential biotechnological applications of microbial surfactants: an overview *Saudi J. Biol. Sci.* **28** 669–79
[185] Pardhi D S, Panchal R R, Raval V H, Joshi R G, Poczai P, Almalki W H *et al* 2022 Microbial surfactants: a journey from fundamentals to recent advances *Front. Microbiol.* **13** 982603

[186] Soberón-Chávez G, González-Valdez A, Soto-Aceves M P and Cocotl-Yañez M 2021 Rhamnolipids produced by *Pseudomonas*: from molecular genetics to the market *Microb. Biotechnol.* **14** 136–46

[187] Wu B, Xiu J, Yu L, Huang L, Yi L and Ma Y 2022 Biosurfactant production by *Bacillus subtilis* SL and its potential for enhanced oil recovery in low permeability reservoirs *Sci. Rep.* **12** 7785

[188] Asgher M, Arshad S, Qamar S A and Khalid N 2020 Improved biosurfactant production from aspergillus niger through chemical mutagenesis: characterization and RSM optimization *SN Appl. Sci.* **2** 966

[189] Thavasi R and Banat I M 2018 Downstream processing of microbial biosurfactants In *Microbial Biosurfactant and their Industrial Application* Ist edn (Boca Raton, FL: CRC Press)

[190] Baumann P and Hubbuch J 2017 Downstream process development strategies for effective bioprocesses: trends, progress, and combinatorial approaches *Eng. Life Sci.* **17** 1142–58

[191] Najmi Z, Ebrahimipour G, Franzetti A and Banat I M 2018 *In situ* downstream strategies for cost-effective bio/surfactant recovery *Biotechnol. Appl. Biochem.* **65** 523–32

[192] Araújo H W C, Andrade R F S, Montero-Rodríguez D, Rubio-Ribeaux D, Alves da Silva C A and Campos-Takaki G M 2019 Sustainable biosurfactant produced by *Serratia marcescens* UCP 1549 and its suitability for agricultural and marine bioremediation applications *Microb. Cell Fact.* **18** 2

[193] Chang J S, Cha D K, Radosevich M and Jin Y 2015 Effects of biosurfactant-producing bacteria on biodegradation and transport of phenanthrene in subsurface soil *J. Environ. Sci. Heal—Part A Toxic Hazard. Subst. Environ. Eng* **50** 611–6

[194] Mateas D J, Tick G R and Carroll K C 2017 *In situ* stabilization of NAPL contaminant source-zones as a remediation technique to reduce mass discharge and flux to groundwater *J. Contam. Hydrol.* **204** 40–56

[195] Zhang Q Q, Yang G F, Sun K K, Tian G M and Jin R C 2018 Insights into the effects of bio-augmentation on the granule-based anammox process under continuous oxytetracycline stress: performance and microflora structure *Chem. Eng. J.* **348** 503–13

[196] Patowary R, Patowary K, Kalita M C and Deka S 2018 Application of biosurfactant for enhancement of bioremediation process of crude oil contaminated soil *Int. Biodeterior. Biodegrad.* **129** 50–60

[197] Rahman K S M, Thahira-Rahman J, Lakshmanaperumalsamy P and Banat I M 2002 Towards efficient crude oil degradation by a mixed bacterial consortium *Bioresour. Technol.* **85** 257–61

[198] Chebbi A, Hentati D, Zaghden H, Baccar N, Rezgui F, Chalbi M *et al* 2017 Polycyclic aromatic hydrocarbon degradation and biosurfactant production by a newly isolated Pseudomonas sp. strain from used motor oil-contaminated soil *Int. Biodeterior. Biodegrad.* **122** 128–40

[199] Imam A, Kanaujia P K, Ray A and Suman S K 2021 Removal of petroleum contaminants through bioremediation with integrated concepts of resource recovery: a review *Indian J. Microbiol.* **61** 250–61

[200] Machado T S, Decesaro A, Cappellaro ÂC, Machado B S, van Schaik Reginato K, Reinehr C O *et al* 2020 Effects of homemade biosurfactant from *Bacillus methylotrophicus* on bioremediation efficiency of a clay soil contaminated with diesel oil *Ecotoxicol. Environ. Saf.* **201** 110798

[201] Zouari O, Lecouturier D, Rochex A, Chataigne G, Dhulster P, Jacques P *et al* 2019 Bio-emulsifying and biodegradation activities of syringafactin producing *Pseudomonas* spp. strains isolated from oil contaminated soils *Biodegradation* **30** 259–72

[202] Diaz De Rienzo M A, Stevenson P S, Marchant R and Banat I M 2016 Effect of biosurfactants on *Pseudomonas aeruginosa* and *Staphylococcus aureus* biofilms in a BioFlux channel *Appl. Microbiol. Biotechnol.* **100** 5773–9

[203] Hage-Hülsmann J, Grünberger A, Thies S, Santiago-Schübel B, Klein A S, Pietruszka J *et al* 2018 Natural biocide cocktails: combinatorial antibiotic effects of prodigiosin and biosurfactants *PLoS One* **13** 1–23

[204] Daverey A, Dutta K, Joshi S and Daverey A 2021 Sophorolipid: a glycolipid biosurfactant as a potential therapeutic agent against COVID-19 *Bioengineered* **12** 9550–60

[205] Baglivo M, Baronio M, Natalini G, Beccari T, Chiurazzi P, Fulcheri E *et al* 2020 Natural small molecules as inhibitors of coronavirus lipid-dependent attachment to host cells: a possible strategy for reducing SARS-COV-2 infectivity? *Acta Biomed.* **91** 161–4

[206] Kumari K, Nandi A, Sinha A, Ghosh A, Sengupta S, Saha U *et al* 2023 The paradigm of prophylactic viral outbreaks measures by microbial biosurfactants *J. Infect. Public Health* **16** 575–87

[207] Thakur P, Saini N K, Thakur V K, Gupta V K, Saini R V and Saini A K 2021 Rhamnolipid the glycolipid biosurfactant: emerging trends and promising strategies in the field of biotechnology and biomedicine *Microb. Cell Fact.* **20** 1–15

[208] Loeto D, Jongman M, Lekote L, Muzila M, Mokomane M, Motlhanka K *et al* 2021 Biosurfactant production by halophilic yeasts isolated from extreme environments in Botswana *FEMS Microbiol. Lett.* **368** 1–10

[209] Lourith N and Kanlayavattanakul M 2009 Natural surfactants used in cosmetics: glycolipids *Int. J. Cosmet. Sci.* **31** 255–61

[210] Sarubbo L A, Silva M da G C, Durval I J B, Bezerra K G O, Ribeiro B G, Silva I A *et al* 2022 Biosurfactants: production, properties, applications, trends, and general perspectives *Biochem. Eng. J.* **181** 108377

[211] Karnwal A, Shrivastava S, Al-Tawaha A R M S, Kumar G, Singh R, Kumar A *et al* 2023 Microbial biosurfactant as an alternate to chemical surfactants for application in cosmetics industries in personal and skin care products: a critical review *Bio. Med. Res. Int.* **2023** 2375223

[212] Adu S A, Twigg M S, Naughton P J, Marchant R and Banat I M 2023 Characterisation of cytotoxicity and immunomodulatory effects of glycolipid biosurfactants on human keratinocytes *Appl. Microbiol. Biotechnol.* **107** 137–52

[213] Resende A H M, Farias J M, Silva D D B, Rufino R D, Luna J M, Stamford T C M *et al* 2019 Application of biosurfactants and chitosan in toothpaste formulation *Colloids Surf. B Biointerfaces* **181** 77–84

[214] Farias J M, Stamford T C M, Resende A H M, Aguiar J S, Rufino R D, Luna J M *et al* 2019 Mouthwash containing a biosurfactant and chitosan: an eco-sustainable option for the control of cariogenic microorganisms *Int. J. Biol. Macromol.* **129** 853–60

[215] Sachdev D P and Cameotra S S 2013 Biosurfactants in agriculture *Appl. Microbiol. Biotechnol.* **97** 1005–16

[216] Köhl J, Kolnaar R and Ravensberg W J 2019 Mode of action of microbial biological control agents against plant diseases: relevance beyond efficacy *Front. Plant Sci.* **10** 1–19

[217] Liu Y, Zeng G, Zhong H, Wang Z, Liu Z, Cheng M *et al* 2017 Effect of rhamnolipid solubilization on hexadecane bioavailability: enhancement or reduction? *J. Hazard. Mater.* **322** 394–401

[218] Mehetre G T, Dastager S G and Dharne M S 2019 Biodegradation of mixed polycyclic aromatic hydrocarbons by pure and mixed cultures of biosurfactant producing thermophilic and thermo-tolerant bacteria *Sci. Total Environ.* **679** 52–60

[219] Rodrigues L, Banat I M, Teixeira J and Oliveira R 2006 Biosurfactants: potential applications in medicine *J. Antimicrob. Chemother.* **57** 609–18

[220] Ochoa-Loza F J, Artiola J F and Maier R M 2001 The biosurfactant developed by *Pseudomonas aeruginosa* is an effective metal complexing agent such as Pb2 ad, Cd2 ad and Zn2 adjacentt *J. Environ. Qual.* **30** 479–85

[221] Kruijt M, Tran H and Raaijmakers J M 2009 Functional, genetic and chemical characterization of biosurfactants produced by plant growth-promoting *Pseudomonas putida* 267 *J. Appl. Microbiol.* **107** 546–56

[222] Fenibo E O, Ijoma G N, Selvarajan R and Chikere C B 2019 Microbial surfactants: the next generation multifunctional biomolecules for applications in the petroleum industry and its associated environmental remediation *Microorganisms* **7** 1–29

[223] Gudiñaand E J and Rodrigues L R 2019 Research and production of biosurfactants for the food industry *Bioprocessing for Biomolecules Production* (Wiley) pp 125–43

[224] Ribeiro B G, Guerra J M C and Sarubbo L A 2020 Potential food application of a biosurfactant produced by *Saccharomyces cerevisiae* URM 6670 *Front. Bioeng. Biotechnol.* **8** 1–13

[225] Ribeiro B G, Guerra J M C and Sarubbo L A 2020 Biosurfactants: production and application prospects in the food industry *Biotechnol. Prog.* **36**

[226] Garcia S N, Osburn B I and Jay-Russell M T 2020 One health for food safety, food security, and sustainable food production *Front. Sustain. Food Syst.* **4** 1–9

[227] Silva I A, Veras B O, Ribeiro B G, Aguiar J S, Campos Guerra J M, Luna J M *et al* 2020 Production of cupcake-like dessert containing microbial biosurfactant as an emulsifier *PeerJ* **2020** 1–23

[228] Augusto P E D 2020 Challenges, trends and opportunities in food processing *Curr. Opin. Food Sci.* **35** 72–8

[229] Market A 2019 Research; renewable, speciality and fine chemicals, world biosurfactant market; opportunities and forecasts, from 2019 to 2026 https://alliedmarketresearch.com/biosurfactantmarket

IOP Publishing

Microbial Surfactants
A sustainable class of versatile molecules
Divya Tripathy and Anjali Gupta

Chapter 3

Producer microorganisms and the metabolic engineering for biosurfactant production

Shruti Roy, Anupam Prakash and Cheshta Sharma

Biosurfactants are a variety of biological molecules that have water and water-phobic components, making them amphiphiles in nature. Biosurfactants can be synthesized by microorganisms such as bacteria, yeast, fungi, including famous cases such as *Penicillium* and *Aspergillus*, highlighting their key role in addressing contemporary challenges, from enhancing oil recovery to ecological conscious strategies in biodegradation. This chapter focuses on the production of biosurfactants and explores the fields of producers and the cutting-edge field of metabolic engineering. Furthermore, this chapter focused on the study of the characteristics of bacteria and yeast and highlighted the remarkable genera and species known for their biosurfactant synthesis. Afterwards, we also explore the potential applications of bacteria, yeasts and yeast-derived surfactants and reveal their importance in different contexts. Finally, the chapter enters the groundbreaking field of metabolic technology, explaining its principles and strategies for optimizing the production process. The accumulated knowledge highlights the biotechnological potential of microbial cell plants and opens new pathways for sustainable and efficient production.

3.1 Microbial sources of surfactants: introduction

Biosurfactants exhibit amphiphilic properties, characterized by the presence of both hydrophilic and hydrophobic moieties. These unique compounds have a remarkable ability to preferentially partition at interfaces, such as liquid/liquid, gas/liquid, or solid/liquid interfaces. This partitioning enables them to perform essential functions like emulsification, foaming, detergency, and dispersing (Rani *et al* 2020). What sets biosurfactants apart is their low toxicity and eco-friendly nature, making them highly valuable for a wide range of potential industrial applications, including bioremediation, healthcare, food processing, and the oil industry. Produced

doi:10.1088/978-0-7503-5989-4ch3

extracellularly by various microorganisms, biosurfactants exhibit superior characteristics compared to chemical surfactants. They are biocompatible, less toxic, highly biodegradable, and exceptionally stable, even under extreme temperature and pH conditions.

The growing appreciation of these advantageous attributes has contributed to the surging demand for biosurfactants in the field of biotechnology, where their multifunctional properties hold great promise for diverse applications. Biosurfactants are a diverse class of biomolecules that possess both hydrophilic and hydrophobic components, making them amphiphilic in nature. These molecules are categorized into two main groups based on their molecular weight: low and high molecular weight biosurfactants. The low molecular weight biosurfactants include glycolipids and lipopeptides, while the high molecular weight biosurfactants consist of polymeric structures. A biosurfactant can exist in various structural forms, which include glycolipids, mycolic acid, polysaccharide-lipid composite, lipoprotein/lipopeptide, phospholipid, or even the microbial cell surface itself. Owing to their biodegradable properties and minimal toxicity, these biosurfactants show great promise as viable alternatives to synthetic surfactants in various remediation technologies (Nguyen *et al* 2008, Tripathy and Mishra 2016).

This chapter delves into the realm of producer microorganisms and the cutting-edge field of metabolic engineering, focusing on their pivotal roles in biosurfactant production. Exploring the prowess of bacteria, we shed light on exemplary genera and species known for their biosurfactant synthesis. By carefully analyzing the advantages and disadvantages of bacterial biosurfactant producers, we gain valuable insights into their potential applications. Subsequently, the chapter unveils yeasts as biosurfactant producers, presenting key genera and species involved in biosurfactant production. Through a detailed examination of the merits and limitations of yeast biosurfactant producers, we gain a nuanced understanding of their suitability for diverse industrial uses.

We explore the biosurfactant-producing abilities of fungi, highlighting notable genera and species responsible for synthesizing these valuable compounds. By delving into the advantages and disadvantages of fungal biosurfactant producers, we discern their significance in different contexts.

Finally, the chapter ventures into the groundbreaking domain of metabolic engineering, elucidating its principles and strategies for optimizing biosurfactant production. The discussion encompasses diverse approaches, such as overexpression of biosurfactant biosynthesis genes, knockout of competing metabolic pathways, culture conditions optimization, and the application of synthetic biology tools. Moreover, we emphasize recent advances in metabolic engineering for biosurfactant production, charting a path toward sustainable and efficient practices in the field.

3.2 Bacteria as biosurfactant producers

In the past, considerable focus has been directed towards the production of surface-active compounds sourced from biological origins due to their promising applications in food processing, pharma, and the oil industry (Venkata Ramana and Karanth 1989). The synthesis of microbial surfactants is influenced not only by the type of producer organism but also by various factors, including nitrogen and carbon availability, temperature, aeration, and trace elements. These elements collectively impact the quantity and type of surfactants produced by the microorganism (table 3.1).

In contrast to chemically synthesized surfactants, which are typically categorized based on the type of polar group they contain, biosurfactants are primarily classified according to their chemical composition and microbial origin. The principal classes of low-mass biosurfactants consist of lipopeptides, glycolipids, and phospholipids, while high-mass biosurfactants include polymeric and particulate surfactants. Most biosurfactants are either anionic or neutral, with their hydrophobic part being derived from long-chain fatty acids or fatty acid derivatives. On the other hand, the hydrophilic component can vary and may include carbohydrates, phosphates, amino acids, or cyclic peptides (Nitschke and Coast 2007).

Table 3.1. Biosurfactant-producing bacteria.

Class of surfactant	Bacteria
Glycolipids	*Pseudomonas aeruginosa*
Rhamnolipids	*Rhodococcus erithropolis*
Trehalose lipids	*Arthobacter* sp.
Lipopeptides	*Bacillus subtilis*
Surfactin/iturin/fengycin	*Pseudomonas fluorescens*
Viscosin	*Bacillus licheniformis*
Lichenysin	*Serratia marcescens*
Serrawettin	
Phospholipids	*Acinetobacter* sp.
	Corynebacterium lepus
Surface-active antibiotics	*Brevibacterium brevis*
Gramicidin	*Paenibacillus polymyxa*
Polymixin	*Myxococcus xanthus*
Antibiotic TA	
Emulsan	*Acinetobacter calcoaceticus*
Alasan	*Acinetobacter radioresistens*
Particulate biosurfactants	*Acinetobacter calcoaceticus*
	Cyanobacteria

3.2.1 Examples of bacterial genera and species that produce biosurfactants

(a) *Pseudomonas aeruginosa*: Among the extensively studied bacteria, *P. aeruginosa* stands out for its impressive biosurfactant production. Its biosurfactant, known as rhamnolipids, exhibits exceptional surface tension reduction and emulsification properties, making it suitable for applications in enhanced oil recovery and bioremediation (Zhang *et al* 2020). The extensively researched glycolipids consist of either one or two rhamnose molecules linked to one or two molecules of β-hydroxydecanoic acid. Among these glycolipids, the most studied ones are those wherein the -OH group of one acid forms a glycosidic linkage with the reducing end of the rhamnose disaccharide, while the -OH group of the second acid is engaged in ester formation (Karanth *et al* 1999). The investigation of glycolipid-containing rhamnose production originated from early research by Jarvis and Johnson in 1949, focusing on *P. aeruginosa* (Jarvis and Johnson 1949). *P. aeruginosa* was found to produce significant glycolipids, namely L-Rhamnosyl-L-rhamnosyl-β-hydroxydecanoyl-β-hydroxydecanoate (referred to as rhamnolipid 1) and L-rhamnosyl-β-hydroxydecanoyl-β-hydroxydecanoate (referred to as rhamnolipid 2) as depicted in figure 3.1 (Edward and Hayashi 1965). These two glycolipids represent the primary products of rhamnose biosynthesis in *P. aeruginosa*.

(b) *Bacillus subtilis*: *B. subtilis* is another prominent biosurfactant producer, known for its production of lipopeptides like surfactin and iturins. These lipopeptides possess potent antimicrobial properties and have found applications in diverse areas such as biocontrol and pharmaceutical formulations

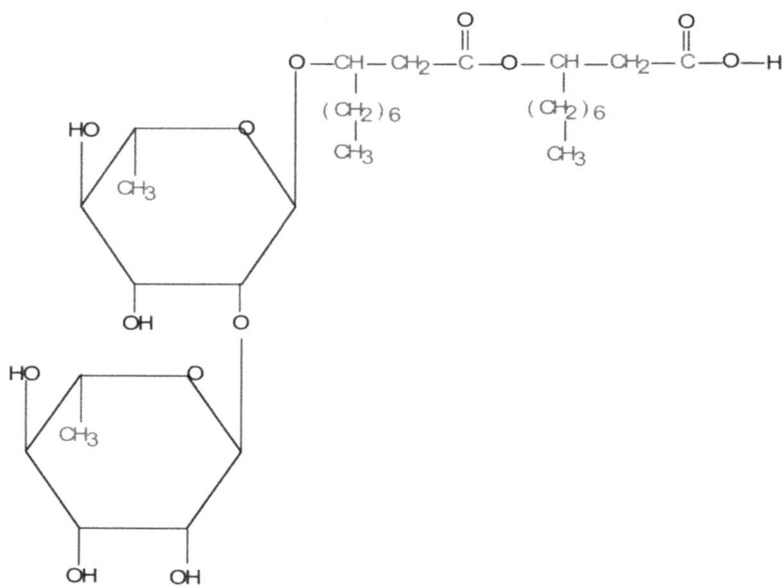

Figure 3.1. Rhamnolipid.

(Choudhary *et al* 2019). Numerous cyclic lipopeptides, such as decapeptide antibiotics (e.g., gramicidins), and lipopeptide antibiotics (e.g., polymyxins), are synthesized, comprising a lipid component connected to a polypeptide chain. The cyclic lipopeptide surfactin, produced by *B. subtilis*, emerges as a highly promising biosurfactant (figure 3.2). Comprising a seven-amino acid ring structure linked to a fatty acid chain through a lactone bond, surfactin remarkably lowers the surface tension from 72 to 27.9 mN m^{-1}, even at a mere 0.005% concentration (Arima *et al* 1968).

(c) *Bacillus licheniformis*: *B. licheniformis* has been associated with foodborne intoxication due to its production of the surfactant lichenysin. This bacterium is capable of synthesizing multiple biosurfactants that exhibit strong stability under varying conditions of temperature, salt concentration, and pH. Notably, these biosurfactants demonstrate structural and physio-chemical similarities to surfactin (McInerney *et al* 1990). The surfactants produced by *B. licheniformis* have a remarkable ability to reduce the surface tension of water to 27 mN m^{-1} and the interfacial tension between water and n-hexadecane to 0.36 mN m^{-1}. Moreover, the combination of these biosurfactants exhibits a synergistic effect, enhancing their overall performance.

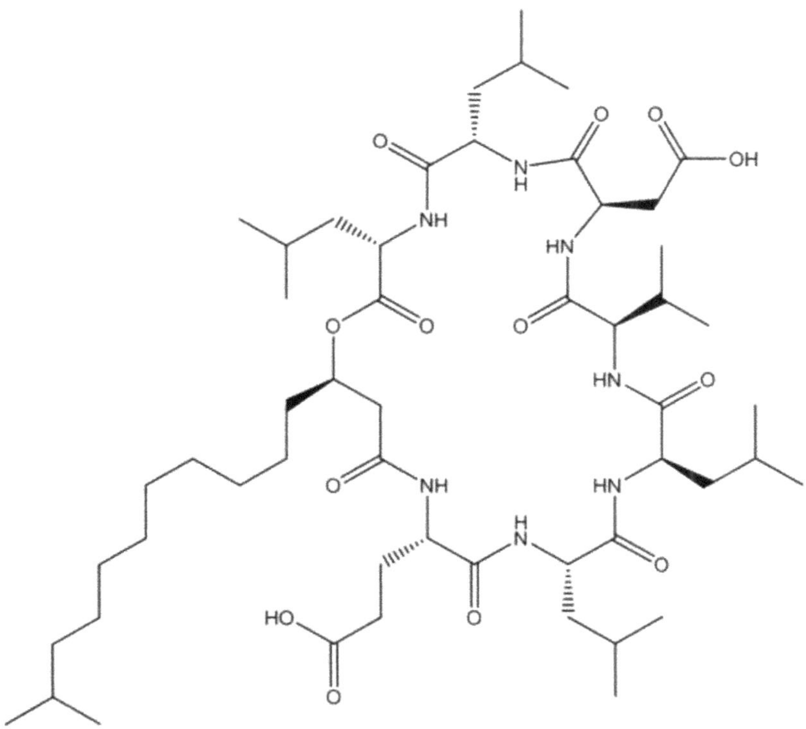

Figure 3.2. Surfactin.

3.2.2 Advantages and disadvantages of bacterial biosurfactant producers

The realm of bacterial biosurfactant producers presents a realm of potential benefits alongside certain challenges. These microorganisms offer a pathway towards environmentally-conscious surfactant production, yet prudent consideration of both advantages and drawbacks is essential.

3.2.2.1 Advantages

(i) Eco-friendly nature: Bacterial biosurfactants are notable for their biodegradability, setting them apart from synthetic counterparts. Their microbial origin ensures minimal ecological impact and enhanced compatibility with living systems.
(ii) Survival in extreme conditions: Many bacterial strains exhibit adaptability in harsh environments, including sites with elevated hydrocarbon levels or extreme salinity. This adaptability positions bacterial biosurfactants as valuable tools in bioremediation processes targeting recalcitrant pollutants.
(iii) Versatile applications: Bacterial biosurfactants boast a wide-ranging scope across industries. Their proficiency in emulsification and antimicrobial activity holds promise in applications such as enhanced oil recovery and medical fields (Santos *et al* 2017).
(iv) Economical substrate Usage: Bacterial biosurfactant producers often exhibit efficient utilization of economical substrates, including agricultural by-products, enhancing cost-effective production pathways (Santos-Merino *et al* 2019).

3.2.2.2 Challenges

(i) Production costs: Large-scale production of bacterial biosurfactants may pose economic challenges. Factors such as fermentation complexities and downstream purification contribute to elevated production expenses (Rufino *et al* 2014).
(ii) Purity and consistency: Ensuring consistent purity and stability of bacterial biosurfactants remains a concern. Variability in production conditions can influence the composition and quality of the biosurfactant, impacting performance (Sivapathasekaran *et al* 2009).
(iii) Regulatory considerations: Regulatory compliance and safety assessments are pivotal considerations in utilizing bacterial biosurfactants. Rigorous evaluations are crucial to meet established standards and ensure safe application.
(iv) Scale-up complexity: Scaling up from laboratory to industrial production necessitates meticulous engineering and optimization. Maintaining product quality and process scalability pose challenges during large-scale implementation.

3.3 Yeasts as biosurfactant producers

Yeasts have emerged as a more prominent focus in biosurfactant studies compared to bacteria, primarily due to concerns regarding the pathogenic and opportunistic characteristics associated with bacterial biosurfactants, which may limit their applicability in the pharmaceutical and food industries (table 3.2). Moreover, certain yeasts exhibit a higher rate of substrate conversion, leading to increased biosurfactant yields compared to bacteria (Sharma and Saharan, 2016).

One of the significant advantages of utilizing yeast as biosurfactant producers lies in their non-toxic and non-pathogenic nature, making them favorable for biotechnological processes. These yeast species have been granted the generally recognized as safe (GRAS) status by the Food and Drug Administration (FDA), further enhancing their appeal for various applications (Rywińska *et al* 2013). Notably, yeasts can efficiently produce biosurfactant from oleaginous substrates, including agro-industrial residues like glycerol, corn-steeping liquor, and residual frying oil. This characteristic renders the industrial production of biosurfactant by yeasts feasible and contributes to cost reduction (Chen *et al* 2018). For instance, green technologies may emerge as valuable tools for lignocellulosic biorefineries, offering alternative and sustainable avenues for short-term biosurfactant production (Marcelino *et al* 2019). Fernandes and Simões (2023) compiled the research articles with primary genera of yeast producers of biosurfactants (figure 3.3).

3.3.1 Examples of yeast genera and species that produce biosurfactants

(a) *Candida* species: Yeast species belonging to the *Candida* genus have garnered significant global attention due to their ability to produce diverse biosurfactants with unique properties. Notably, *Candida* species such as *Candida lipolytica, Candida antarctica, Candida sphaerica*, and *Candida bomicola* have emerged as particularly interesting subjects for researchers, given their potential industrial applications. The glycolipids synthesized primarily by yeast, such as *Candida*, consist of a dimeric carbohydrate sophorose linked to a long-chain hydroxyl fatty acid via a glycosidic linkage (figure 3.4) (Vatsa *et al* 2021). Typically, sophorolipids are found in both

Table 3.2. Biosurfactant-producing yeast.

Class of surfactant	Yeast
Sophorolipids mannosylerythritol lipids	*Candida bombicola*, *Candida apicola* *Candida antartica* *Pseudozyma* sp.
Liposan	*Candida lipolytica*
Lipomanan	*Candida tropicalis*

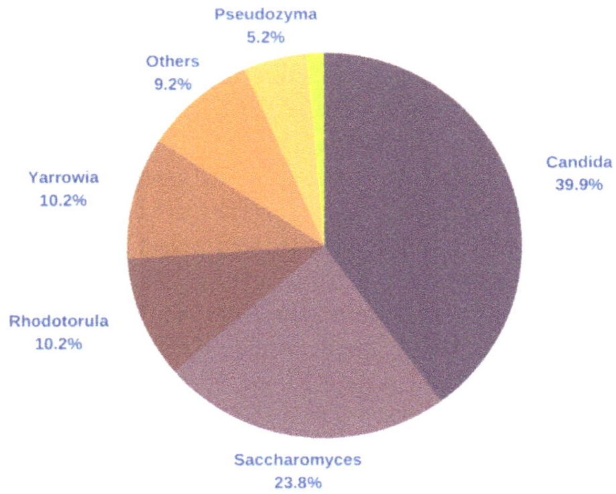

Figure 3.3. Research articles concerning the primary genera of yeast producers of biosurfactants during the past five years (2018–22).

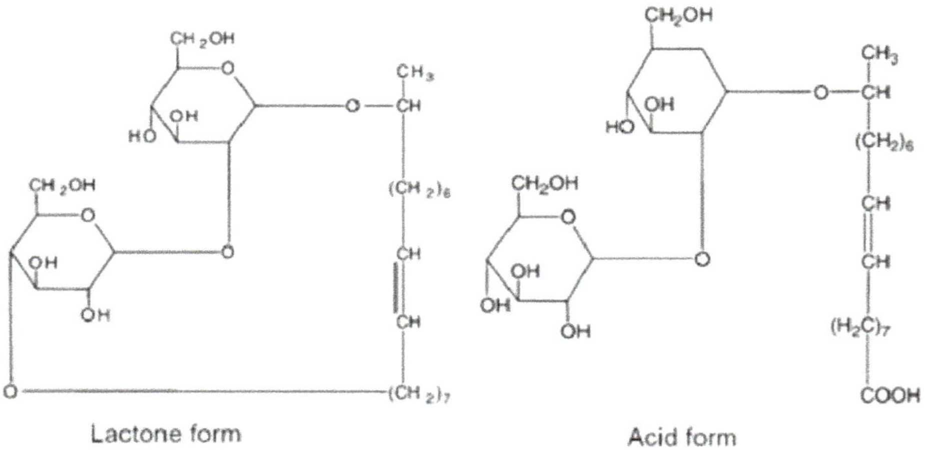

Figure 3.4. The arrangement of lactonized and free-acid forms of sophorolipids.

free-acid and macrolactone forms. The lactone form of sophorolipids has proven crucial for various applications (Gaur *et al* 2019). These biosurfactants are composed of a combination of at least six to nine diverse hydrophobic sophorolipids.

(b) *Saccharomyces cerevisiae:* Biosurfactants produced by *S. cerevisiae* have gained widespread recognition as safe (Generally Recognized as Safe—GRAS), allowing their metabolites to be applied in medical, food, and other relevant fields. Additionally, their biosurfactants offer superior functionality compared to conventional surfactants. In the structural analysis, nuclear

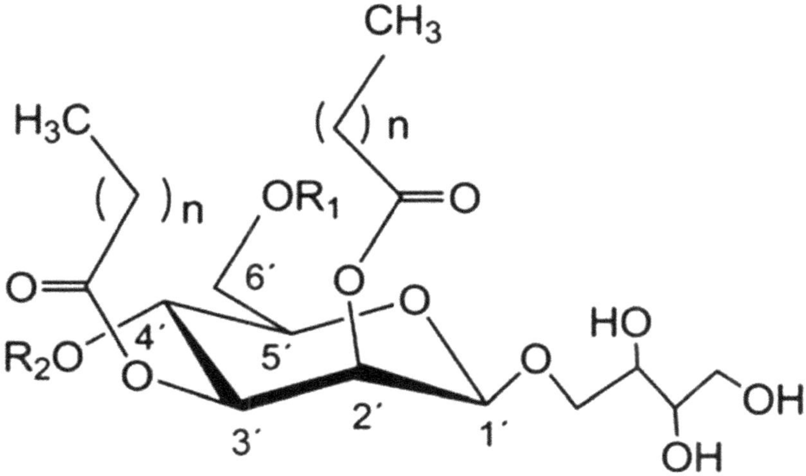

Figure 3.5. Mannosylerythritol lipids.

magnetic resonance (NMR) and Fourier transform infrared spectroscopy (FT-IR) techniques were utilized to characterize the biosurfactant produced by *S. cerevisiae*. The results indicated a glycolipid structure, with a significant proportion of linoleic acid (approximately 50.58%) in the fatty acid profile. Remarkably, the biosurfactant exhibited non-toxicity towards fibroblast and macrophage cell lines, with cell inhibition remaining below 15%. These findings open up promising avenues for utilizing the biosurfactant in various applications while ensuring safety and effectiveness (Ribeiro *et al* 2020).

(c) *Pseudozyma:* Mannosylerythritol lipids (MELs) are a type of glycolipid biosurfactants known for their surface-active properties. MELs consist of 4-O-β-D-mannopyranosyl-meso-erythritol as the hydrophilic group and can have a fatty acid and/or an acetyl group as the hydrophobic portion (figure 3.5). These biosurfactants are primarily produced by *Pseudozyma* sp. as a major component, while *Ustilago* sp. synthesizes them as a minor component (Arutchelvi *et al* 2008). Despite being known for more than five decades, MELs have recently garnered renewed attention due to their favorable environmental characteristics, mild production requirements, structural diversity, self-assembling attributes, and versatile biochemical functions.

3.3.2 Advantages and disadvantages of yeast biosurfactant producers

Yeasts have garnered considerable attention as noteworthy generators of biosurfactants, showcasing substantial potential across multiple industries. However, akin to any biological process, yeast-mediated biosurfactant production presents a set of distinct advantages and limitations. In this segment, we delve into the prominent strengths and challenges linked with east biosurfactant producers, drawing insights from empirical research.

3.3.2.1 Advantages:

(i) Diverse array of biosurfactants: Yeasts have demonstrated their proficiency in yielding an assortment of biosurfactant types, encompassing glycolipids, and lipopeptides, contributing to their adaptability in a multitude of applications (Makkar *et al* 2011).

(ii) Biocompatibility: Biosurfactants originating from yeasts generally exhibit biocompatibility and a reduced level of toxicity when contrasted with their synthetic counterparts, making them a suitable choice for deployment in pharmaceuticals and cosmetics (Fernandes and Simões 2023).

(iii) Robust stability and functionality: Yeast-derived biosurfactants often manifest stability across a wide spectrum of pH and temperature conditions, retaining their operational efficacy even under rigorous environmental settings—a desirable trait for specific industrial and environmental contexts (Nitschke and Pastore, 2002).

(iv) Versatile substrate utilization: Yeasts showcase the capability to metabolize a diverse range of substrates for biosurfactant production, including renewable and waste materials, thus bolstering sustainability and cost-effectiveness (Lotfabad *et al* 2018).

3.3.2.2 Limitations

(i) Restricted yields: Biosurfactant yields from yeast-based processes can at times be comparatively lower than those from other microorganisms, thereby potentially impacting their viability for large-scale commercial applications (Santos *et al* 2016).

(ii) Genetic manipulation complexities: Manipulating yeast strains to enhance biosurfactant production can prove intricate due to inherent limitations in transformation efficiency and genetic tools, thereby potentially impeding endeavors in metabolic engineering.

(iii) Financial overheads: Factors such as the cost associated with yeast strain cultivation, maintenance, and subsequent downstream processing can collectively contribute to escalated production costs, consequently influencing the economic competitiveness of yeast-derived biosurfactants (Santos *et al* 2016).

(iv) Substrate constraints: Specific yeast species may exhibit preferences for certain substrates, potentially limiting the scope of available feedstock and application possibilities (Fernandes *et al* 2020).

3.4 Fungi as biosurfactant producers

Fungal-derived biosurfactants exhibit remarkable chemical structure versatility, enabling their application in a wide array of sectors, including personal care, food, agriculture, pharmaceuticals, biomedicine, materials engineering, bioenergies, and environmental remediation. The biochemical composition and molecular weight directly influence the versatility of these biosurfactants, granting them physicochemical and biological properties of significant biotechnological interest. Notably, low molecular weight biosurfactants typically possess molar masses ranging from 0.5 to 1.5 kDa, while high molecular weight counterparts, also known as Exopolysaccharides (EPS) or bio-emulsifiers, can reach impressive sizes of up to 500 kDa. This broad range of molecular weights contributes to the diverse functionalities and potential applications of fungal biosurfactants across multiple industries (da Silva *et al* 2021). Table 3.3 furnishes a comprehensive outline of frequently utilized genera in biosurfactant production, along with commonly identified biosurfactants encompassing glycolipids, lipopeptides, enamides, and similar compounds.

3.4.1 Examples of fungal genera and species that produce biosurfactants

(a) *Penicillium:* The biosurfactant produced by *Penicillium*, a prominent fungal genus, falls within the category of glycolipids. These surface-active agents possess notable properties such as effective emulsification and foaming, rendering them highly valuable across diverse applications. The glycolipids synthesized by *Penicillium* comprise hydrophilic sugar components coupled with hydrophobic fatty acid chains. This specific molecular arrangement facilitates a reduction in surface tension at liquid interfaces (Sanches *et al* 2021). Leveraging their amphiphilic nature, these biosurfactants can interact with both aqueous and hydrophobic substances, thereby showcasing potential in domains like enhanced oil recovery, bioremediation, and pharmaceuticals. The distinctive biosurfactant attributes produced by

Table 3.3. Multiple categories of fungal-derived biosurfactants.

Biosurfactant class	Fungi
Glycolipid	*Aspergillus niger*
Monoglucosyloxyoctadecenoic	*Penicillium citrinum*
Fusaroside	*Fusarium* sp.
α,β-Trehalose	*Fusarium fujikuroi*
Glycoprotein	*Aspergillus ustus* MSF3
Lipopeptide	*Fusarium SP BS-8*
	Penicillium chrysogenum SNP5
Enamide	*Fusarium proliferatum*
Sap-Pc protein	*Penicillium chrysogenum MUT 5039*
Cerato-platanins	*Aspergillus terreus MUT 271*
Monoglycerides	*Exophiala dermatitidis SK80*
Polymeric biosurfactant	*Curvularia lunata IM 2901*

Penicillium underscore its significance within the realm of sustainable biotechnological progress.

(b) *Aspergillus:* The fungal genus *Aspergillus* takes center stage in the realm of biosurfactant production, notably excelling in the creation of lipopeptide biosurfactants. These compounds exhibit significant structural variety and possess the capacity to reduce surface tension and interfacial tension at the respective surfaces and interfaces. Lipopeptides find versatile applications across multiple sectors, such as the food and cosmetic industries, where they are valued for their roles in emulsification and de-emulsification, dispersion, foaming, moisturization, and distribution. Additionally, they hold promise as agents that can lower viscosity, solubilize and mobilize hydrocarbons, and sequester metals, thus positioning them for potential environmental and bioremediation applications. Furthermore, their aptitude for creating pores and destabilizing biological membranes allows for applications as agents with antimicrobial, hemolytic, antiviral, antitumor, and insecticidal properties. Moreover, lipopeptides exert influence at the surface level and can modulate enzyme activities, thereby potentially enhancing the function of specific enzymes to improve microbial processes or inhibiting other enzymes, presenting their potential as antifungal agents (Femina Carolin *et al* 2021).

(c) *Fusarium*: *Fusarium* sp. has recently gained attention for its capacity to produce distinctive biosurfactants, marked by their exceptional attributes. These surface-active agents exhibit remarkable proficiency in mitigating both surface and interfacial tension, establishing their adaptability for diverse applications. The biosurfactants hailing from *Fusarium* manifest compelling traits including emulsification, dispersion, and foaming, positioning them favorably within sectors like the food and cosmetic industries. Moreover, their potential as modifiers of viscosity, facilitators of hydrocarbon solubilization, and agents for sequestering metals accentuate their pertinence within environmental and bioremediation landscapes (Qazi *et al* 2013) Additionally, these biosurfactants display properties that span antimicrobial, antiviral, antitumor, and insecticidal realms, underpinned by their ability to create pores and perturb biological membranes. *Fusarium's* adeptness in biosurfactant production, encompassing enzymatic modulation and surface reactivity, accentuates their far-reaching utility and potential contributions to diverse technological domains (Muneer *et al* 2013).

3.4.2 Advantages and disadvantages of fungal biosurfactant producers

The utilization of fungal organisms as biosurfactant producers offers distinct advantages for various applications, although these benefits come hand in hand with certain limitations. This section explores the merits and drawbacks of fungal biosurfactant production, drawing upon authentic examples from the literature.

3.4.2.1 Advantages

(i) Chemical structure versatility: Fungal-derived biosurfactants display an impressive array of chemical structures, affording them a wide spectrum of applications across diverse sectors, including personal care, agriculture, pharmaceuticals, and environmental remediation (Makkar *et al* 2011).

(ii) Biodegradability: Fungal biosurfactants stand out for their eco-friendliness, serving as environmentally-conscious alternatives to conventional synthetic surfactants. Their natural origin underscores their reduced environmental impact and long-term ecological viability (Thavasi *et al* 2011).

(iii) Low toxicity: Fungal biosurfactants commonly exhibit lower toxicity compared to their synthetic counterparts. This attribute renders them particularly suitable for applications involving direct human contact, such as in personal care products (Bharali *et al* 2011).

(iv) Performance under extreme conditions: A notable advantage of fungal biosurfactants lies in their robust performance across a wide range of pH, temperature, and salinity conditions. This adaptability positions them as valuable assets for applications in demanding industrial and environmental contexts (Banat *et al* 2010).

(v) Bioemulsification properties: Specific fungal-derived biosurfactants possess inherent bioemulsification properties, enabling effective dispersion and stabilization of oil–water mixtures. This quality holds relevance for applications such as oil spill mitigation and bioremediation (Das *et al* 2014).

3.4.2.2 Disadvantages

(i) Production challenges: Scaling up fungal biosurfactant production can present complexities and financial considerations. The optimization of culture conditions, the selection of appropriate fungal strains, and the upscaling of production processes pose challenges that could impact overall commercial viability (Sarachat *et al* 2017).

(ii) Yield variation: The yields of fungal biosurfactants may exhibit considerable variation due to factors such as strain variability, cultivation conditions, and fermentation methodologies. Such variability might influence consistent and predictable biosurfactant production, which could impact industrial applications (Thavasi *et al* 2011).

(iii) Extraction and purification complexity: Extracting and purifying fungal biosurfactants can entail intricate procedures that demand specialized techniques. The attainment of high-purity biosurfactants tailored for specific applications may prove to be a complex endeavor (Shah *et al* 2019).

(iv) Limited biosynthesis understanding: Despite progress, the biosynthesis pathways of fungal-derived biosurfactants remain incompletely elucidated for many species. This knowledge gap hampers the ability to engineer strains for enhanced biosurfactant production and properties (Satpute *et al* 2010).

(v) Competition with established surfactants: Fungal biosurfactants face formidable competition from established synthetic surfactants across various industries. Gaining market traction and establishing themselves as credible alternatives could pose challenges due to the prevailing use of synthetic products (Banat *et al* 2010).

3.5 Metabolic engineering for biosurfactant production

The application of metabolic engineering has emerged as a powerful and innovative approach to amplify biosurfactant production in microorganisms. By strategically modifying cellular metabolic pathways and genetic elements, metabolic engineering aims to optimize biosurfactant yields, enhance product properties, and unlock new possibilities for sustainable and economically feasible biosurfactant production. This section delves into the core principles of metabolic engineering and presents various strategies utilized to boost biosurfactant synthesis.

3.5.1 Principles of metabolic engineering

Metabolic engineering involves targeted modifications to cellular metabolic pathways, redirecting precursor fluxes towards desired biosynthetic pathways. By identifying crucial enzymes and regulatory points, metabolic engineers can fine-tune cellular metabolism to favor increased biosurfactant production (Stephanopoulos *et al* 1998). This interdisciplinary field draws insights from genetics, molecular biology, systems biology, and synthetic biology to enable the rational design of microbial cell factories for optimal biosurfactant synthesis.

3.5.2 Strategies for biosurfactant production using metabolic engineering

3.5.2.1 Overexpression of biosurfactant biosynthesis genes

A fundamental strategy in metabolic engineering entails the overexpression of genes involved in biosurfactant synthesis. By augmenting the expression of biosurfactant biosynthesis genes, such as those encoding critical enzymes or regulatory components, the metabolic flux toward biosurfactant production can be significantly enhanced (Sivapuratharasan *et al* 2022). This approach has demonstrated successful outcomes in various microorganisms, leading to substantial improvements in biosurfactant yields.

3.5.2.2 Knockout of competing metabolic pathways

Microbial cells often possess multiple competing metabolic pathways that can divert precursor molecules away from biosurfactant production. Metabolic engineering aims to alleviate this issue by eliminating or downregulating these competing pathways through genetic knockout or repression. This redirection of metabolic flux can effectively channel resources towards biosurfactant synthesis, thereby further elevating production efficiency.

In the realm of metabolic engineering, computational modeling has traditionally served as a guiding tool for experimental endeavors, foreseeing the impact of genetic alterations on metabolic processes. However, these modeling approaches often involve intricate kinetic techniques (Fell, 1996) demanding precise enzyme kinetic data, which, regrettably, remains largely elusive. Similarly, metabolic control analysis, necessitating flux control coefficient measurements derived from experimentation, faces a similar scarcity of data. A promising alternative, known as constraint-based modeling (CBM), circumvents these challenges by dissecting the behavior of genome-scale metabolic networks, relying solely on elementary physical and chemical constraints (Price *et al* 2003). These comprehensive network models have gained traction across numerous researchers. The efficacy of constraint-based modeling in investigating large-scale microbial networks has been emphatically established, spanning studies in metabolic engineering and various other domains (Tepper *et al* 2010).

Distinct CBM methodologies cater to diverse genetic manipulations feasible in engineering microbial strains. A frequently employed approach, OptKnock, explores gene knockout combinations to facilitate the hyper-production of targeted metabolites (Burgard *et al* 2003). An analogous technique, OptStrain, not only accommodates gene knockouts but also assimilates novel enzyme-encoding genes from disparate species into a designated microbial genome (Pharkya *et al* 2004). A recent stride in this domain, OptReg, has surfaced, pursuing modifications encompassing both up- and down-regulation of metabolic enzymes, in conjunction with gene knockouts, to fulfill specific bio-production objectives (Pharkya and Maranas, 2006).

3.5.2.3 Optimization of culture conditions

Environmental factors exert a crucial influence on biosurfactant production. As part of metabolic engineering, optimizing culture conditions, such as temperature, pH, nutrient availability, and oxygen supply, is employed to create an ideal micro-environment for enhanced biosurfactant synthesis. A comprehensive understanding of the impact of these factors on biosurfactant production aids in maximizing yields and improving the scalability of the process (Dong *et al* 2022).

3.5.2.4 Use of synthetic biology tools

The integration of synthetic biology tools into metabolic engineering has revolutionized the field, offering powerful genetic manipulation and pathway construction capabilities. Synthetic biology facilitates the design and synthesis of custom DNA sequences, genetic circuits, and engineered promoters, allowing precise control over biosurfactant biosynthesis pathways. This level of control enhances predictability and efficiency in biosurfactant production (Santos-Merino *et al* 2019).

3.6 Advances in metabolic engineering for biosurfactant production

Recent progress in metabolic engineering has yielded promising outcomes in biosurfactant production. From leveraging advanced omics technologies to identify novel biosurfactant-producing genes to utilizing machine learning algorithms for pathway optimization, these advancements are propelling the field forward. Metabolic pathway manipulation has gained significant prominence within this domain, facilitated by expanding insights into the intricate processes of biosurfactant biosynthesis and the pathogenic tendencies of diverse microorganisms. Metabolic engineering emerges as a pivotal approach, encompassing the dual objectives of diminishing pathogenic attributes and crafting robust non-pathogenic biosurfactant producers. These dual facets hold paramount significance in ensuring both economic viability and safety in the biosurfactant production landscape (Chong and Li 2017). Furthermore, the fusion of metabolic engineering with synthetic biology tools is paving the way for tailor-made biosurfactant-producing microorganisms, unlocking new avenues for sustainable and economically viable biosurfactant production. Metabolic engineering presents a transformative approach to unleash the full potential of biosurfactant production. Through targeted modifications and harnessing the power of genetic engineering, researchers are poised to revolutionize biosurfactant synthesis, charting a path toward a greener and more sustainable future.

3.7 Conclusions

In summary, the comprehensive investigation into microbial biosurfactants, spanning a spectrum of microorganisms such as bacteria, yeasts, and fungi, including noteworthy instances like *Penicillium* and *Aspergillus*, has unveiled a captivating realm rich with biotechnological potential. These remarkable compounds, characterized by their diverse structural variations and multifaceted attributes, offer a wide array of promising applications.

Collectively, the accumulated insights emphasize biosurfactants' pivotal role in addressing contemporary challenges, ranging from enhanced oil recovery to ecologically conscious strategies in bioremediation. The utilization of metabolic engineering to harness the capabilities of microorganisms for escalated biosurfactant production emerges as a beacon of innovation, presenting opportunities for sustainable solutions in the realm of surfactants.

Nonetheless, amidst the potential, obstacles persist. The intricacies tied to the large-scale production of biosurfactants, coupled with concerns regarding purity, stability, and regulatory compliance, underscore the need for a comprehensive and holistic approach to successful integration.

As we navigate the ever-evolving landscape of biotechnology, the exploration of biosurfactants continues to be an ongoing journey. With each stride, we inch closer to unlocking their full potential in reshaping industries, bolstering environmental remediation endeavors, and even shaping the forefront of medical and pharmaceutical advancements.

As we conclude this exploration, it becomes unmistakably clear that biosurfactants are more than mere chemical compounds; they embody agents of transformation, embodying the amalgamation of scientific acumen, ecological responsibility, and technological progress. The road ahead is illuminated by the immense potential of these dynamic entities, guiding us towards an era characterized by sustainability and ingenuity.

References

Arima K, Kakinuma A and Tamura G 1968 Surfactin, a crystalline peptide lipid surfactant produced by *Bacillus subtilis*: isolation, characterization and its inhibition of fibrin clot formation *Biochem. Biophys. Res. Commun.* **31** 488–94

Arutchelvi J I *et al* 2008 Mannosylerythritol lipids: a review *J. Ind. Microbiol. Biotechnol.* **35** 1559–70

Banat I M, Franzetti A, Gandolfi I, Bestetti G, Martinotti M G, Fracchia L and Marchant R 2010 Microbial biosurfactants production, applications, and future potential *Appl. Microbiol. Biotechnol.* **87** 427–44

Bharali P, Konwar B K and Thakur A J 2011 Fungal biosurfactants of potential importance in biomedical applications *Front. Microbiol.* **2** 1–8

Burgard A P *et al* 2003 Optknock: a bilevel programming framework for identifying gene knockout strategies for microbial strain optimization *Biotechnol. Bioeng.* **84** 647–57

Chen C, Sun N, Li D, Long S, Tang X, Xiao G and Wang L 2018 Optimization and characterization of biosurfactant production from kitchen waste oil using *Pseudomonas aeruginosa Environ. Sci. Pollut. Res.* **25** 14934–43

Chong H and Li Q 2017 Microbial production of rhamnolipids: opportunities, challenges and strategies *Microb. Cell Fact.* **16** 137

Choudhary D K, Kasotia A, Jain S and Vaishnav A 2019 Surfactins and iturins: microbial cyclic lipopeptides for sustainable agriculture *Microbial Cyclic Di-nucleotide Signaling* (Singapore: Springer) pp 123–36

da Silva A F, Banat I M, Giachini A J *et al* 2021 Fungal biosurfactants, from nature to biotechnological product: bioprospection, production and potential applications *Bioprocess. Biosyst. Eng.* **44** 2003–34

Das P, Mukherjee S and Sen R 2014 Antimicrobial potential of a lipopeptide biosurfactant derived from a marine *Bacillus circulans J. Appl. Microbiol.* **117** 373–83

Dong H *et al* 2022 Optimization and characterization of biosurfactant produced by indigenous *Brevibacillus borstelensis* isolated from a low permeability reservoir for application in MEOR *RSC Advances* **12** 2036–47

Edward J R and Hayashi J A 1965 Structure of rhamnolipid from *Pseudomonas aeruginosa Arch. Biochem. Biophys.* **111** 415–21

Fell D A 1996 *Understanding the Control of Metabolism* (London: Portland Press)

Femina Carolin C, Senthil Kumar P and Tsopbou Ngueagni P 2021 A review on new aspects of lipopeptide biosurfactant: types, production, properties and its application in the bioremediation process *J. Hazard. Mater.* **407** 124827

Fernandez C W, See C R and Peter 2020 Decelerated carbon cycling by ectomycorrhizal fungi is controlled by substrate quality and community composition *New Phytologist* **226** 569–82

Fernandes N A T and Simões L A 2023 Dias Dr biosurfactants produced by yeasts: fermentation, screening, recovery, purification, characterization, and applications *Fermentation* **9** 207

Gaur V K, Regar R K, Dhiman N, Gautam K, Srivastava J K, Patnaik S, Kamthan M and Manickam N 2019 Biosynthesis and characterization of sophorolipid biosurfactant by *Candida* spp.: application as food emulsifier and antibacterial agent *Bioresour. Technol.* **285** 121314

Jarvis F G and Johnson M J 1949 A glycolipid produced by *Pseudomonas aeruginosa J. Am. Oil Chem. Soc.* **71** 4124–6

Karanth N G K, Deo P and Veena A 1999 Microbial production of biosurfactants and their importance *Curr. Sci.* **77** 116–26

Lotfabad T B, Abassi H, Ahmadkhaniha R, Roostaazad R, Masoomi F, Zahiri H S and Noghabi K A 2018 Sophorolipids production by *Candida bombicola* ATCC 22214 and its potential application in microbial enhanced oil recovery *Biotechnol. Progr.* **34** 645–57

Makkar R S, Cameotra S S and Banat I M 2011 Advances in utilization of renewable substrates for biosurfactant production *AMB Express* **1** 1–19

Marcelino P R F, Peres G F D, Terán-Hilares R, Pagnocca F C, Rosa C A, Lacerda T M, dos Santos J C and da Silva S S 2019 Biosurfactants production by yeasts using sugarcane bagasse hemicellulosic hydrolysate as new sustainable alternative for lignocellulosic biorefineries *Ind. Crops Prod.* **129** 212–23

McInerney M J, Javaheri M and Nagle D P 1990 Properties of the biosurfactant produced by *Bacillus liqueniformis* strain JF-2 *I. J. Microbiol. Biotechnol.* **5** 95–102

Muneer A Q, Subhan M, Fatima N, Ishtiaq Ali M and Ahmed S 2013 Role of biosurfactant produced by *Fusarium* sp. BS-8 in enhanced oil recovery (EOR) through sand pack Column *Int. J. Biosci., Biochem. Bioinform.* **3** 6

Nguyen T T, Youssef N H, McInerney M J and Sabatini D A 2008 Rhamnolipid biosurfactant mixtures for environmental remediation *Water Res.* **42** 1735–43

Nitschke M and Coast S G 2007 Biosurfactants in food industry *Trends Food Sci. Technol.* **18** 252–9

Nitschke M and Pastore G M 2002 Biosurfactant production by Candida species using renewable substrates *Biotechnol. Progr.* **18** 1433–7

Pharkya P and Maranas C D 2006 An optimization framework for identifying reaction activation/ inhibition or elimination candidates for overproduction in microbial systems *Metab. Eng.* **8** 1–13

Pharkya P *et al* 2004 Optstrain: a computational framework for redesign of microbial production systems *Genome Res.* **14** 2367–76

Price N D *et al* 2003 Genome-scale microbial in silico models: the constraints-based approach *Trends Biotechnol.* **21** 62–169

Qazi M A, Subhan M, Fatima N, Ali M I and Ahmed S 2013 Role of biosurfactant produced by *Fusarium sp.* BS-8 in enhanced oil recovery (EOR) through sand pack column. *IJBBB* **3** 598

Rani M, Weadge J T and Jabaji S 2020 Isolation and characterization of biosurfactant-producing bacteria from oil well batteries with antimicrobial activities against food-borne and plant pathogens *Front. Microbiol.* **11** 64

Ribeiro B G, Guerra J M C and Sarubbo L A 2020 Potential food application of a biosurfactant produced by *Saccharomyces cerevisiae* URM 6670 *Front. Bioeng. Biotechnol.* **8** 434

Rufino R D, Luna J M, de Campos Takaki G M and Sarubbo L A 2014 Characterization and properties of the biosurfactant produced by *Candida lipolytica* UCP 0988 *Electron. J. Biotechnol.* **17** 1–5

Rywińska A, Juszczyk P, Wojtatowicz M, Robak M, Lazar Z, Tomaszewska L and Rymowicz W 2013 Glycerol as a promising substrate for yarrowia lipolytica biotechnological applications *Biomass Bioenergy* **48** 148–66

Sanches M A, Luzeiro I G, Alves Cortez A C, Simplício de Souza É, Albuquerque P M, Chopra H K and Braga de Souza J V 2021 Production of biosurfactants by ascomycetes *Int. J. Microbiol.* **2021** 6669263

Santos D K F, Rufino R D, Luna J M, Santos V A and Sarubbo L A 2016 Biosurfactants: multifunctional biomolecules of the 21st century *Int. J. Mol. Sci.* **17** 401

Santos D K F, Rufino R D, Luna J M, Santos V A and Sarubbo L A 2017 Biosurfactants: multifunctional biomolecules of the 21st century *Int. J. Mol. Sci.* **18** 1–22

Santos-Merino M, Singh A K and Ducat D C 2019 New applications of synthetic biology tools for cyanobacterial metabolic engineering *Front. Bioeng. Biotechnol.* **7** 33

Sarachat S, Pechyen C and Reungsang A 2017 Current trends in microbial biosurfactants in environmental and industrial applications *J. Surf. Sci. Technol.* **33** 167–86

Satpute S K, Banat I M, Dhakephalkar P K, Banpurkar A G and Chopade B A 2010 Biosurfactants, bioemulsifiers and exopolysaccharides from marine microorganisms *Biotechnol. Adv.* **28** 436–50

Shah S, Sharma S and Shah A 2019 Purification and characterization of lipopeptide biosurfactant produced by *Bacillus subtilis* SBS8 isolated from silty-clay soil of the Gujarat region in India *Biochem. Cell Arch.* **19** 15–22

Sharma D and Saharan B S 2016 Functional characterization of biomedical potential of biosurfactant produced by *Lactobacillus helveticus Biotechnol. Rep.* **11** 27–35

Sivapathasekaran C, Mukherjee S, Samanta R and Sen R 2009 High-performance liquid chromatography purification of biosurfactant isoforms produced by a marine bacterium *ABC* **395(3)** 845–54

Sivapuratharasan V, Lenzen C, Michel C, Muthukrishnan A B, Jayaraman G and Blank L M 2022 Metabolic engineering of *Pseudomonas taiwanensis* VLB120 for rhamnolipid biosynthesis from biomass-derived aromatics *Metab. Eng. Commun.* **15** e00202

Stephanopoulos G, Aristos A A and Høiriis Nielsen J 1998 *Metabolic Engineering Principles and Methodologies* (San Diego: Academic)

Tepper N and Shlomi T 2010 Predicting metabolic engineering knockout strategies for chemical production: accounting for competing pathways *Bioinformatics* **26** 536–43

Thavasi R, Jayalakshmi S, Balasubramanian T and Banat I M 2011 Biosurfactant production by *Bacillus* sp. MIG isolated from the marine environment *Acta Biol. Indica* **1** 173–82

Tripathy D B and Mishra A 2016 Sustainable biosurfactants *Sustain. Inorg. Chem.* **1** 175–92

Vatsa P, Sanchez L, Clement C, Baillieul F, Dorey S, Riollet C and Cordelier S 2021 Yeast-based production of glycolipid biosurfactants *Glycolipids* (New York: Academic) ch 2 pp 29–45

Venkata Ramana K and Karanth N G 1989 Production of biosurfactants by the resting cells of *Pseudomonas aeruginosa* CFTR-6 *Biotechnol. Lett.* **11** 437–42

Zhang Y, Xu Y, Zou W, Jiang L, Yin H and Shen Z 2020 Rhamnolipids: production, microbial rhamnolipid synthesis, and potential applications *Appl. Microbiol. Biotechnol.* **104** 1173–88

Chapter 4

Kinetics of microbial growth and biosurfactant production

Ashish Kumar Agrahari, Pratichi Singh and Payal Gulati

This chapter describes the microbial growth, product formation and consumption of substrate. The microbes present in fed culture show the growth pattern involving four phases, namely lag, log, exponential and death phase. The growth of these microbes depends upon the availability of the nutrients, oxygen availability, pH, light and temperature. The kinetics of biosurfactant production represent the temporal and quantitative aspects of the synthesis of biosurfactants by microorganisms. This chapter also includes an overview of the microbial and biosurfactant measurement techniques such as optical density (OD) measurement, colony cdunt methods, Dry weight measurement, and surface tension measurement. Moreover, biosurfactants' various industrial, agricultural and environmental applications are also described in it. This chapter underscores the significance of these techniques and their potential for diverse scientific and industrial applications.

4.1 Kinetics of microbial growth

This section presents the brief of growth and death of microbes. Kinetics is a branch of science that controls the rates and phenomena of any processes including biological, chemical and physical. In microbiology branch, kinetics study includes every dynamic process of microbial life such as growth, survival, death, product formation, acclimatization, mutations, and interactions. Microbe(s) growth kinetics depicts the correlation between particular growth rate (μ) of microbe(s) population versus concentration of substrate (s).

Growth kinetics can be defined as the autocatalytic reaction where growth rate is equivalent to the cell concentration. In the cell, it is based on the correlation between energy production and product synthesis. It is classified as non- growth-associated and growth-associated. Cell division leads to cell growth for different microorganisms. There are several methods by which cell division occurs in microorganisms

doi:10.1088/978-0-7503-5989-4ch4

such binary fission in bacteria, intra-cellular division in viruses and budding in yeast [1].

The bacterial growth involves cell elongation which includes enlargement in cell wall, cell membrane and enhancement in cell volume. Cell division starts with replication of cellular DNA followed by formation of two copies of chromosomal DNA which divides equally in each cell. When the cell begins to elongate, FtsZ, a protein moves towards the midpoint of the cell and surrounds the chromosome by forming a ring at the periphery of the midpoint of the chromosome. After, nucleoids are separated in the elongated cells and septum formation is completed, which in turn divides the elongated cell into two equal proportion daughter cells, as shown in figure 4.1. The entire cell cycle of active culture *Escherichia coli* bacteria takes 20 min.

The time required to complete a fission cycle by bacteria is known as doubling time which remains constant if environmental conditions remain favourable [2]. For industrial purposes, bacteria are cultured in three types of cultures including batch, fed-batch and continuous [3, 4].

In batch culture, a fixed quantity of media is added and a closed system is maintained for growth of the desired microbe. In this system, when nutrients are consumed then all the cellular product is collected from the medium before replacing it with a fresh batch. Therefore, in this system both growth and yield are affected by availability and concentration of the nutrients. Nutrients concentration and growth rate are directly proportional to each other, i.e., higher concentration nutrients result in increased cellular yield and vice-versa [5]. When microorganisms are cultured in batch culture then following growth phases are observed in the bacterial growth such as lag, log, stationary and death phase [6].

4.1.1 Growth curve phases

The bacterial growth curve represents the count of alive cells in a bacterial population over a period of time. The growth curve includes following phases as indicated in a graph, represented in figure 4.2.

4.1.1.1 Lag phase

This phase lasts for 1–4 h. This is the initial phase where cells adapt themselves to the culture medium. In this period bacteria do not divide, instead they mature by synthesizing RNA, enzymes and other molecules which are required to utilize nutrients present in the medium [1]. Length of this phase is dependent upon the environmental conditions, culture media and bacterial species. This phase length can be long or short, depending upon the health of the bacteria.

4.1.1.2 Exponential (log) phase

This phase lasts for 8 h and is also known as logarithmic phase. In this phase, bacteria enter a new phase where chromosomal replication, cellular growth and reproduction take place. This phase involves cell doubling and new bacteria appeared per unit time is equivalent to the current population. When growth is

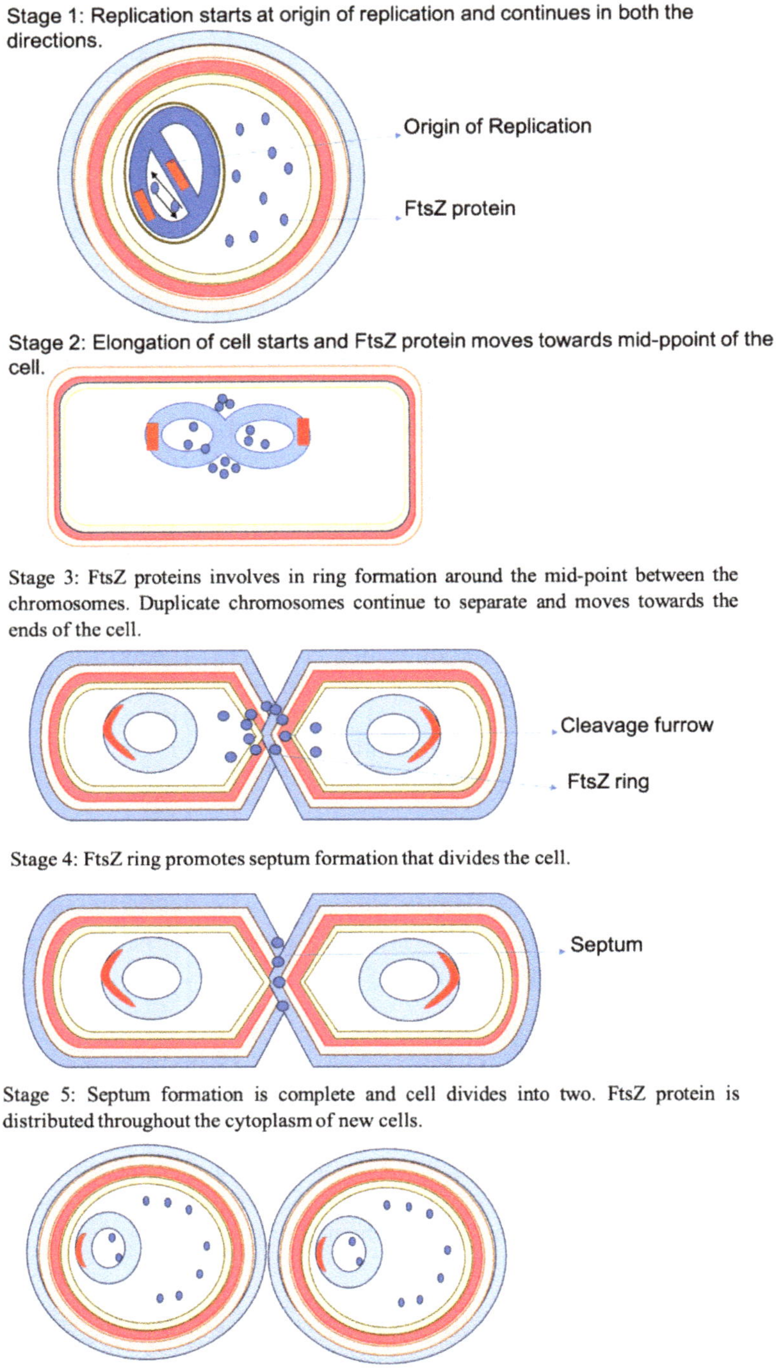

Figure 4.1. Binary fission of bacterium.

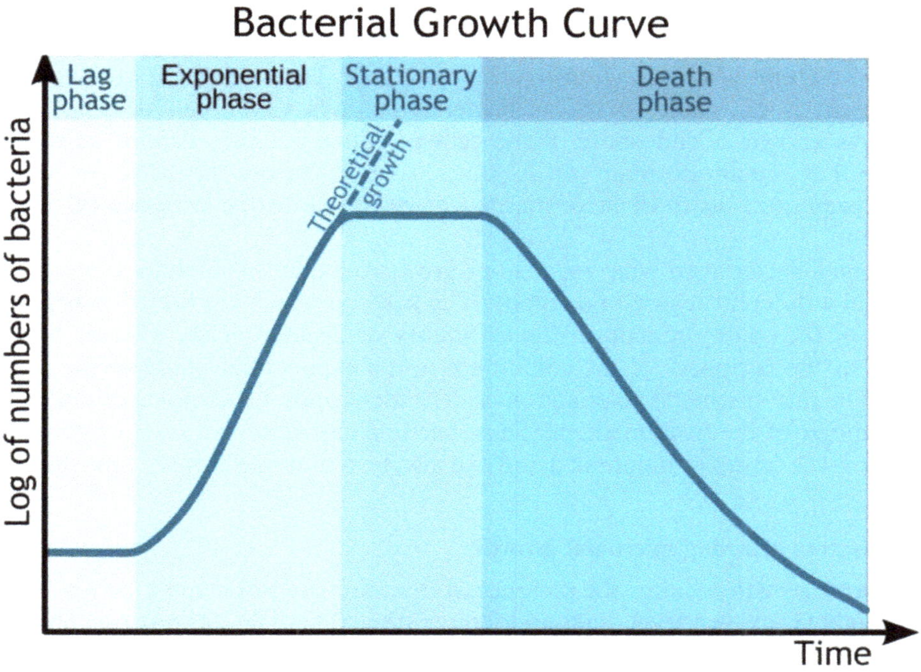

Figure 4.2. Bacterial growth curve.

not limited then doubling occurs at a constant rate where both number of cells and population rate increase with period of time. Therefore, for exponential growth period, the plot natural log of cell number versus time gives straight line and slope of this graph represents the rate of growth of organism. This gives the quantitative count for cell division per unit time. During this log phase, the metabolic rate of every single cell is at its peak and this is sometimes desired for laboratory and industrial purposes [6]. But this exponential growth continues for a specified time period unless the nutrients of medium are exhausted and waste is accumulated. For example: cyanobacteria doubling time period is four times in a single day.

4.1.1.3 Stationary phase

This phase occurs due to depletion of nutrients from the medium and results in accumulation of waste, i.e., formation of inhibitory products such as organic acids take place. In the stationary phase the number of dying cells is equivalent to the newly formed cells and, therefore, the population size remains constant. This is represented by a straight horizontal linear part of the curve, as depicted in figure 4.2. The metabolic rate of living cells is decreased [7].

4.1.1.4 Death (decline) phase

In this phase, all the nutrients are exhausted and toxic products as well as autocatalytic enzymes are accumulated, consequently, bacterial population reaches a point where cells die at a faster rate in comparison to the alive cells which are still alive, metabolizing and reproducing. Mostly, all the cells die in this phase but there are some cultures where a few survivors may remain. The case where a few survivors are present is especially for those bacterial cultures which can develop resting structures known as endospore. Some cells are alive but they cannot be cultured because they remain dormant [6]. At times, some survivors stay alive for about a month even after death of large number of cells, due to the presence of released nutrients.

Colonial bacteria growing on a solid growth medium also show these growth phases in a different region of a colony. The bacteria which are buried in the oldest section of the colony remain in the stationary or decline phase, whereas bacteria present at the periphery of the colony are in the exponential phase of the growth curve. The rate at which bacteria grow is dependent upon the removal of old culture and addition of the fresh medium. Thus, bacteria can grow fast or slow depending upon how the set-up is maintained and can also be maintained longer time duration.

4.1.2 Factors affecting microbial growth

Microbe(s) growth requires the presence of optimum environmental factors such as nutrients, pH, oxygen level, light and temperature. These factors vary according to the species, required for their appropriate growth [8].

4.1.2.1 Nutrients

When culture medium is substrate sufficient then growth of bacteria takes place at a faster rate and conversely, when culture medium is substrate deficient then growth rate of bacteria decreases. Thus, growth rate of bacteria is dependent upon the availability of the nutrients, i.e., substrate. Specific bacteria have specific requirements of nutrients, such as chemo-heterotrophic bacteria use organic carbon for energy and structure. Therefore, sources of carbon are required where glucose and fructose serve the purpose. Requirement of other sources includes: nitrogen for making of amino acids, proteins, and lipids; sulfur and phosphorus for making components of bacteria. For proper functioning of enzymes and other processes elements required are potassium, calcium, magnesium, iron, manganese, cobalt and zinc. The plot of substrate concentration versus growth rate is shown in figure 4.3.

4.1.2.2 pH

It is an important factor for the microbial growth as pH affects the ionic properties of the bacterial cell. Several bacteria are responsible for production of organic acids which lower the pH of the medium that hampers the bacterial growth. Moreover, some media constituents are affected by the lower pH. Thus, optimum pH is essential for appropriate growth of the organisms. Mostly bacteria grow at a neutral pH but there are some bacteria which can survive at acidic or basic pH, for example *Lactobacillus lactis* requires an acidic pH for proper growth. This bacterium is used for industrial purpose. Relationship of pH and bacterial growth is demonstrated in figure 4.4.

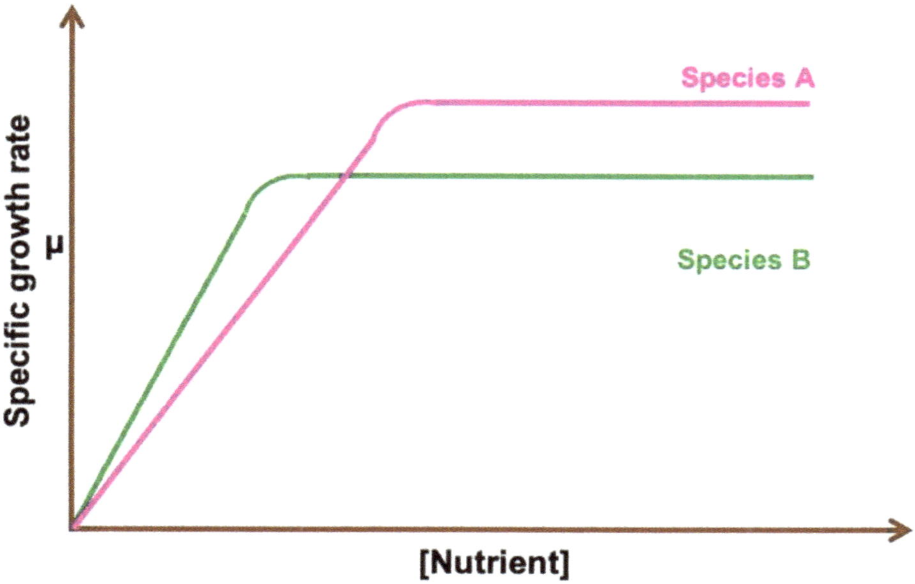

Figure 4.3. Substrate concentration versus growth rate.

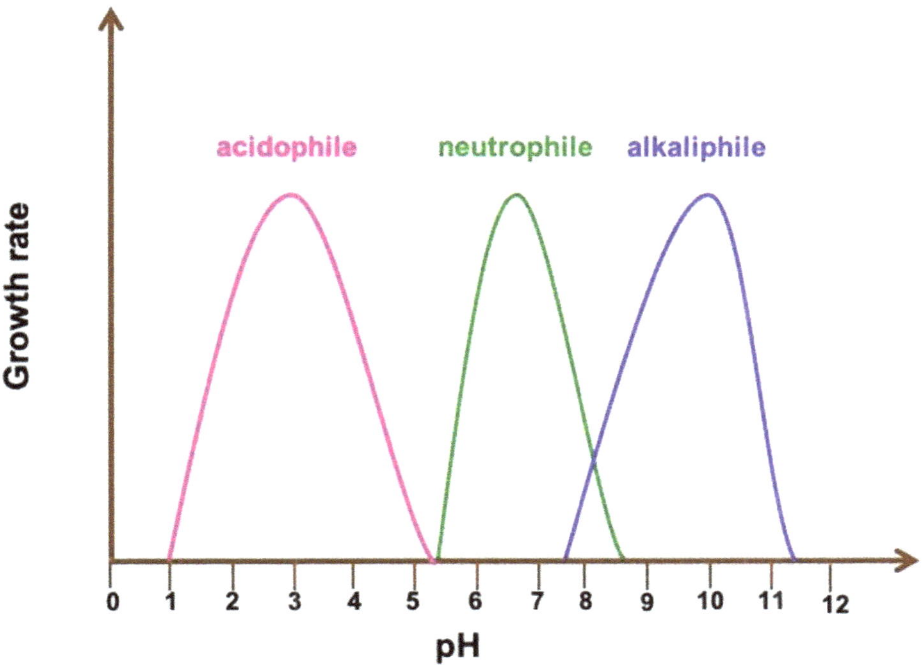

Figure 4.4. Relationship of pH versus bacterial growth.

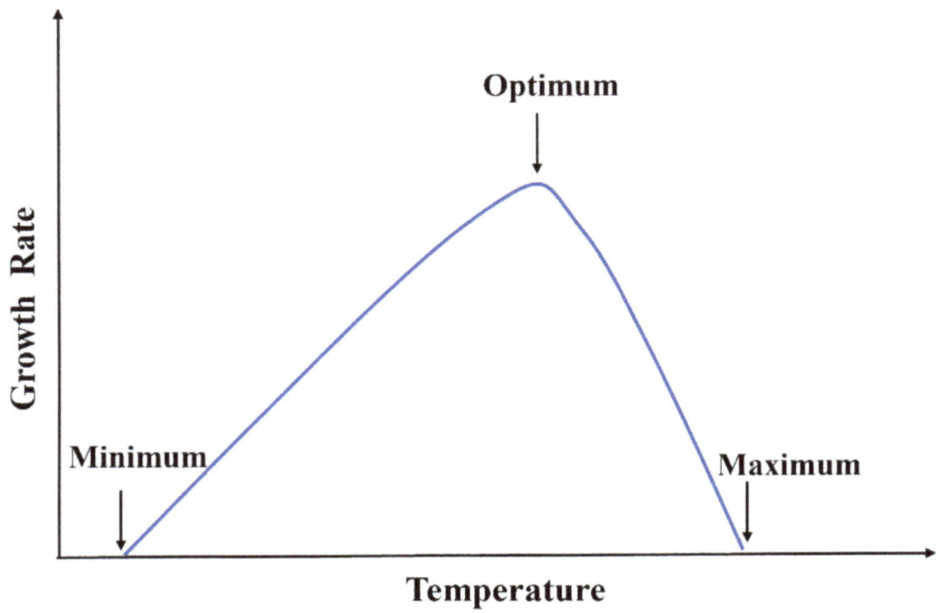

Figure 4.5. Relationship of temperature versus bacterial growth rate.

4.1.2.3 Temperature

One of the important factors for bacterial growth is temperature where bacterial growth at lowest temperature is called minimum temperature and at highest temperature is known as maximum temperature. No bacterial growth takes place below and above minimum and maximum temperature, respectively. As below minimum temperature, the cell membrane of bacteria will become solid and stiff, no nutrients will transport inside the cell and no growth is promoted. In contrast, above maximum temperature, proteins and enzymes of the cell are denatured, thus bacterial growth will stop. The relationship of temperature and bacterial growth rate is shown in figure 4.5.

Temperature is an important environmental factor that affects the growth of organisms. Several pathogens grow at 37 °C and bacteria are classified in three categories on the basis of optimal temperature range[1].

- Mesophile: optimal temperature range of mesophile bacteria is 25 °C–40 °C.
- Psychrophile: optimal temperature range of psychrophile bacteria is below 20 °C.
- Thermophile: optimal temperature range of thermophile bacteria is 55 °C–80 °C. For example: *Bacillus stearothermophilus*.

4.1.2.4 Oxygen availability

Bacteria is broadly categorized into three groups on the basis of oxygen requirement.

[1] https://microbenotes.com/environmental-factors-affecting-microbial-growth/ (accessed 15 September 2023).

- Aerobic bacteria: Oxygen is essential for this bacterial growth and they are also known as obligate aerobes. For example, *Pseudomonas aeruginosa.*
- Facultative anaerobic bacteria: these bacteria can grow in the presence of as well as in the absence of oxygen. Mostly pathogenic bacteria lie in this category. For example, *E. coli.*
- Anaerobes: These are obligate anaerobes which cannot grow in the presence of oxygen. For example, *Clostridium tetani.*

Another group apart from these is called microaerophilic which requires a trace amount of oxygen. For example, *Helicobacter pylori.*

4.1.2.5 Light

Bacteria which require light for growth are phototrophic bacteria. Some bacteria can also grow in the absence of light. The presence of electromagnetic radiation such as UV rays can hamper the growth of bacteria. A unique bacterium such as photochromogenic mycobacterium, makes pigments in the presence of light.

4.2 Kinetics of biosurfactant production

The kinetics of biosurfactant production represent the temporal and quantitative aspects of the synthesis of biosurfactants by microorganisms. It is a fundamental aspect of bioprocess engineering that plays a pivotal role in optimizing the production of these valuable molecules. Biosurfactants, with their unique surface-active properties, have a wide range of applications, and understanding their production kinetics is essential for efficient and sustainable processes.

The kinetics of biosurfactant production encompass various key aspects:

- Temporal dynamics: Biosurfactant production kinetics elucidates how the concentration of biosurfactants changes over time throughout the microbial cultivation process. It tracks the transition between growth phases, biosurfactant synthesis, and the eventual decline in production as cells enter the stationary phase or undergo lysis [9].
- Growth phase influence: These kinetics are closely associated with the growth phases of microorganisms. The exponential growth phase is typically marked by the highest rates of biosurfactant production. The specific kinetics may vary depending on whether biosurfactant production is primarily growth-associated or non-growth-associated [9].
- Factors affecting kinetics: Numerous factors impact the kinetics of biosurfactant production, including the availability of nutrients, environmental conditions (e.g., temperature, pH), and the presence of inducers or stressors. The choice of microbial strain and its genetic characteristics also significantly influence the kinetics [10].
- Monitoring and optimization: To optimize biosurfactant production processes, it is imperative to continuously monitor critical parameters such as cell density, biosurfactant concentration, and nutrient levels. Process optimization often involves adjusting environmental conditions, nutrient

concentrations, and inoculum size to enhance biosurfactant production rates and yields [11].

- Applications: Understanding biosurfactant production kinetics is essential for tailoring biosurfactant production processes to meet the requirements of specific applications. Biosurfactants have diverse uses in industries such as bioremediation, enhanced oil recovery, food processing, and pharmaceuticals [9].
- Future directions: Ongoing research focuses on advancing the kinetics of biosurfactant production. This includes the development of genetically engineered microbial strains optimized for biosurfactant synthesis. Innovations in bioprocess engineering and bioreactor design contribute to more precise control of biosurfactant production kinetics [12].

4.2.1 Biosurfactant production process

The biosurfactant production process is a multi-stage biotechnological endeavor that involves the cultivation of microorganisms capable of synthesizing biosurfactants under controlled conditions. This process can be broken down into several key steps, each of which is crucial for the successful production of biosurfactants.

4.2.1.1 Microorganism selection

The journey to biosurfactant production begins with the careful selection of a suitable microorganism. Various microorganisms, including bacteria, yeasts, and fungi, have demonstrated the ability to produce different types of biosurfactants. The choice of microorganism depends on factors such as the desired biosurfactant type, environmental conditions, and the specific application of the biosurfactant.

4.2.1.2 Nutrient medium formulation

Once a microorganism is selected, a nutrient medium is prepared. This medium serves as the growth medium for the microorganism and provides the necessary carbon and nitrogen sources, minerals, vitamins, and trace elements essential for cell growth and biosurfactant production. The composition of the medium is tailored to meet the metabolic requirements of the chosen microorganism [9].

4.2.1.3 Inoculation and fermentation

The selected microorganism is inoculated into the nutrient medium, typically in a bioreactor or fermenter. Here, environmental conditions such as temperature, pH, and oxygen levels are carefully controlled to optimize the growth of the microorganism. During this phase, the microorganisms consume nutrients, grow, and metabolize to produce biosurfactants as a secondary metabolite [9].

4.2.1.4 Biosurfactant production

Biosurfactant production occurs primarily during the growth phase of the microorganism. Depending on the kinetics involved (growth-associated or non-growth-associated), biosurfactants are synthesized and excreted into the surrounding medium. The rate and yield of biosurfactant production depend on factors like the specific growth rate, the availability of carbon sources, and the presence of inducers or stressors [9].

4.2.1.5 Harvesting and cell separation

Once biosurfactant production reaches its peak, the culture is harvested. This involves separating the microbial cells from the medium, often through centrifugation or filtration. The cells can be recycled for further use in subsequent fermentations or discarded [13].

4.2.1.6 Biosurfactant purification

The harvested broth may contain impurities such as residual nutrients, cell debris, and other contaminants. To obtain pure biosurfactants suitable for commercial applications, purification steps are undertaken. Techniques like precipitation, solvent extraction, chromatography, and ultrafiltration are employed to isolate and concentrate the biosurfactants [12].

4.2.1.7 Formulation and application

The purified biosurfactants can be further formulated into various products, including detergents, emulsifiers, and foaming agents. These products find applications in a wide range of industries, including agriculture, food processing, cosmetics, pharmaceuticals, and environmental remediation [14].

Understanding and optimizing each stage of the biosurfactant production process are essential for achieving high yields and maintaining product quality. Researchers and industrial practitioners continuously work to improve microorganism strains, fermentation conditions, and purification methods to enhance the efficiency and sustainability of biosurfactant production processes.

4.2.2 Biosurfactant production kinetics

Understanding the kinetics of biosurfactant production is essential for optimizing the yield, quality, and efficiency of biosurfactant production processes. The kinetics of biosurfactant production can be categorized into two main types: growth-associated and non-growth-associated production kinetics. Each of these kinetics has distinct characteristics that influence the production of biosurfactants.

4.2.2.1 Growth-associated production

Growth-associated biosurfactant production is characterized by the production of biosurfactants that are directly linked to the growth of microorganisms. This type of production is often observed when biosurfactants serve as essential components for microbial growth or when biosurfactants are produced as a consequence of normal metabolic activities during exponential growth. Key features of growth-associated biosurfactant production include:

- Specific growth rate dependency: The rate of biosurfactant production is directly proportional to the specific growth rate of the microorganisms. Faster-growing microorganisms tend to produce biosurfactants at higher rates [15].

- Exponential phase production: Biosurfactants are typically synthesized and excreted during the exponential (log) phase of microbial growth. This phase is characterized by rapid cell division and high metabolic activity [16].
- Nutrient-dependent: Growth-associated biosurfactant production is heavily influenced by the availability of nutrients, especially carbon sources. Adequate carbon sources are essential because they serve as precursors for biosurfactant synthesis [15].
- Examples: Rhamnolipids produced by *Pseudomonas* species and sophorolipids produced by yeast species are examples of biosurfactants that are often produced in a growth-associated manner [17].

4.2.2.2 Non-growth-associated production

Non-growth-associated biosurfactant production is characterized by the production of biosurfactants that are not directly linked to the growth of microorganisms. Instead, biosurfactants are produced as secondary metabolites under specific conditions, such as stress or the presence of inducers. Key features of non-growth-associated biosurfactant production include:

- Stress or inducer-dependent: Biosurfactant production in this mode is often triggered by stressful conditions or the presence of specific inducers. Stressors can include factors like limited nutrient availability, high salinity, or the presence of hydrophobic substrates [16].
- Stationary phase production: Non-growth-associated biosurfactant production is frequently observed during the stationary phase of microbial growth when cell growth has ceased, and cells enter a maintenance phase [15].
- Nutrient-independent: Unlike growth-associated production, non-growth-associated production can occur even when nutrient levels are limited. Biosurfactant production in this mode is driven by factors other than nutrient availability [17].
- Examples: Surfactin produced by *Bacillus subtilis* and lipopeptides produced by various bacterial species are examples of biosurfactants often produced in a non-growth-associated manner [15].

Understanding whether biosurfactant production follows growth-associated or non-growth-associated kinetics is crucial for process design and optimization. It allows for the development of strategies that can enhance biosurfactant production under specific conditions, leading to more efficient and cost-effective processes.

4.2.3 Factors affecting biosurfactant production

The composition of the biosurfactant along with its emulsifying activity depends upon the culture conditions and producer strain. Therefore, certain parameters like, type of carbon and nitrogen source, their C:N ratio, nutritional constraints and physical–chemical factors including pH, divalent cations, temperature and aeration affects the biosurfactant production in addition to the type of polymer generated [18].

4.2.3.1 Substrate concentration

Source of carbon: Carbon substrate plays a major role in estimation of quality and quantity of biosurfactant production [19]. Certain reported carbon sources for biosurfactant production are glucose, glycerol, sucrose, crude oil and diesel [20].

Source of nitrogen: It is essential for biosurfactant production medium as microbial growth requires proteins and enzymes, which are a nitrogen source. Several nitrogen sources that have been employed for biosurfactant production are yeast extract, ammonium sulfate, ammonium nitrate, urea, urea peptone, meat extract and malt extract. Mostly yeast extract is used as a nitrogen source for biosurfactant production and its usage amount is dependent upon the organism and culture medium. The preferred nitrogen source for biosurfactant production is different for different species such as *Arthrobacter paraffineus* requires ammonium salt and urea, whereas *P. aeruginosa* requires nitrogen [21].

Hence, production cost of biosurfactants strongly depends upon the carbon and nitrogen sources in the media. Thus, recently new novel resources have come into use which can be used as carbon and nitrogen source in media. These resources are by-products or remnant waste products of the food industry, such as from a distillery, molasses, frying oil, vegetable oil or plant oil, which can be used singly or with the stabilized resources. These products reduce the cost of biosurfactant production [22–24]. The cheapest source of biosurfactant production from *Pseudomonas*, *Bacillus*, and *Candida* species involves the use of vegetable oil and hydrocarbon-based substrates [25].

4.2.3.2 pH

pH is also an important environmental factor in addition to temperature, agitation and oxygen availability. These factors all together play an important role in cellular growth and activity. There are certain reports which present pH range for biosurfactant production. Optimum pH medium is required for sophorolipid synthesis through *Torulopsis bombicola* [26]. At pH range from 6 to 6.5, rhamnolipid synthesis was maximum in *Pseudomonas* spp. But decreased abruptly above $pH = 7$ [27]. In contrast, it was found penta- and disaccharide lipid was produced in pH range from 6.5 to 8, in *Nocardia corynbacteroides* [28]. Moreover, critical micelle concentration and surface tension of biosurfactant production maintained stability over a broad range of pH [29].

4.2.3.3 Temperature

Temperature promotes changes in the biosurfactant composition in the *A. paraffineus* and *Pseudomonas* sp. strain DSM-2874 [30]. A thermophilic *Bacillus* species increased in number and generated biosurfactant above 40 °C [31]. There was no significant change found in biosurfactant properties when they were given heat treatment. The properties which remained intact were surface tension, interfacial tension and emulsification efficiency, all of which remained stable even after autoclaving at 120 °C for 5 min [32].

4.2.3.4 Aeration

Important factors for biosurfactant production include aeration and agitation, as both facilitate the transportation of oxygen from gaseous phase to the aqueous phase. This function is linked with the physiological function of microbes which includes emulsification. It has been reported that synthesis of biological emulsifiers can increase the solubility of water insoluble substrates. As a result, nutrient transportation is facilitated to microorganisms. Adamczak *et al* reported 45.5g l^{-1} synthesis value of surfactant was attained when air flow rate was maintained at 1 vvm and dissolved oxygen concentration at 50% saturation [21].

4.2.3.5 Agitation

Agitation is the procedure through which microbial cells are maintained in culture medium with homogenous suspension along with nutrients. The major role of agitator is to ensure proper mixing of nutrients and enhance the transportation of oxygen in the culture medium. It also prevents the formation of clumps, thereby increasing the metabolic rate. An increment in the agitation speed leads to a fall in yield of biosurfactant production because of the shear stress in *Nocardia erythropolis* [33, 34]. Wang and Wang reported biosurfactant production in the case of *A. calcoaceticus* RAG-1 and found that cell-bound polymer/dry-cell ratio decreases as the shear stress increases [35]. In contrast, in the case of yeast, biosurfactant production increases as the agitation and aeration increase [36].

4.3 Measurement of microbial growth and biosurfactant production

4.3.1 Optical density (OD) measurement

OD measures the value of absorbance. It is the ratio of light striking the material to the intensity transmitted. It is abbreviated as OD and equals the negative of log (base 10) of the reciprocal of transmittance, given by the equation:

$$O.\,D = -\log_{10}(T)$$

Transmission value lies between 0 and 1. The typical method employed for determination of microbial growth is the estimation of OD at 600 nm wavelength. This technique involves the absorbance as a detection mode and generally analysis of a portion of the sample through which light can pass easily, more precisely it can be said to be a suspension of microorganisms. The scattering of light phenomenon is utilized to measure the growth of microorganisms. The presence of particles such as microorganisms in a solution causes scattering of light, the more the particles the more is the scattering of light by them. Therefore, if the bacteria and yeast population is replicating in the medium then it will increase the scattering of light and the greater is the measured absorbance intensity. Consequently, it is concluded that absorbance detection mode is utilized to measure the scattering of light in lieu of measuring simply the absorbance of light by the molecules. At OD = 600 nm wavelength, only scattering of light is measured and not the absorbance value[2].

[2] https://bmglabtech.com/en/blog/measure-microbial-growth-using-the-od600/#:~:text=The%20most%20common%20way%20to,through%20a%20suspension%20of%20microorganisms (accessed 19 September 2023).

4.3.2 Colony count method

A colony counter is an instrument and more often used for counting the colonies of bacteria or any other microorganism on a plate which is composed of a jelly substance suitable for growth of microbes. Colony count method involves the spreading and pouring plate approach, in which a diluted bacterial sample is spread over agar plates and incubated at 37 °C in an incubator overnight. Next morning, the number of colonies can be counted and multiply with the dilution factor. These results are given as colony forming units (CFUs) [37]. A CFU gives the precise value of how many bacterial cells are present in a specific sample and this approach detects the viable CFU. The major purpose of this method is to check the count of cells present in the sample and their propensity to continue to grow in specific factors such as nutritional medium and temperature.

There are several types of colony counters but broadly these are categorized as manual and automated. In manual or semi-automatic method, skilled personnel are required to analyse the difference between different types of colonies and encircle them on the outside surface of the plate with a pen. Therefore, this strategy is useful in keeping track of colonies but is laborious, chaotic and error prone. The automatic bacterial colony counter is an effective strategy as it involves an image processing technique such as thresholding, grey scaling and filtering etc.

4.3.3 Dry weight measurement

In this measurement technique, a small aliquot of about 1 g or 2 g wet tissue is homogenized and dried at 105 °C to attain a constant weight. This weight of the dried homogenized sample corresponds to the dry weight of the sample. Percentage moisture calculation is done by estimating the quantity of weight lost during the sample drying.

Calculation of percentage of dry weight is given by the equation:

$$\text{Dry Wt.\%} = \frac{(\text{Beaker Wt.} + \text{Dry Wt.}_{\text{sample}}) - (\text{Beaker Wt.})}{(\text{Beaker Wt.} + \text{Wet Wt.}_{\text{sample}}) - (\text{Beaker Wt.})} \times 100$$

There are some microorganisms such as filamentous organisms which cannot be directly measured. Therefore, they are first filtered from the culture medium and then dried and weighed. This dry weight measurement technique is appropriate for broth culture but growth of the organism cannot be continued as this process requires the demise of the organism [38].

4.3.4 Surface tension measurement

Surface tension, represented by γ, is the energy of any air–fluid interface. It is defined as energy per unit surface area or force per unit length. It is a free form of energy where surface free energy is measured by contact angle method in the case of solids and by surface tension in the case of liquids [39]. Surface tension depends on many factors including temperature, measurement time, substances of the apparatus and fluid viscosity. Upon formation of the new surface, active chemicals are diffused and aligned to it. In the duration of this process, there is rapid and continuous change in surface tension. To track these changes, dynamic surface tension technique is employed. At equilibrium condition of this process, static surface tension is measured by obtaining maximum force at a fluid–gas interface on a sample.

An essential criterion utilized in choosing the biosurfactant producing micro-organism is the capability to minimize surface tension to a value of at least 40 mN m^{-1} [40]. Cell free supernatant shows a huge reduction of the surface tension from 72.1 to 25.14 mN m^{-1}, which distinctly satisfies the essential criterion. Likewise, a research group of Thavasi *et al* suggested reduction in surface tension in five strains of bacteria that synthesize biosurfactants.

A Du Noiiy tensiometer is the instrument employed for surface tension measurement [41]. This method measures the pull experienced by the platinum ring on the liquid surface which is under investigation. The pull experienced is transferred by the lever which is present in a right-angle direction to a twisted steel wire and its torsion is recorded on a dial.

4.4 Application of microbial growth and biosurfactant production

The global production rate of surfactants has increased to 17 million metric tons. The expected future growth rate is 3%–4% every year [19]. The chemically produced surfactants are of petroleum nature and are non-biodegradable, as a result they are toxic to the environment. These compounds get accumulated and their by-products including synthesis process are hazardous to the environment. Thus, in order to protect the environment, scientist teams are working on regulation of the environment and are promoting an interest in synthesis of surfactants which are of microbial origin [42]. Biosurfactants have many applications in petroleum, medicine, industry and agriculture.

4.4.1 Industrial application

- Biosurfactant use in the food industry:
  Biosurfactants are utilized in the food processing industry as food preparation ingredients and anti-adhesive agents. The major role of biosurfactants in food preparation is that they cause formation and stabilization of emulsion as they have potential to reduce the surface and interface tension. They also play several other roles such as curbing aggregation of fat molecules, promoting stabilization in aerated drinks, enhancing the shelf-life of starch-based products, changing rheology-based things of wheat dough and enhancing texture as well as consistency of fat-based compounds [43].
- Biosurfactant use in the cosmetics industry:
  Due to advantages of biosurfactants in comparison to the chemically synthesized surfactants, they are utilized as emulsifier, foam-forming, solubilizers, cleaners, wetting and antimicrobial agents. They are also used in mediation of enzymatic actions, repelling insects, ant-acids, bath products, acne pads, anti-dandruff agent, contact lens solutions, baby products, mascara, lipsticks, toothpaste, dentine cleansers [44].

- Biosurfactant use in petroleum:
  The biosurfactants and bio-emulsifiers are known as novel molecules and also known as versatile by-product in the microbes-based technology. They are used as a bio-fouling and bio-corrosion degradation agent of hydrocarbons and enzymes in oil reservoirs and upgradation of petroleum, respectively [45]. The important role of biosurfactant is in extraction, transportation, upgradation and refining of petroleum.
- Biosurfactant use in oil recovery:
  In this field, microorganisms and their metabolic products are exploited to enhance the oil production in marginally producing reservoirs and the mechanism involved for this process is acidification of solid phase. Microbes such as *Bacillus subtilis*, *P. aeruginosa*, and *T. bombicola* consume hydrocarbons and crude oil as a carbon source and are therefore employed for cleaning of oil spills [46].

4.4.2 Environmental applications

- Use of biosurfactants in agriculture:
  In order to increase the solubilization of bio-hazardous chemical components like PAH, mobilizing reagents (surfactants) are employed. Thus, apparent solubility of hydrophobic organic contaminants (HOCs) is enhanced. Surfactants assist the microbe's adsorption to polluted soil, therefore diffusion path length is decreased between the adsorption and bio-uptake site. Another major role of surfactants is to improve the hydrophilicity of the heavy soil in order to increase the wettability and unform distribution of fertilizers in the soil. *Pseudomonas* produces rhamnolipid biosurfactant which has antimicrobial activity, which is evidenced by no adverse effects on humans or the environment. Another biosurfactant such as fengycins has anti-fungal activity and is utilized to control plant diseases [47].
- Use of biosurfactants in detergents:
  Almost every chemically synthesized surfactant used in modern day laundry detergents exerts toxic effects on the living organism present in fresh water. Therefore, general awareness has been promoted regarding environmental risk and hazards caused by chemical surfactants and the need to replace them with eco-friendly and natural substitutes. Cyclic lipopeptide (CLP) is one of the biosurfactants which is stable over a wide pH range from 7 to 12 and does not even show any effect on surface activity at high temperature [48]. They usually display excellent emulsion forming ability with vegetable derived oils and also show good stability as well as compatibility with laundry detergents, and therefore can be incorporated in them for their formation.
- Biosurfactants use in biopesticides:
  To curb the arthropods pest, several chemicals and pesticides are employed, which shows undesirable effects. Generation of pesticide-resistant arthropods population and cost of pesticides has led to the search for new eco-friendly and cheap pest control agents. Biosurfactants such as lipopeptide, synthesised

by many bacteria show pest control ability against species such as the fruit fly, *Drosophila melanogaster* [13].

4.4.3 Medical and pharmaceutical applications

- Antimicrobial activity:
  Structural diversity of biosurfactants means they display several functions. Biosurfactants exhibit toxic effects on the permeability of the cell membrane, which is similar to the effect of detergents. Biosurfactants exhibit several activities such as anti-bacterial, anti-fungal, antiviral and they also act as anti-adhesive agents to pathogens which makes them useful in treating diseases. These properties also make them useful in therapeutics and as pro-biotic agents [44]. An example includes production of biosurfactants by marine species such as *Bacillus circulans*, which show antimicrobial activity against gram positive and negative pathogens.
- Anti-cancer activity:
  Microbial extracellular glycolipids cause cellular differentiation and proliferation in the human promyelocyte leukemic cell line. Exposure of PC 12 cells to mannosyl erythritol lipids increases the acetylcholine activity, which interferes with G1 phase of cell cycle resulting in over-growth of partial cellular differentiation and neurites [43].
- Anti-adhesive agents:
  Rodrigues *et al* have reported that biosurfactants prevent the adhesion of pathogens to a solid surface or infection positions through an example such as pre-coating of vinyl urethral catheter with surfactant solution resulting in reduced production of biofilm by microorganisms like *Salmonella typhimurium*, *Salmonella enterica*, *E. coli* and *Proteus mirabilis* [49]. Muthusamy *et al* described that silicon rubber coated with surfactant of *Streptococcus thermophilus* prevented the adhesion of species *Candida albicans*. Similarly, coating of *Limosilactobacillus fermentum* and *Lactobacillus acidophilus* surfactants on glass reduces 77% biofilm of *Enterococcus faecalis* [47].
- Gene delivery:
  A research group of Gharaei-Fathabad suggested that development of efficient and safe technology for inserting synthetic nucleotides inside mammalian cells is serious for fundamental sciences and clinical use.

4.5 Conclusion

Microbial growth involves the growth of cellular components which results in increase in cell count. The concept and kinetic study of microbial growth and death provide an approach to solve problems involved in industries and in fermentation process to enhance productivity. The growth rate of microbes depends on nutrients, oxygen availability, light, pH and temperature. This basic knowledge of kinetics provides enhancement in disease control, products of industry and safety measures of the environment. Synthesis of laboratory-based biosurfactants and their applications are discussed in this chapter. At a plant level, synthesis of biosurfactants remains a challenge in comparison to those synthesized in the laboratory due to availability of nutrients, and micro-nutrients including effects of environmental factors.

References

[1] Sakthiselvan P, Meenambiga S S and Madhumathi R 2019 Kinetic studies on cell growth *Cell Growth* (London: InTech Open) p 13

[2] Björklund M 2019 Cell size homeostasis: metabolic control of growth and cell division *Biochim. Biophys. Acta. (BBA)-Mol. Cell Res.* **1866** 409–17

[3] Son S, Tzur A, Weng Y, Jorgensen P, Kim J, Kirschner M W *et al* 2012 Direct observation of mammalian cell growth and size regulation *Nat. Methods* **9** 910–2

[4] Ram Y, Dellus-Gur E, Bibi M, Karkare K, Obolski U, Feldman M W *et al* 2019 Predicting microbial growth in a mixed culture from growth curve data *Proc. Natl. Acad. Sci.* **116** 14698–707

[5] Buckley D H, Stahl D A, Martinko J M and Madigan M T 2015 *Brock, biología de los microorganismos* 14th edn (Pearson)

[6] Hogg S 2013 *Essential Microbiology* (New York: Wiley)

[7] Di Caprio F 2021 Cultivation processes to select microorganisms with high accumulation ability *Biotechnol. Adv.* **49** 107740

[8] Lerner B W 2003 *World of Microbiology and Immunology* (Gale Group Inc.)

[9] Banat I M, Franzetti A, Gandolfi I, Bestetti G, Martinotti M G, Fracchia L *et al* 2010 Microbial biosurfactants production, applications and future potential *Appl. Microbiol. Biotechnol.* **87** 427–44

[10] Singh A, Van Hamme J D and Ward O P 2007 Surfactants in microbiology and biotechnology: part 2. application aspects *Biotechnol. Adv.* **25** 99–121

[11] Ron E Z and Rosenberg E 2001 Natural roles of biosurfactants: minireview *Environ. Microbiol.* **3** 229–36

[12] Satpute S K, Banat I M, Dhakephalkar P K, Banpurkar A G and Chopade B A 2010 Biosurfactants, bioemulsifiers and exopolysaccharides from marine microorganisms *Biotechnol. Adv.* **28** 436–50

[13] Mulligan C N 2005 Environmental applications for biosurfactants *Environ. Pollut.* **133** 183–98

[14] Marchant R and Banat I M 2012 Microbial biosurfactants: challenges and opportunities for future exploitation *Trends Biotechnol.* **30** 558–65

[15] Franzetti A, Gandolfi I, Bestetti G, Smyth T J P and Banat I M 2010 Production and applications of trehalose lipid biosurfactants *Eur. J. Lipid Sci. Technol.* **112** 617–27

[16] Perfumo A, Rancich I and Banat I M 2010 Possibilities and challenges for biosurfactants use in petroleum industry *Biosurfactants* (New York: Springer) pp 135–45

[17] Sekhon Randhawa K K and Rahman P K S M 2014 Rhamnolipid biosurfactants—past, present, and future scenario of global market *Front. Microbiol.* **5** 454

[18] Salihu A, Abdulkadir I and Almustapha M N 2009 An investigation for potential development on biosurfactants *Biotechnol. Mol. Biol. Rev.* **3** 111–7

[19] Rahman P K S M and Gakpe E 2008 Production, characterisation and applications of biosurfactants-review *Biotechnology* **7** 360–70

[20] Desai J D and Banat I M 1997 Microbial production of surfactants and their commercial potential *Microbiol. Mol. Biol. Rev.* **61** 47–64

[21] Adamczak M and Bednarski W O 2000 Influence of medium composition and aeration on the synthesis of biosurfactants produced by candida antarctica *Biotechnol. Lett.* **22** 313–6
[22] Patel P, Modi A, Minipara D and Kumar A 2021 Microbial biosurfactants in management of organic waste *Sustainable Environmental Clean-up* (Amsterdam: Elsevier) pp 211–30
[23] de Almeida P F, Moreira R S, de Almeida R C C, Guimaraes A K, Carvalho A S, Quintella C *et al* 2004 Selection and application of microorganisms to improve oil recovery *Eng. Life Sci.* **4** 319–25
[24] De Lima C J B, Ribeiro E J, Servulo E F C, Resende M M and Cardoso V L 2009 Biosurfactant production by *Pseudomonas aeruginosa* grown in residual soybean oil *Appl. Biochem. Biotechnol.* **152** 156–68
[25] Sivapathasekaran C and Sen R 2017 Origin, properties, production and purification of microbial surfactants as molecules with immense commercial potential *Tenside Surfactants Deterg.* **54** 92–107
[26] Göbbert U, Lang S and Wagner F 1984 Sophorose lipid formation by resting cells of Torulopsis bombicola *Biotechnol. Lett.* **6** 225–30
[27] Guerra-Santos L, Käppeli O and Fiechter A 1984 *Pseudomonas aeruginosa* biosurfactant production in continuous culture with glucose as carbon source *Appl. Environ. Microbiol.* **48** 301–5
[28] Powalla M, Lang S and Wray V 1989 Penta-and disaccharide lipid formation by Nocardia corynebacteroides grown on n-alkanes *Appl. Microbiol. Biotechnol.* **31** 473–9
[29] Abu-Ruwaida A S, Banat I M, Haditirto S and Khamis A 1991 Nutritional requirements and growth characteristics of a biosurfactant-producing Rhodococcus bacterium *World J. Microbiol. Biotechnol.* **7** 53–60
[30] Drouin C M and Cooper D G 1992 Biosurfactants and aqueous two-phase fermentation *Biotechnol. Bioeng.* **40** 86–90
[31] Banat I M 1993 The isolation of a thermophilic biosurfactant producing Bacillus sp *Biotechnol. Lett.* **15** 591–4
[32] Abu-Ruwaida A S, Banat I M, Haditirto S, Salem A and Kadri M 1991 Isolation of biosurfactant-producing bacteria, product characterization, and evaluation *Acta Biotechnol.* **11** 315–24
[33] Margaritis A, Zajic J E and Gerson D F 1979 Production and surface-active properties of microbial surfactants *Biotechnol. Bioeng.* **21** 1151–62
[34] Margaritis A, Kennedy K and Zajic J E 1980 Application of an air lift fermenter in the production of biosurfactants *Dev. Ind. Microbiol.* **21** 95
[35] Wang S and Wang D I C 1990 Mechanisms for biopolymer accumulation in immobilized Acinetobacter calcoaceticus system *Biotechnol. Bioeng.* **36** 402–10
[36] Spencer J F T, Spencer D M and Tulloch A P 1979 Extracellular glycolipids of yeasts *Econ. Microbiol.* **3** 523–40
[37] Elizabeth O T, Fehintola O T, Seun A J, Bolanle F R and Oyenike A T 2021 Concept and kinetics of microbial growth and death: a review *Himal J. Agric* **2** 5
[38] Black J G and Black L J 2018 *Microbiology: Principles and Explorations* (New York: Wiley)
[39] Buckton G 1988 The assessment, and pharmaceutical importance, of the solid/liquid and the solid/vapour interface: a review with respect to powders *Int. J. Pharm.* **44** 1–8
[40] Bodour A A and Miller-Maier R M 1998 Application of a modified drop-collapse technique for surfactant quantitation and screening of biosurfactant-producing microorganisms *J. Microbiol. Methods* **32** 273–80

[41] Frobisher M Jr 1926 Relations of surface tension to bacterial phenomena *J. Infect. Dis.* **38** 66–91

[42] Benincasa M 2007 Rhamnolipid produced from agroindustrial wastes enhances hydrocarbon biodegradation in contaminated soil *Curr. Microbiol.* **54** 445–9

[43] Muthusamy K, Gopalakrishnan S, Ravi T K and Sivachidambaram P 2008 Biosurfactants: properties, commercial production and application *Curr. Sci.* **94** 736–47

[44] Gharaei-Fathabad E 2011 Biosurfactants in pharmaceutical industry: a mini-review *Am. J. Drug Discov. Dev.* **1** 58–69

[45] De Almeida D G, de Soares Da Silva R C F, Luna J M, Rufino R D, Santos V A, Banat I M *et al* 2016 Biosurfactants: promising molecules for petroleum biotechnology advances *Front. Microbiol.* **7** 1718

[46] Das K and Mukherjee A K 2007 Crude petroleum-oil biodegradation efficiency of bacillus subtilis and *Pseudomonas aeruginosa* strains isolated from a petroleum-oil contaminated soil from North-East India *Bioresour. Technol.* **98** 1339–45

[47] Traudel K and Merten S 2017 Possible food and agricultural application of microbial surfactants: an assessment *Biosurfactants and Biotechnology* (Routledge) pp 183–210

[48] Mukherjee A K 2007 Potential application of cyclic lipopeptide biosurfactants produced by bacillus subtilis strains in laundry detergent formulations *Lett. Appl. Microbiol.* **45** 330–5

[49] Rodrigues L, Banat I M, Teixeira J and Oliveira R 2006 Biosurfactants: potential applications in medicine *J. Antimicrob. Chemother.* **57** 609–18

IOP Publishing

Microbial Surfactants
A sustainable class of versatile molecules
Divya Tripathy and Anjali Gupta

Chapter 5

Bioreactors for biosurfactant production and downstream processing surfactant extraction

Pramilaa Kumar, Soghra Nashath Omer, Venkat Kumar Shanmugam and Dharmaraj Senthilkumar

Recent years have seen a major increase in interest in the biotechnological production of biosurfactants because of their many useful and environmentally friendly applications. Because they provide exact control over environmental conditions and scalability, bioreactors are essential to the culture of microorganisms for the generation of biosurfactants. In order to synthesize biosurfactants, this abstract examines the application of several bioreactor types, such as stirred tanks, bubble columns, and solid-state fermentation, highlighting the benefits and drawbacks of each. It also covers downstream processing methods, emphasizing surfactant extraction in particular since it is essential for isolating and purifying biosurfactants from the fermentation broth. The chapter highlights the significance of effective downstream processing techniques to guarantee the economic viability and purity of biosurfactant products, as well as the importance of optimizing bioreactor design and operational parameters for increased biosurfactant yields. With implications for a range of commercial applications, this thorough analysis advances knowledge of the bioproduction of biosurfactants from fermentation to extraction.

5.1 Introduction

These days, surfactants are widely used in many different industries. Cosmetics, medicines, petroleum, petrochemicals, food, drinks, environmental control, and agrochemicals are a few industries that have used this substance (Kosaric 1992). The current estimate for the global surfactant industry is approximately US$9.4 billion annually, and by the end of the century, demand for these products is predicted to rise at a rate of 35% (Desai and Banat 1997).

doi:10.1088/978-0-7503-5989-4ch5

Yet, there is now more interest in these microbe-produced substances as chemical surfactant substitutes due to the development of biosurfactants, which may lower surface and interfacial tension with high specificity, low toxicity, and biodegradability (Tripathy and Mishra 2016).

Protein-like substances, glycolipids, lipopeptides, protein complexes, glycolipids, polysaccharide phospholipids, fatty acids, and neutral lipids are examples of microbial molecules with high surface activity that are known as biosurfactants (Rodrigues *et al* 2006).

However, biosurfactants are still not able to hold a market share against surfactants produced chemically. This might be brought on by their high production costs in comparison to the ineffective bioprocessing techniques now in use, low strain yield, and the requirement to use expensive substrates (Cameotra and Makkar 1998). The low yields and high production costs of microbial surfactants have prevented their commercial manufacture, even though they have several commercially desirable features and definite benefits over their synthetic counterparts. Furthermore, it is challenging to create biosurfactants at a reasonable cost for several reasons, including the rarity of overproducing bacterial strains and the generally poor productivity of those that are found, necessitating the use of a complicated medium (Fiechter 1992). Figure 5.1 represents application and production of biosurfactants.

Because they make it possible to cultivate bacteria, fungi, and animal cells efficiently, bioreactors are the foundation of innumerable biological production processes for high-value goods in the pharmaceutical and biotechnology industries. Nevertheless, the research (Mandenius 2016, Palomares and Ramírez 2019) emphasizes that aeration, one of the primary bioreactor design parameters, has not seen

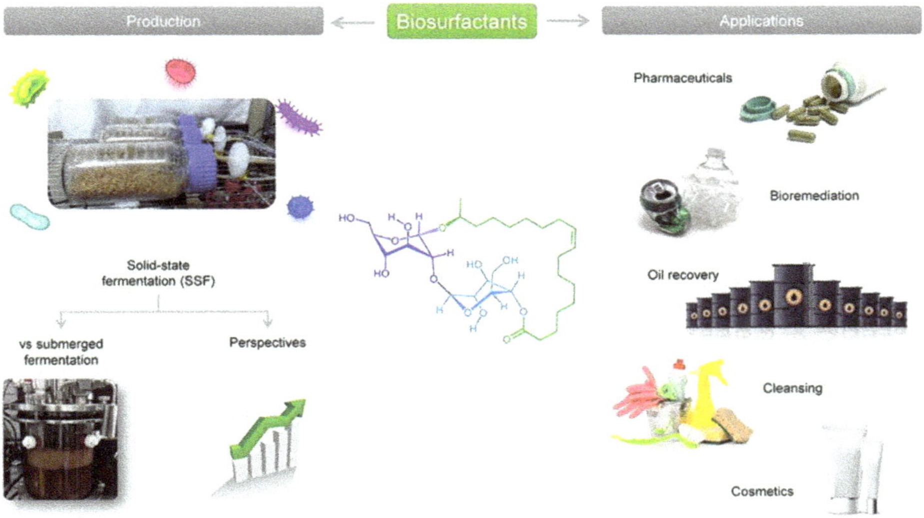

Figure 5.1. Schematic representation of biosurfactants production and application. Reprinted from Banat et al (2021), copyright (2021), with permission from Elsevier.

much innovation since the twentieth century, despite technical advancements in process control, sensor technology, and vessel material: in submerged cultures, bubble dispersion and stirring are still used for gas delivery. Conventional bioreactor types are unable to combine high gas intake with low-stress mixing conditions because of this lack of innovation Although stirred-tank reactors frequently achieve excessive volumetric mass transfer coefficients (kLa = 360–1080 unit), the shear force caused by the stirrer's tip along with air bubbles and their rupture, is a disadvantage in many processes, particularly with shear-sensitive cultures of cells or microorganisms that produce biosurfactants due to more production of foam (Chisti 2000).

Membrane aeration is an enhanced technique for transferring gases in a bioprocess with respect to the propensity for foam formation. Commonly used in water treatment, bioreactors made of membranes combine a number of benefits for the process, including a substantial decrease in total shear stress (Henzler 2000), low energy consumption in the oxygen transfer (Syron and Casey 2008), linear process scaling by expanding the membrane area (Côté *et al* 1988, Drioli *et al* 2005), and the lack of the production of foam due to purely diffusion of oxygen (De Kronemberger *et al* 2007, Coutte *et al* 2010). Membrane oxygenators have emerged as a viable, albeit less often used industrial alternative for *ex situ* without bubbles aeration of cell culture media since the research (Knazek *et al* 1972, Frahm *et al* 2009, Charcosset 2012).

Membrane oxygenators function by bringing oxygen, carbon dioxide, blood, or a medium into touch with one another through a gas-permeable membrane (Djeljadini *et al* 2021). Despite the excellent volumetric efficiency offered by external membrane modules, substantial flow rates are required to exchange the oxygen-poor medium within the bioreactor. In addition, creating dissolved oxygen gradients and complicating process design and operation are the outcomes of circulating the fluid through the external loop. Moreover, the likelihood of contamination is raised by the external loop (Carstensen *et al* 2012). Membrane aeration has been limited to mammalian cell culture due to the previously noted restrictions in external membrane modules, as well as issues with scalability and membrane fouling (Furusaki 1991, Catapano *et al* 2004, Qi *et al* 2008, Frahm *et al* 2009).

Fungi and bacteria have higher growth rates than cell cultures, despite the fact that cell cultures have a larger oxygen demand per cell (Guarino *et al* 2004, Wagner *et al* 2011). As a result, at a certain point, the oxygen requirement of those cultivations is typically exceeded. *In situ*, membrane aeration, as opposed to external membrane modules, can offer options for fermentations requiring more oxygen. Moreover, foaming-prone systems like biosurfactant synthesis benefit from the *in situ* membrane aeration approach.

Rhamnolipids are an important family of biosurfactants that fall under the glycolipid class and are primarily generated by the genus *Burkholderia* and *Pseudomonas* (Abdel-Mawgoud *et al* 2010). Since *Pseudomonas aeruginosa*, the naturally occurring producer, is an invasive human infectious agent, rhamnolipid synthesis has been created using recombinant non-pathogenic Pseudomonas *putida KT2440* (Tiso *et al* 2017, Wittgens *et al* 2011). In addition to the usual difficulties in microbial cultivation, processes involved in the generation of biosurfactants are particularly vulnerable to significant foaming. Aeration causes surface-active molecules (biosurfactants) to adsorb at the gas–liquid interfaces enclosing the gas bubbles, generating an amphiphile-enriched layer that stabilizes the foam when gas bubbles are injected into the microbial fermentation (Junker 2007).

5.1.1 Bioreactors for biosurfactant production

The process of producing biosurfactants in a bioreactor is intricate and requires careful observation of all relevant parameters. The first step in the process is to select an appropriate culture method (batch, or fed-batch, continuous), which differs based on the product, microorganism type, and kind of bioreactor (Blank *et al* 1984, Bernard and Payton 1995, Brück *et al* 2020). Each culture method has different substrates, biomass kinetics, and ultimate product or byproduct concentrations, according to Atlić *et al* (2011).

For instance, researchers (Brumano *et al* 2017) reported a sophisticated process using *Aureobasidium pullulans LB 83* to produce surfactin in a stirred-tank bioreactor. In order to assess the effects of sucrose concentrations (20–80 g l^{-1}) and aeration rates (0.1–1.1 min^{-1}), their team employed a centrally organized face model. In an oil spreading test, their findings showed a rise in tense activity of 8.05 cm and effectiveness of 0.0838 cm h^{-1} when every parameter was applied at high levels.

Continuous cultures: One technique that is becoming more and more important is the ongoing culture of microorganisms (Amutha and Gunasekaran 2001). Consistent microbial growth in continuous culture, or a uniform level of growth in a steady environment, is the main characteristic of this methodology. The particular rate of development of microorganisms can be efficiently controlled within limitations in continuous culture conditions, such as bioreactors (Kiran *et al* 2010, Samad *et al* 2015).

The continuous bioreactor may come in a variety of forms. For instance:

(a) Auxostat: An auxostat is a device that maintains the computation at a constant level by monitoring the steady media rate of flow using inputs from an analysis of a microbial growth environment (Adamberg *et al* 2003).
(b) Retentostat: This is a modified chemostat that recycles biomass into the bioreactor by connecting a filtration unit to the sewage line.
(c) Turbidostats: A turbidostat is a continually growing system that receives input from the culture vessel's dilution rate and its optical density (Klok *et al* 2013).

Batch processes: In the fermentation industry, batch fermentation is frequently used to produce a wide range of microbial products, such as hormones, medications, vitamins, and secondary metabolic products (Kebbouche-Gana *et al* 2013). In the batch method, substrates and microorganisms are added to a bioreactor in batches to synthesize products (Vanavil *et al* 2013). A straightforward method for managing the fermentation process and maintaining regulated conditions within a bioreactor is the batch technique. On the other hand, acid concentration, microbial biomass, and byproduct concentration may all experience competitive variations during the fermentation process. A continuously agitated container with numerous additional fittings, such as a sparger, an insulating jacket to control fluctuations in temperature, and a gas sprayer, make up a batch bioreactor (Zhang *et al* 2014, Behrens *et al* 2016). Even though batch fermentation is a simple process, it comes with a lot of costs and requires a lot of time for pre-and post-run procedures, such as cleaning, filling, and emptying bioreactors. Such procedures may require a lot more time in some circumstances than the growth of the microbial biomass (Probert and Gibson 2002).

Fed-batch process: Batch fermentation is adapted in the batch procedure. In the bioprocessing industry, it is the most widely used fermentation method of operation. After adding nutrition to the fermenter, microorganisms are cultured and inoculated using the batch method for a certain amount of time in order to provide nourishment for them. Only when the volume of the fermentation broth reaches 75% of the bioreactor volume is the fermentation stopped (Rodriguez-Contreras *et al* 2013). The consistent medium substrate feed flow in fed-batch systems allows the intended secondary metabolites to reach extremely high concentrations and levels. The feed substrate level can be controlled at the target level, which is frequently rather low, with this culturing technique (Sarilmiser *et al* 2015). This will enable the avoidance of unfavorable alterations, such as those that lead to substrate inhibition or modifications in cellular metabolism when substrate concentrations are high. Additionally, fed-batch systems can be used when a lot of biomass is needed (Yao *et al* 2015, Xu *et al* 2020).

5.1.2 Downstream processing for surfactant extraction

Petroleum is the primary source of chemically produced surfactants. Because they typically do not biodegrade, they continue to be hazardous to the environment. Once more, the production procedures and byproducts of synthetic surfactants may be toxic to the environment, and they may also bioaccumulate.

Applications of naturally occurring surfactants are potential substitutes for chemically synthesized ones due to growing awareness of the need to preserve this ecosystem (Benincasa 2007, Sayed *et al* 2012a, 2012b, Tawfik 2015, Tawfik *et al* 2016). Many biosurfactants have remarkably short alkyl chains compared to standard chemically synthesized surfactants, which typically have alkyl chains of ten or more carbon atoms. These structures help these biosurfactants become soluble in water. When compared with artificial surfactants with comparable structures, they can have impressively low critical micelle concentrations (CMCs) despite their water solubility. For instance, the concentrations of the CMC of non-ionic (low pH) multicomponent mono-rhamnolipid mixtures, in which the heptyl chain congener is the most abundant, vary from less than 1 mM to less than 10 mM based on the ionic strength of the solution (Lebron-Paler 2008). The rhamnolipids become deprotonated and the CMC value rises as the pH rises. Studies in science have demonstrated that biosurfactants are superior to their chemical counterparts in many ways.

- Biodegradability: These surface-active chemicals are readily broken down by microorganisms due to their biological origin, low toxicity, and straightforward chemical structure. As a result, they do not linger in the environment.
- Biological compatibility and digestibility: Because of their biological origin, they have an innate quality of compatibility that allows them to be used without restriction in food additives, cosmetics, medications, and agricultural sectors.
- Minimal toxicity: Compared to surfactants manufactured from chemicals, biosurfactants show lower levels of toxicity. For instance, *Rhodococcus* species 413A glycolipids were 50% less hazardous than Tween 80 in naphthalene solubilization experiments (Kanga *et al* 1997).
- Surface and interface activity: According to Mulligan, a good surfactant may reduce the interfacial pressure of water and hexadecane from 40 to 1 mN m^{-1} and the surface tension of water from 72 to 35mN m^{-1}. The interfacial tension between water and hexadecane can be reduced by surfactants to less than 1 mN M^{-1} and the surface tension of water to 25 mN m^{-1}, respectively (Muthusamy *et al* 2008).
- Supply of raw materials: A wide range of inexpensive raw materials can be used to make biosurfactants. The carbon sources, which include carbohydrates, lipids, and hydrocarbons, can be combined or used separately (Kosaric 2001).
- Acceptable manufacturing economics: One benefit of mass production is that biosurfactants can also be made from industrial waste and byproducts.
- Physiological factors: Biosurfactants are active in a wide range of salinities, pHs, and temperatures.
- Specificity: Because biosurfactants are complex organic compounds with certain functional groups, they frequently behave in a specific way. This characteristic makes it crucial for the detoxification of particular contaminants, the creation of particular cosmetics, the de-emulsification of industrial emulsions, and various food and medicinal uses.
- Environmental control: spills of oil, the handling of industrial emulsions, the detoxification and biological degradation of industrial effluents, and the bioremediation of contaminated soil can all be successfully handled by biosurfactants.

With soaps included, the global production of surfactants reached 17 million metric tonnes in 2000, and 3%–4% annual growth rates are predicted going forward (Rahman and Gakpe 2008).

5.2 Types of bioreactors for biosurfactant production

Biosurfactants, or biologically derived surfactants, are used in food, pharmaceutical, and environmental remediation industries. Bioreactors are crucial equipment for producing these surfactants. The synthesis of biosurfactants can be done in a variety of bioreactor types. Here are a few typical kinds and links to pertinent sources for more details.

5.2.1 Stirred tank bioreactors

Stirred-tank bioreactors are a popular choice for producing biosurfactants. Their ability to mix and aerate effectively is essential for the growth of microorganisms and the creation of biosurfactants. The yield of biosurfactants can be affected by the impeller design and agitation speed selected.

5.2.1.1 Features and components

The primary parts consist of a cylindrical tank with a motor-driven central shaft that houses one or more paddles, agitators, or impellers. Impellers come in a variety of forms, such as naval fans, rustom discs, concave blades, and Rushton turbines. The ratio of height to diameter, liquid height to width, or aspect ratio affects the number. Typically, the surface texture aspect ratio (STR) is 3–5 (Su *et al* 2019). A sparger is used to supply air to the culture media through a flow rate controller. Either a single-hole tube or a multi-hole ring serves as the sparger. The STRs are more widely available on the market, and they might achieve more effective mass and gas transfers to the developing cells with adjustable control regimes.

5.2.1.2 Advantages and disadvantages

- Effective mixing: The uniform distribution and optimal generation of biosurfactants are ensured by the efficient mixing of nutrients and bacteria in stirred-tank bioreactors.
- Controllable agitation: These microbes that produce biosurfactants and are sensitive to shear can be used since the speed of the agitation can be changed to regulate shear forces.
- Scalability: The generation of biosurfactants on an industrial or laboratory scale is made possible by the easily scalable nature of these bioreactors, which come in a range of sizes.
- Temperature and pH control: Precise temperature and pH control are provided by stirred-tank bioreactors, which are necessary to sustain microbial development and the synthesis of biosurfactants (Pandey *et al* 2008).
- High energy consumption: The agitation and aeration processes in stirred-tank bioreactors necessitate a large energy input, which can raise operating costs and have an adverse effect on the environment (Doran 1995).
- Foam development: In stirred-tank bioreactors, agitation can result in the development of foam, particularly in microbial cultures that generate biosurfactants. It is possible for foam to escape the bioreactor, which could cause contamination problems as well as the loss of growth medium and biosurfactants.
- Shear stress: Stirred-tank bioreactors with high-speed agitation may produce shear stress, which may be harmful to microorganisms that are sensitive to shear and may lower the production of biosurfactants.
- Limitations on heat transfer: Heat transfer can be difficult in large-scale stirred-tank bioreactors, which can result in temperature gradients inside the reactor. Microorganisms' ability to proliferate and produce biosurfactants

can be impacted by temperature fluctuations (Biotechnology for Biofuels and Bioproducts n.d.).

5.2.2 Application in biosurfactant production

For the production of biosurfactants, stirred-tank bioreactors are widely utilized, particularly when regulated manufacturing conditions and significant yields are needed. Stirred-tank bioreactors can be used to grow a variety of microorganisms, including *Bacillus* species, *Candida* species, and *Pseudomonas* species, in order to make biosurfactants. The generated biosurfactants can be used for a number of purposes, such as:

- Enhanced oil recovery (EOR): By lowering the interfacial tension between water and oil and facilitating oil displacement, biosurfactants can be utilized to boost oil recovery from reservoirs (Cameotra and Makkar 2004).
- Bioremediation: By making hydrophobic pollutants more soluble and encouraging microbial breakdown, biosurfactants can help remove hydrophobic chemicals from polluted soil and water, such as petroleum and hydrocarbon (Rivardo *et al* 2011).
- Food and pharmaceutical industries: Biosurfactants can enhance foaming and emulsification in food processing, and they can be utilized in drug delivery systems in the pharmaceutical business (Gudina *et al* 2013).
- Cosmetics and personal care products: Due to their emulsifying and moisturizing qualities, certain biosurfactants are utilized in cosmetic and personal care products (Banat *et al* 2014).

Since stirred-tank bioreactors allow for the controlled culture of microorganisms for a variety of industrial and environmental functions, these applications demonstrate the versatility and significance of these bioreactors in the manufacture of biosurfactants.

5.2.3 Membrane bioreactors

A special kind of bioreactor that combines membrane filtration and microbial cultivation is called a membrane bioreactor (MBR). They have several uses, one of which is the synthesis of biosurfactants. Here's an overview of the features and components, advantages and disadvantages, and their applications in biosurfactant production.

5.2.3.1 Features and components

- Membrane units: Membrane units that are submerged or side-stream make up MBRs. The purpose of these modules, which are usually composed of ceramic or polymeric membranes, is to physically separate the suspended solids and activated sludge biomass from the treated water.
- Biological reactor: MBRs come with a biological reactor that uses microorganisms to break down organic materials in wastewater. This reactor is

comparable to traditional activated sludge procedures. The membranes allow only treated water to pass through, leaving the activated sludge from the MBR inside the system (Judd 2010).

- Membrane cleaning system: To avoid fouling and preserve membrane function, MBRs are fitted with membrane cleaning systems. Chemical cleaning, air scouring, and backwashing are common cleaning techniques (Le-Clech 2010).
- System for pumping permeates: In order to sustain a pressure gradient across the membranes and allow water to pass through them while trapping the biomass inside, MBRs use a permeate pumping mechanism (Stephenson *et al* 2000).
- Observation and control: MBRs use control and monitoring systems to make the treatment process as efficient as possible. Sensors for monitoring characteristics including biomass concentration, membrane fouling, and water quality are frequently included in these systems.
- Recirculation of sludge: Sludge recirculation is frequently used in MBRs to boost biological treatment, maintain high biomass concentrations in the bioreactor, and optimize membrane performance (Aslam *et al* 2017).

5.2.3.2 Advantages and disadvantages

- High-quality treated water produced by MBRs is free of bacteria, pathogens, and suspended particles, making it acceptable for reuse or release into the environment (Le-Clech *et al* 2006).
- Compact footprint: MBRs are suited for applications where space is at a premium since they have a lower footprint than traditional wastewater treatment systems (Judd 2008).
- Reduced sludge production: Because of their high biomass concentration and longer solids retention period, MBRs usually create less sludge than traditional activated sludge processes (Ding *et al* 2019).
- Flexible design: MBRs offer process design flexibility by being able to be tailored to suit a range of applications and wastewater types (Stephenson *et al* 2000).
- Efficient removal of nutrients: MBRs can be designed to efficiently remove nutrients such as phosphate and nitrogen, adhering to strict discharge guidelines.
- Membrane fouling: Membrane fouling can lower filtration effectiveness and raise maintenance costs, making it a serious problem for MBRs (Le-Clech *et al* 2006).
- High energy consumption: The energy needed for membrane filtration and aeration makes MBRs more energy-intensive than other wastewater treatment methods.
- Initial investment costs: MBRs often require a larger initial capital investment than traditional treatment systems, which may prevent them from being widely used (Judd 2010).

- Maintenance of membranes: Cleaning and replacement of membranes regularly are required, which raises operating expenses.
- Hazard of membrane damage: Improper or harsh cleaning techniques may cause damage to the membranes, increasing the expense of replacement (Jeon *et al* 2016).

5.2.3.3 Application in biosurfactant production

- Enhanced productivity: By consistently eliminating extra biomass and encouraging the development of microorganisms that produce biosurfactants, MBRs can maintain higher numbers of cells and biosurfactant levels in the culture medium. Consequently, the output of biosurfactants is raised (e Silva *et al* 2014).
- Higher product purity: The effective separation of biosurfactants from the culture medium made possible by the use of MBRs leads to improved product quality. As a result, fewer downstream processing steps are required (Nayarisseri *et al* 2018).
- Decreased environmental impact: MBRs help to reduce the amount of wastewater produced and make it possible to repurpose treated culture media, which results in a more environmentally friendly and sustainable process for producing biosurfactants.
- Scalability: MBRs are well suited for large-scale biosurfactant synthesis in industry since they are readily scalable from laboratory to industrial settings (Deziel *et al* 1996).

5.2.4 Bubble column bioreactors

One kind of bioreactor that uses gas sparging to produce bubbles and keep aeration and mixing within a liquid-filled column is called a bubble column bioreactor (BCR). They are useful in many different bioprocesses, including the synthesis of biosurfactants, and have distinctive properties.

5.2.4.1 Features and components

Pneumatically mixed reactors known as BCRs were designed for sensitive cell cultures, including those of plants, mammals, and filamentous fungi. A cylindrical container having an aspect ratio of 2:1 or 3:1—and maybe as high as 6:1 in certain situations—is the BCR.

To aid in bubble discharge and foam rupturing, the top diameter of the BCRs may be increased. Gas is dispensed as bubbles from the bottom spargers into the phase of liquid or solid–liquid suspension (slurry). The BCRs lack an internal component to extend the liquid-to-air contact period, in contrast to the STRs (Mansi Sharma *et al* 2015, p 1). Consequently, the BCRs exhibit superior heat and mass transmission along with the convenience of either adding or removing the catalyst/biocatalyst to give a greater interfacial surface area of around 30% (Degaleesan *et al* 2001). BCRs have been used in the metallurgical, petrochemical, chemical, and biochemical sectors as reactors or multi-reactors (Angeles *et al* 2014). In order to expose the culture to light sources for photosynthetic reactions, fluid circulation is produced by varying gas flow rates from a dark, central section to the outer area (Sharma *et al* 2015).

5.2.4.2 Advantages and disadvantages

- High mass transfer: The synthesis of surfactants and aerobic microbial growth are facilitated by the effective gas–liquid mass transfer offered by bubble column bioreactors.
- Scalability: They are well suited for large-scale surfactant synthesis since they can be readily scaled from laboratory to industrial sizes.
- Reduced shear stress: Compared to stirred-tank bioreactors, bubble column bioreactors usually have less shear stress, which makes them appropriate for microorganisms that are sensitive to shear.
- Simple design: BCRs require less maintenance because to its comparatively simple design, which has fewer moving elements (Ghribi and Ellouze-Chaabouni 2011).
- Inadequate mixing: Compared to stirred-tank bioreactors, BCRs may not mix as well, which could cause an unequal distribution of nutrients and oxygen.
- Foam formation: When surfactant-producing bacteria are present, foaming is a common problem in BCRs. The foam can exit the reactor and cause losses.
- Limited heat transfer: When precise temperature control is needed, bubble column bioreactors may not be the best option due to their less efficient heat transfer.
- Complex control: In order to maintain appropriate gas–liquid ratios and avoid problems like flooding or channeling, BCRs may need complex control systems in order to operate at optimal circumstances (Stanbury *et al* 2013).

5.2.4.3 Application in biosurfactant production

Biosurfactant generation and extraction can be carried out in BCRs. As the microorganisms develop, they produce surfactants, which can aid in the extraction of hydrophobic substances. The resulting surfactants can help with surfactant extraction procedures by making hydrophobic substances easier to solubilize and extract from an aqueous phase (Ali Mansoori *et al* 2008).

5.2.5 Packed bed bioreactors

5.2.5.1 Features and components

A growth media compartment and a packed-bed compartment with macroporous molecules to immobilize the cells make up a traditional packed-bed bioreactor (PBR). Using the fixed bed, an oxygen-dense medium is recycled.

The fresh medium is combined with the departing culture media, which contains depleted nutrients like oxygen and glucose. This replenishes the concentration of vital nutrients in the culture medium (Sen *et al* 2017). PBRs have been employed in the generation of many cell lines and biologics, including stem cell antileukemic medicines, monoclonal antibodies, and retroviral vectors mostly utilized in gene therapy. Low shear forces, a low concentration of loosely suspended cells, and a high cell density are necessary for these cultures (Chaudhuri and Al-Rubeai 2005). PBRs have increased mass transfer and contact time, easier operation, and improved product quality control. Reusable biocatalysts, such as stable enzymes, can be utilized repeatedly to continuously recover products (Chaudhuri and Al-Rubeai 2005). By raising the inflow rate and nutrient broth concentrations, the concentration of nutrients can be increased; however, because of unfavorable mixing circumstances, controlling parameters like pH can be difficult (Su *et al* 2019).

5.2.5.2 Advantages and disadvantages

- Efficient mass transfer: The high surface area of gas–liquid contact in BCRs facilitates efficient mass transfer and oxygen transfer, both of which are important for the development of microorganisms and the synthesis of biosurfactants (Md 2012).
- Scalability: The excellent scalability of these bioreactors facilitates the transfer from laboratory-to-industrial-scale biosurfactant synthesis.
- Decreased shear stress: BCRs are suited for microorganisms that are sensitive to shear because they generally produce less shear stress than stirred-tank bioreactors (Gong *et al* 2021).

5.2.5.3 Application in biosurfactant production

In order to produce biosurfactants, BCRs are utilized, particularly in situations where effective oxygen transfer and aeration are essential for microbial development. They are used in the culture of microorganisms that are known to create biosurfactants, like *Pseudomonas* and *Bacillus* strains (Thavasi *et al* 2007).

5.3 Downstream processing for surfactant extraction

5.3.1 Centrifugation

Depending on the pace at which the particles in a mixture settle, a centrifuge is used in the centrifugation process to separate the particles from a liquid. Numerous fields, including as chemistry, biology, and industry, frequently employ it. In this post, I'll give a brief synopsis of centrifugation's workings, benefits, and drawbacks, as well as some other resources for more learning.

5.3.1.1 Principles and mechanisms

The principles of sedimentation, which divide components or particles in a mixture according to their size and density, are the basis of centrifugation. The sample is subjected to centrifugal force by the centrifuge, which causes the less dense particles to settle at different levels or remain suspended in the sample, while the denser particles migrate outward and settle at the bottom (Chaffey 2003).

The biosurfactant generated by the fermentation process needs to be separated from the cells and broth and then purified in order to be properly characterized.

Centrifugation at 5000–10 000 g for 10–20 min can be used to separate the cells (Nayak *et al* 2009, Shavandi *et al* 2011, Luna *et al* 2013, e Silva *et al* 2014). The pellet of cells is subsequently removed and dried out to determine the dry weight of the cells for the experiment, and the resulting supernatant can be further purified to characterize the biosurfactant.

5.3.1.2 Advantages and disadvantages

- Quick separation: Centrifugation is a quick separation method that can handle big sample volumes in a short amount of time.
- High separation efficiency: It is capable of efficiently separating particles with minute size and density variations.
- Versatility: There are many uses for centrifugation, such as particle separation, protein purification, and cell separation.
- Usability: Contemporary centrifuges offer exact control over the separation process and are simple to use (Rickwood 2001).
- Cost: The acquisition and upkeep of centrifuges can be costly.
- Limited scale: The size of the centrifuge rotor and the number of centrifugation tubes that are available determine the maximum amount of sample that can be treated.
- Hazard of sample damage: Difficult biological samples may occasionally be harmed by strong centrifugal pressures (BeClood.com n.d.).

5.3.1.3 Application in surfactant extraction

The extraction and purification of surfactants can be accomplished using centrifugation. Emulsions and micellar systems may contain surfactants, and centrifugation is a useful technique for separating the surfactant-rich phase from other constituents (Abrar and Trathnigg 2010, Najjar and Abu-Shamleh 2020).

5.3.2 Filtration

Solids and liquids can be separated and purified using the separation technique known as filtration, which is widely employed in many different industries. Filtration can be used to remove surfactants from liquid mixtures in the context of surfactant extraction. An outline of the filtration process's benefits and drawbacks, as well as its uses in surfactant extraction, is provided below.

5.3.2.1 Principles and mechanisms

- Using a porous material or filter, filtration is the process of removing particles, including surfactants, from a fluid combination. The following fundamental ideas and procedures underpin the process:
- Sieving: The pore diameters of filtration media allow larger particles, such as surfactant molecules, to stay in the medium while allowing smaller particles to flow through.
- Depth filtration: This type of filtration allows for a better loading capacity since particles are held throughout the filter medium rather than simply at the surface.

- Cake filtration: Particles build up on the filter media and create a 'cake' that can serve as an extra filter barrier to help hold on to smaller particles.
- Crossflow filtration: This technique reduces clogging and extends the life of the filter by having the fluid to be filtered flow parallel to the filter medium (Jornitz 2019, Svarovsky 2000).

5.3.2.2 Advantages and disadvantages

Filtration has the following benefits for surfactant extraction:

- An extremely easy and affordable way to extract surfactants from liquid solutions.
- Highly effective in eliminating surfactants and particulates, producing a refined and cleaner product.
- Versatility in selecting pore diameters and filter media according to the particular needs of the surfactant extraction procedure.

Filtration has several drawbacks when it comes to surfactant extraction:

- Possible blockage of the filter medium, which over time may result in a decrease in filtration effectiveness.
- Restrictions on the use of surfactants with small molecular sizes due to the possibility of the molecules passing through filter pores.
- Requirement for routine filter media replacement and maintenance.

5.3.2.3 Application in surfactant extraction

In surfactant extraction procedures, filtration is frequently employed, especially for separating and purifying synthetic and biosurfactants from reaction mixes or fermentation broths. It is frequently used to eliminate contaminants, cells, or other particles following the initial step of surfactant synthesis or extraction. Filtration can aid in the recovery of the surfactant from the microbial culture medium in the synthesis of biosurfactants (Santos *et al* 2017).

5.3.3 Precipitation

A common separation technique in several industries, including as chemistry, medicine, and surfactant manufacturing, is precipitation. When specific requirements are satisfied, it involves the production of solid particles (precipitate) from a solution.

5.3.3.1 Principles and mechanisms

The basis of precipitation is the idea that when a substance's solubility in a solution is surpassed, it forms crystals or solid particles that separate from the fluid. Numerous processes, including cooling, the addition of a precipitating agent, or pH modification, might cause this (Ghasem and Henda 2015).

By acidifying the supernatant to pH 2.0 with hydrochloric acid and precipitating the biosurfactant with methanol, it is possible to purify the biosurfactant from the supernatant. After that, the precipitate can be separated by centrifuging it. After that, the precipitate is cleaned with methanol and dried at 37 °C. This process, which produces pure biosurfactant for more research, is detailed in Luna *et al* (2013).

5.3.3.2 Advantages and disadvantages

- Simplicity: The process of precipitation is a direct and somewhat simple method of separation.
- High purity: Because the precipitate and solution are separated, high-purity goods can be produced.
- Cost-effective: Precipitation is frequently economical and appropriate for extensive operations.
- Selectivity: It might not be very selective and could cause contaminants to co-precipitate.
- Difficulties with redissolution: In some applications, it can be a disadvantage that some precipitates are difficult to redissolve.
- Waste generation: In order to manage the precipitated solids, precipitation may produce waste or call for extra treatment

5.3.3.3 Application in surfactant extraction

The extraction and purification of surfactants are frequently accomplished via precipitation. By modifying the pH, adding certain salts to the solution, or chilling it down, surfactants can separate from a mixture. The residual solution can then be separated from the precipitated surfactants using centrifugation or filtration (Stellner and Scamehorn 1986).

5.3.4 Extraction with solvents

Solvent extraction is a widely employed separation method that is employed for the retrieval and refinement of diverse substances, including surfactants. Solvents are used in surfactant extraction to help isolate the surfactants from a combination.

5.3.4.1 Principles and mechanisms

The target molecule (surfactants) must be differently soluble in two immiscible phases, usually a solvent and the feed solution, in order for solvent extraction to take place. The surfactants can move from the feed solution into the solvent phase by coming into touch with one another. The chemical affinity of the surfactants for the solvent drives this transfer (Liquid–liquid extraction 2023).

5.3.4.2 Advantages and disadvantages

- Selectivity: The ability to extract solvents with a high degree of selectivity makes it possible to separate surfactants from other combination ingredients.
- High recovery: The result is frequently a high surfactant recovery from the feed solution.
- Versatility: Solvent extraction is adaptable to many types and mixes of surfactants.

- Selecting the right solvent: Selecting the right solvent is essential and might need to be optimized for a particular surfactant.
- Environmental issues: Using organic solvents for solvent extraction may give rise to safety and environmental issues.
- Multiple steps: To reach the necessary level of purification, complex combinations may need more than one extraction step.

5.3.4.3 Application in surfactant extraction

In the surfactant sector, solvent extraction is a useful method for isolating and purifying surfactants. It is frequently used to extract various surfactant components or eliminate contaminants from intricate mixes. After being extracted, the surfactants can be separated from the aqueous phase in the organic phase (Yazdi 2011).

5.4 Conclusion and future perspectives

In conclusion, bioreactors are a crucial component of the process of producing biosurfactants. A variety of bioreactor models, such as bubble column, and stirred-tank bioreactors, provide alternatives for maximizing biosurfactant yield and microbial development. The particular microorganisms and feedstock, as well as the required biosurfactant properties, determine the type of bioreactor to be used. The manufacturing of biosurfactants has been improved and is now more economical and sustainable thanks in large part to bioreactors.

Extraction of surfactants and processing downstream: Solvent extraction is one of the downstream processing procedures that is essential for isolating and purifying biosurfactants. High-purity surfactants can be isolated and contaminants removed using the flexible surfactant recovery technique known as solvent extraction. This stage is crucial to guaranteeing the functionality and quality of the finished product.

Future perspectives

- Bioreactor optimization: It is projected that bioreactor design and operation will continue to innovate and improve over time. To improve the generation of biosurfactants, researchers will probably concentrate on maximizing the parameters of the bioreactor, including pH, temperature, oxygen supply, and nutrient availability. The manufacture of biosurfactants will become more reliable and efficient with the advancement of automation and control systems.
- Sustainability and feedstock selection: As sustainability becomes more and more important, there will be a greater emphasis on employing inexpensive and renewable feedstocks to produce biosurfactants. As substitute substrates, researchers will look into industrial and agricultural byproducts including leftover cooking oil and agro-industrial residues. This fits perfectly with the worldwide movement towards circular and green economies.
- Process integration: The integration of downstream processing steps with upstream bioreactor activities will be given top importance. This will minimize the impact on the environment, optimize resource use, and lower

total processing costs. Researchers investigate creative process designs and the creation of continuous production procedures.

- Advanced separation techniques: In order to increase the purity and effectiveness of surfactant extraction, researchers will keep investigating advanced separation techniques like chromatography and membrane technology. There will also be interest in the creation of sustainable and ecologically friendly extraction solvents.
- Biotechnological approaches: Customized biosurfactants with improved characteristics can be produced thanks to emerging biotechnological techniques like metabolic engineering and synthetic biology. This will increase the commercial feasibility of biosurfactants and broaden their variety of uses.
- In summary, downstream processing techniques and bioreactors are essential to the synthesis and purification of biosurfactants. Future studies will concentrate on enhancing bioreactor efficiency, feedstock sustainability, and the creation of cutting-edge separation techniques due to the growing demand for biodegradable and sustainable surfactants. Biosurfactants are an interesting sector for continued discovery and development because of their significant promise for tackling industrial and environmental concerns.

References

Abdel-Mawgoud A M, Lépine F and Déziel E 2010 Rhamnolipids: diversity of structures, microbial origins and roles *Appl. Microbiol. Biotechnol.* **86** 1323–36

Abrar S and Trathnigg B 2010 Separation of nonionic surfactants according to functionality by hydrophilic interaction chromatography and comprehensive two-dimensional liquid chromatography *J. Chromatogr.* A **1217** 8222–9

Adamberg K, Kask S, Laht T-M and Paalme T 2003 The effect of temperature and pH on the growth of lactic acid bacteria: a pH-auxostat study *Int. J. Food Microbiol.* **85** 171–83

Ali Mansoori G, Bastami T R, Ahmadpour A and Eshaghi Z 2008 Environmental application of nanotechnology *Annu. Rev. Nano Res.* **2** 439–93

Amutha R and Gunasekaran P 2001 Production of ethanol from liquefied cassava starch using co-immobilized cells of *Zymomonas mobilis* and *Saccharomyces diastaticus J. Biosci. Bioeng.* **92** 560–4

Angeles M J, Leyva C, Ancheyta J and Ramírez S 2014 A review of experimental procedures for heavy oil hydrocracking with dispersed catalyst *Catal. Today* **220–222** 274–94 (Int. Symp. on Advances in Hydroprocessing of Oil Fractions (ISAHOF 2013))

Aslam M, Charfi A, Lesage G, Heran M and Kim J 2017 Membrane bioreactors for wastewater treatment: a review of mechanical cleaning by scouring agents to control membrane fouling *Chem. Eng. J.* **307** 897–913

Atlić A, Koller M, Scherzer D, Kutschera C, Grillo-Fernandes E, Horvat P, Chiellini E and Braunegg G 2011 Continuous production of poly([R]-3-hydroxybutyrate) by cupriavidus necator in a multistage bioreactor cascade *Appl. Microbiol. Biotechnol.* **91** 295–304

Banat I M, Carboué Q, Saucedo-Castañeda G and de Jesús Cázares-Marinero J 2021 Biosurfactants: The green generation of speciality chemicals and potential production using Solid-State fermentation (SSF) technology *Bioresource Technol.* **320** 124222

Banat I M, Satpute S K, Cameotra S S, Patil R and Nyayanit N V 2014 Cost effective technologies and renewable substrates for biosurfactants' production *Front. Microbiol.* **5** 697
BeClood.com n.d. Centrifugation: its advantages and its limitations https://suezwaterhandbook.com/processes-and-technologies/liquid-sludge-treatment/centrifugation/centrifugation-its-advantages-and-its-limitations (accessed 11 May 2023)
Behrens B, Helmer P O, Tiso T, Blank L M and Hayen H 2016 Rhamnolipid biosurfactant analysis using online turbulent flow chromatography-liquid chromatography-tandem mass spectrometry *J. Chromatogr.* A **1465** 90–7
Benincasa M 2007 Rhamnolipid produced from agroindustrial wastes enhances hydrocarbon biodegradation in contaminated soil *Curr. Microbiol.* **54** 445–9
Bernard A and Payton M 1995 Fermentation and growth of *Escherichia coli* for optimal protein production *Curr. Protoc. Protein Sci.* **00**
Biotechnology for Biofuels and Bioproducts n.d BioMed Cent. https://biotechnologyforbiofuels.biomedcentral.com/ (accessed 11 May 2023)
Blank L L, Grosso L J and Benson Jr J A 1984 A survey of clinical skills evaluation practices in internal medicine residency programs *Acad. Med.* **59** 401–6
Brück H, Coutte F, Delvigne F, Dhulster P and Jacques P 2020 Optimization of biosurfactant production in a trickle-bed biofilm reactor with genetically improved bacteria *25th National Symp. for Applied Biological Science (NSABS))*
Brumano L P, Antunes F A F, Souto S G, Dos Santos J C, Venus J, Schneider R and da Silva S S 2017 Biosurfactant production by *Aureobasidium pullulans* in stirred tank bioreactor: new approach to understand the influence of important variables in the process *Bioresour. Technol.* **243** 264–72
Cameotra S S and Makkar R S 2004 Recent applications of biosurfactants as biological and immunological molecules *Curr. Opin. Microbiol.* **7** 262–6
Cameotra S S and Makkar R S 1998 Synthesis of biosurfactants in extreme conditions *Appl. Microbiol. Biotechnol.* **50** 520–9
Carstensen F, Apel A and Wessling M 2012 *In situ* product recovery: submerged membranes vs. external loop membranes *J. Membr. Sci.* **394** 1–36
Catapano G, Hornscheidt R, Wodetzki A and Baurmeister U 2004 Turbulent flow technique for the estimation of oxygen diffusive permeability of membranes for the oxygenation of blood and other cell suspensions *J. Membr. Sci.* **230** 131–9
Chaffey N, Alberts B, Johnson A, Lewis J, Raff M, Roberts K and Walter P 2003 Molecular biology of the cell, 4th edn (Book review) *Ann. Bot.* **91** 401
Chang H N and Furusaki S 1991 Membrane bioreactors: present and prospects *Bioreactor Systems and Effects, Advances in Biochemical Engineering/Biotechnology* (Berlin: Springer) pp 27–64
Charcosset C 2012 *Membrane Processes in Biotechnology and Pharmaceutics* (Amsterdam: Elsevier)
Chaudhuri J and Al-Rubeai M (ed) 2005 *Bioreactors for Tissue Engineering: Principles, Design and Operation* (Dordrecht: Springer)
Chisti Y 2000 Animal-cell damage in sparged bioreactors *Trends Biotechnol.* **18** 420–32
Côté P, Bersillon J-L, Huyard A and Faup G 1988 Bubble-free aeration using membranes: process analysis *J. Water Pollut. Control Fed.* **60** 1986–92
Coutte F, Lecouturier D, Ait Yahia S, Leclère V, Béchet M, Jacques P and Dhulster P 2010 Production of surfactin and fengycin by *Bacillus subtilis* in a bubbleless membrane bioreactor *Appl. Microbiol. Biotechnol.* **87** 499–507

De Kronemberger F A, Anna L M M S, Fernandes A C L B, De Menezes R R, Borges C P and Freire D M G 2007 Oxygen-controlled biosurfactant production in a bench scale bioreactor *Biotechnology for Fuels and Chemicals* ed W S Adney, J D McMillan, J Mielenz and K T Klasson (Totowa, NJ: Humana Press) pp 401–13

Degaleesan S, Dudukovic M and Pan Y 2001 Experimental study of gas-induced liquid-flow structures in bubble columns *AIChE* J **47** 1913–31

Desai J D and Banat I M 1997 Microbial production of surfactants and their commercial potential *Microbiol. Mol. Biol. Rev.* **61** 47–64

Deziel E, Paquette G, Villemur R, Lepine F and Bisaillon J 1996 Biosurfactant production by a soil pseudomonas strain growing on polycyclic aromatic hydrocarbons *Appl. Environ. Microbiol.* **62** 1908–12

Ding S, Deng Y, Bond T, Fang C, Cao Z and Chu W 2019 Disinfection byproduct formation during drinking water treatment and distribution: a review of unintended effects of engineering agents and materials *Water Res.* **160** 313–29

Djeljadini S *et al* 2021 Porous PVDF monoliths with templated geometry *Adv. Mater. Technol.* **6** 2100325

Doran P M 1995 *Bioprocess Engineering Principles* (Amsterdam: Elsevier)

Drioli E, Curcio E and Di Profio G 2005 State of the art and recent progresses in membrane contactors *Chem. Eng. Res. Des.* **83** 223–33

e Silva N M P R, Rufino R D, Luna J M, Santos V A and Sarubbo L A 2014 Screening of *Pseudomonas* species for biosurfactant production using low-cost substrates *Biocatal. Agric. Biotechnol.* **3** 132–9

Fiechter A 1992 Biosurfactants: moving towards industrial application *Trends Biotechnol.* **10** 208–17

Frahm B, Brod H and Langer U 2009 Improving bioreactor cultivation conditions for sensitive cell lines by dynamic membrane aeration *Cytotechnology* **59** 17–30

Ghasem N and Henda R 2015 *Principles of Chemical Engineering Processes: Material and Energy Balances* (Routledge) 2nd edn

Ghribi D and Ellouze-Chaabouni S 2011 Enhancement of *Bacillus subtilis* lipopeptide biosurfactants production through optimization of medium composition and adequate control of aeration *Biotechnol. Res. Int.* **2011** 653654

Gong Z, Yang G, Che C, Liu J, Si M and He Q 2021 Foaming of rhamnolipids fermentation: impact factors and fermentation strategies *Microb. Cell Fact.* **20** 77

Guarino R D, Dike L E, Haq T A, Rowley J A, Pitner J B and Timmins M R 2004 Method for determining oxygen consumption rates of static cultures from microplate measurements of pericellular dissolved oxygen concentration *Biotechnol. Bioeng.* **86** 775–87

Gudina E J, Pereira J F, Costa R, Coutinho J A, Teixeira J A and Rodrigues L R 2013 Biosurfactant-producing and oil-degrading *Bacillus subtilis* strains enhance oil recovery in laboratory sand-pack columns *J. Hazard. Mater.* **261** 106–13

Henzler H-J 2000 Particle stress in bioreactors *Influence of Stress on Cell Growth and Product Formation, Advances in Biochemical Engineering/Biotechnology* ed K Schügerl, G Kretzmer, H J Henzler, P M Kieran, G Kretzmer, P E MacLoughlin, D M Malone, W Schumann, P A Shamlou and S S Yim (Berlin: Springer) pp 35–82

Jeon S, Rajabzadeh S, Okamura R, Ishigami T, Hasegawa S, Kato N and Matsuyama H 2016 The effect of membrane material and surface pore size on the fouling properties of submerged membranes *Water* **8** 602

Jornitz M W 2019 *Filtration and Purification in the Biopharmaceutical Industry* (CRC Press) 3rd edn

Judd S 2010 *The MBR Book: Principles and Applications of Membrane Bioreactors for Water and Wastewater Treatment* (Amsterdam: Elsevier)

Judd S 2008 The status of membrane bioreactor technology *Trends Biotechnol.* **26** 109–16

Junker B 2007 Foam and its mitigation in fermentation systems *Biotechnol. Prog.* **23** 767–84

Kanga S A, Bonner J S, Page C A, Mills M A and Autenrieth R L 1997 Solubilization of naphthalene and methyl-substituted naphthalenes from crude oil using biosurfactants *Environ. Sci. Technol.* **31** 556–61

Kebbouche-Gana S, Gana M L, Ferrioune I, Khemili S, Lenchi N, Akmouci-Toumi S, Bouanane-Darenfed N A and Djelali N-E 2013 Production of biosurfactant on crude date syrup under saline conditions by entrapped cells of *Natrialba* sp. strain E21, an extremely halophilic bacterium isolated from a solar saltern (Ain Salah, Algeria) *Extremophiles* **17** 981–93

Kiran G S, Sabu A and Selvin J 2010 Synthesis of silver nanoparticles by glycolipid biosurfactant produced from marine *Brevibacterium casei* MSA19 *J. Biotechnol.* **148** 221–5

Klok A J, Verbaanderd J A, Lamers P P, Martens D E, Rinzema A and Wijffels R H 2013 A model for customising biomass composition in continuous microalgae production *Bioresour. Technol.* **146** 89–100

Knazek R A, Gullino P M, Kohler P O and Dedrick R L 1972 Cell Culture on artificial capillaries: an approach to tissue growth *in vitro Science* **178** 65–7

Kosaric N 2001 Biosurfactants and their application for soil bioremediation *Food Technol. Biotechnol.* **39** 295–304

Kosaric N 1992 Biosurfactants in industry *Pure Appl. Chem.* **64** 1731–7

Lebron-Paler A 2008 Solution and Interfacial characterization of rhamnolipid biosurfactant from *P. aeruginosa* ATCC 9027 *PhD Dissertation* University of Arizona, Tucson, AZ

Le-Clech P 2010 Membrane bioreactors and their uses in wastewater treatments *Appl. Microbiol. Biotechnol.* **88** 1253–60

Le-Clech P, Chen V and Fane T A G 2006 Fouling in membrane bioreactors used in wastewater treatment *J. Membr. Sci.* **284** 17–53

Liquid–Liquid Extraction 2023 (Wikipedia) https://en.wikipedia.org/wiki/Liquid%E2%80%93liquid_extraction

Luna J M, Rufino R D, Sarubbo L A and Campos-Takaki G M 2013 Characterisation, surface properties and biological activity of a biosurfactant produced from industrial waste by *Candida sphaerica* UCP0995 for application in the petroleum industry *Colloids Surf.* B **102** 202–9

Mandenius C-F 2016 *Bioreactors: Design, Operation and Novel Applications* (New York: Wiley)

Md F 2012 Biosurfactant: production and application *J. Pet. Environ. Biotechnol.* **3** 124

Muthusamy K, Gopalakrishnan S, Ravi T K and Sivachidambaram P 2008 Biosurfactants: properties, commercial production and application *Curr. Sci.* **94** 736–47

Najjar Y and Abu-Shamleh A 2020 Harvesting of microalgae by centrifugation for biodiesel production: a review *Algal Res.* **51** 102046

Nayak A S, Vijaykumar M H and Karegoudar T B 2009 Characterization of biosurfactant produced by *Pseudoxanthomonas sp.* PNK-04 and its application in bioremediation *Int. Biodeterior. Biodegrad.* **63** 73–9

Nayarisseri A, Singh P and Singh S K 2018 Screening, isolation and characterization of biosurfactant producing *Bacillus subtilis* strain ANSKLAB03 *Bioinformation* **14** 304–14

Palomares L A and Ramírez O T 2019 Hydrodynamic stress and heterogeneities in animal cell culture *Comprehensive Biotechnology* (Elsevier) 3rd ednvol 2 pp 108–18
Pandey A, Soccol C R and Larroche C 2008 *Current Developments in Solid-State Fermentation* (Berlin: Springer Science & Business Media)
Probert H M and Gibson G R 2002 Investigating the prebiotic and gas-generating effects of selected carbohydrates on the human colonic microflora *Lett. Appl. Microbiol.* **35** 473–80
Qi H N, Goudar C T, Michaels J D, Henzler H-J, Jovanovic G N and Konstantinov K B 2008 Experimental and theoretical analysis of tubular membrane aeration for mammalian cell bioreactors *Biotechnol. Prog.* **19** 1183–9
Rahman P K and Gakpe E 2008 Production, characterisation and applications of biosurfactants-Review *Biotechnology* **7** 360–70
Rickwood D 2001 Centrifugation techniques *Encyclopedia of Life Sciences* (Wiley)
Rivardo F, Martinotti M G, Turner R J and Ceri H 2011 Synergistic effect of lipopeptide biosurfactant with antibiotics against *Escherichia coli* CFT073 biofilm *Int. J. Antimicrob. Agents* **37** 324–31
Rodrigues L R, Teixeira J A, van der Mei H C and Oliveira R 2006 Physicochemical and functional characterization of a biosurfactant produced by *Lactococcus lactis* 53 *Colloids Surf. B Biointerfaces* **49** 79–86
Rodriguez-Contreras A, Koller M, de Sousa Dias M M, Calafell M, Braunegg G and Marqués-Calvo M S 2013 Novel poly [(R)-3-hydroxybutyrate]-producing bacterium isolated from a bolivian hypersaline lake *Food Technol. Biotechnol.* **51** 123
Samad A, Zhang J, Chen D and Liang Y 2015 Sophorolipid production from biomass hydrolysates *Appl. Biochem. Biotechnol.* **175** 2246–57
Santos D K F, Resende A H M, de Almeida D G, Soares da Silva R, de cf, Rufino R D, Luna J M, Banat I M and Sarubbo L A 2017 *Candida lipolytica* UCP0988 biosurfactant: potential as a bioremediation agent and in formulating a commercial related product *Front. Microbiol.* **8** 767
Sarilmiser H K, Ates O, Ozdemir G, Arga K Y and Oner E T 2015 Effective stimulating factors for microbial levan production by *Halomonas smyrnensis* AAD6T *J. Biosci. Bioeng.* **119** 455–63
Sayed G H, Ghuiba F M, Abdou M I, Badr E A A, Tawfik S M and Negm N A M 2012a Synthesis, surface and thermodynamic parameters of some biodegradable nonionic surfactants derived from tannic acid *Colloids Surf. Physicochem. Eng. Asp.* **393** 96–104
Sayed G H, Ghuiba F M, Abdou M I, Badr E A A, Tawfik S M and Negm N A M 2012b Synthesis, surface, thermodynamic properties of some biodegradable vanillin-modified polyoxyethylene surfactants *J. Surfactants Deterg.* **15** 735–43
Sen P, Nath A and Bhattacharjee C 2017 Packed-bed bioreactor and its application in dairy, food, and beverage industry *Current Developments in Biotechnology and Bioengineering* (Amsterdam: Elsevier) pp 235–77
Sharma M, Mandal G, Mandal S, Bhattacharjee H and Mukhopadhyay R 2015 Functional role of lysine 12 in *Leishmania major* AQP1 *Mol. Biochem. Parasitol.* **201** 139–45
Sharma M, Thukral N, Soni N K and Maji S 2015 Microalgae as future fuel: real opportunities and challenges *J. Thermodyn. Catal.* **6** 1
Shavandi M, Mohebali G, Haddadi A, Shakarami H and Nuhi A 2011 Emulsification potential of a newly isolated biosurfactant-producing bacterium, *Rhodococcus* sp. strain TA6 *Colloids Surf. B Biointerfaces* **82** 477–82

Stanbury P F, Whitaker A and Hall S J 2013 *Principles of Fermentation Technology* (Amsterdam: Elsevier)

Stellner K L and Scamehorn J F 1986 Surfactant precipitation in aqueous solutions containing mixtures of anionic and nonionic surfactants *J. Am. Oil Chem. Soc.* **63** 566–74

Stephenson T, Brindle K, Judd S and Jefferson B 2000 *Membrane Bioreactors for Wastewater Treatment* (IWA Publishing)

Su R, Sujarani M, Shalini P and Prabhu N 2019 A review on bioreactor technology assisted plant suspension culture *Asian J. Biotechnol. Bioresour. Technol.* **5** 1–13

Svarovsky L (ed) 2000 *Solid-liquid Separation* 4th edn (Oxford : Butterworth-Heinemann)

Syron E and Casey E 2008 Membrane-aerated biofilms for high rate biotreatment: performance appraisal, engineering principles, scale-up, and development requirements *Environ. Sci. Technol.* **42** 1833–44

Tawfik S M 2015 Synthesis, surface, biological activity and mixed micellar phase properties of some biodegradable gemini cationic surfactants containing oxycarbonyl groups in the lipophilic part *J. Ind. Eng. Chem.* **28** 171–83

Tawfik S M, Abd-Elaal A A and Aiad I 2016 Three gemini cationic surfactants as biodegradable corrosion inhibitors for carbon steel in HCl solution *Res. Chem. Intermed.* **42** 1101–23

Thavasi R, Jayalakshmi S, Balasubramanian T and Banat I M 2007 Biosurfactant production by corynebacterium kutscheri from waste motor lubricant oil and peanut oil cake *Lett. Appl. Microbiol.* **45** 686–91

Tiso T, Zauter R, Tulke H, Leuchtle B, Li W-J, Behrens B, Wittgens A, Rosenau F, Hayen H and Blank L M 2017 Designer rhamnolipids by reduction of congener diversity: production and characterization *Microb. Cell Fact.* **16** 14

Tripathy D B and Mishra A 2016 Sustainable biosurfactants *Sustain. Inorg. Chem.* **1** 175–92

Vanavil B, Perumalsamy M and Rao A S 2013 Biosurfactant production from novel air isolate NITT6L: screening, characterization and optimization of media *J. Microbiol. Biotechnol.* **23** 1229–43

Wagner B A, Venkataraman S and Buettner G R 2011 The rate of oxygen utilization by cells *Free Radic. Biol. Med.* **51** 700–12

Wittgens A *et al* 2011 Growth independent rhamnolipid production from glucose using the non-pathogenic *Pseudomonas putida* KT2440 *Microb. Cell Fact.* **10** 80

Xu N, Liu S, Xu L, Zhou J, Xin F, Zhang W, Qian X, Li M, Dong W and Jiang M 2020 Enhanced rhamnolipids production using a novel bioreactor system based on integrated foam-control and repeated fed-batch fermentation strategy *Biotechnol. Biofuels* **13** 80

Yao S, Zhao S, Lu Z, Gao Y, Lv F and Bie X 2015 Control of agitation and aeration rates in the production of surfactin in foam overflowing fed-batch culture with industrial fermentation *Rev. Argent. Microbiol.* **47** 344–9

Yazdi A S 2011 Surfactant-based extraction methods *TrAC, Trends Anal. Chem.* **30** 918–29

Zhang Q, Li Y and Xia L 2014 An oleaginous endophyte *Bacillus subtilis* HB1310 isolated from thin-shelled walnut and its utilization of cotton stalk hydrolysate for lipid production *Biotechnol. Biofuels* **7** 152

IOP Publishing

Microbial Surfactants
A sustainable class of versatile molecules
Divya Tripathy and Anjali Gupta

Chapter 6

Microbial surfactants from waste resources

Vishal Sharma, Dharmaraj Senthilkumar, Achyut Pandey and Shruti Mishra

Biosurfactants, derived from microbial sources, have become integral components of various everyday products, including detergents, cosmetics, and food additives. Their widespread use across industries such as petroleum, pharmaceuticals, agriculture, and environmental remediation demonstrates their economic and environmental advantages over synthetic surfactants. Notable classes of biosurfactants include beta-amyloids, produced primarily by aerophilic bacteria, and well-known compounds like rhamnolipids, surfactins, sophorolipids, emulsans, and mannosylerythritol lipids. This chapter emphasizes the extensive applications of biosurfactants across various industries, including nanotechnology product formulation, petroleum, personal healthcare product manufacturing, pharmaceuticals, agriculture, agrochemicals, and food processing. Additionally, biosurfactants play a crucial role in environmental remediation, offering a natural and environmentally benign alternative to chemical surfactants. Their sustainable nature, coupled with the potential for cost-effective production through biotechnological methods, makes them an attractive choice. However, challenges related to soil adsorption, limited bioavailability, insolubility, and hydrophobicity toward pollutants exist. These hurdles can be overcome through optimization of growth and production conditions, the use of renewable substrates, and efficient downstream processing, ultimately leading to the development of economically viable biosurfactants. Further research is needed to fine-tune the production conditions, capitalizing on the societal and economic benefits they offer.

6.1 Introduction

These days, bacterial surfactants are widely used in human lifestyles as an opulent part of everyday goods like detergents, food additives, and cosmetics. The petroleum, pharmaceutical, medical, agricultural, and environmental industries all make extensive use of them. Substituting synthetic surfactants with biodegradable microbial surfactants can enhance the economics and mitigate environmental

doi:10.1088/978-0-7503-5989-4ch6

problems (Bhardwaj 2013). A class of molecules known as surfactants combines hydrophobic and hydrophilic moieties in a single molecule. They disperse across two immiscible fluids, lower surface/interfacial tensions (ST/IFT), and make non-polar chemicals soluble in polar solvents (Liu *et al* 2015, Udoh and Vinogradov 2019). They exhibit qualities including lubricating, detergency, and solubilization. They can also froth and stabilize (Sarubbo *et al* 2014, Pradhan and Bhattacharyya 2018).

Surfactants can be generated by chemical or biological means. Since the biologically derived surfactants are made from living things, particularly microbes, they are referred to as biosurfactants. These compounds are generated either by the surface chemistry of the cells or as metabolic by-products (Karanth *et al* 1999). Aerophilic bacteria in water-based media with a carbon-containing feedstock, such as hydrocarbons, carbohydrates, fats, and oil, are the primary source of beta-amyloids. These microbes are primarily from the fungus *Fusarium*, the bacterium *Pseudomonas*, and the yeast genera *Candida* and *Pseudozyma* (Rocha e Silva *et al* 2014). Rhamnolipids, surfactins, sophorolipids, emulsans, and mannosylerythritol lipids are the most prevalent biosurfactants. The physiological roles of these surface-active compounds vary depending on which class of biosurfactant a given biosurfactant belongs to. These functions include enabling biosurfactants producing bacteria to grow on water-immiscible substrates, ensuring exponential biomass increase, exhibiting antimicrobial activities against potential predators, helping them survive unfriendly environmental conditions, and viral and cell desorption for survival (Md Badrul Hisham 2012). Low- and high- molecular weight (LMW and HMW) biosurfactants are the two general categories into which biosurfactants fall. Whereas the HMW biosurfactants are more of an emulsion-stabilizing agent, the LMW biosurfactants reduce ST and ITF. The LMW biosurfactants include glycolipids, lipopeptides, and phospholipids, whereas the HMW biosurfactants are particulate and polymeric (Ron and Rosenberg 2001, Shekhar *et al* 2015). Biosurfactants are classified into several groups based on their chemical composition: glycolipids (mannosylerithritol lipids, rhamnolipids, sophorolipids, trehalolipids), lipopeptides (surfactin, lichenysin, iturin, fengycin, serrawettin), fatty acids/phospholipids/neutral lipids (phosphatidylethanolamine, spiculisporic acid), polymeric biosurfactants (emulsan, alasan, biodispersan, liposan), and particulate biosurfactants (vesicles, whole-cell) (Muthusamy *et al* 2008, Pacwa-Płociniczak *et al* 2011, Stancu 2015). Studies using spectroscopy and chromatography enable this categorization. Research utilizing these studies demonstrated that the hydrophobic moiety of biosurfactants is a long-chain fatty acid, whereas the hydrophilic component may consist of cyclic peptide, alcohol, amino acid, carbohydrate, phosphate, or carboxylic acid (Santos *et al* 2016).

6.1.1 Importance of microbial surfactants

Important substances called biosurfactants have the potential to take the place of manufactured surfactants. Petroleum, organic chemicals, medicines, cosmetics, food and beverage, bioremediation procedures, petrochemicals, biological control, and other industrial sectors are just a few of the industries in which they find numerous uses (Pardhi *et al* 2022).

Biosurfactants, or microbial surfactants, are naturally occurring substances that are produced by microbes. Because of their special qualities and eco-friendliness, they are essential in many applications. The following are some of the major functions of microbial surfactants.

Environmental sustainability: Compared to synthetic surfactants, microbial surfactants are more ecologically friendly since they are non-toxic and biodegradable. This is especially crucial for sectors of the economy where environmental damage and pollution are major issues (Tripathy and Mishra 2016).

Enhanced oil recovery: Microbial surfactants are utilized in the petroleum industry to boost reservoir oil recovery efficiency. They have the ability to lessen the tension that exists between water and oil, which facilitates the removal of oil from porous rock formations.

Microbial surfactants facilitate the process of bioremediation in polluted settings. They facilitate the breakdown of hydrophobic pollutants, such as oil and other organic contaminants, and aid in their solubilization.

Agriculture: Microbial surfactants have the potential to enhance the application and dispersal of fertilizers and pesticides on plant surfaces. As a result, these items work better and need less, which might have a positive influence on the environment overall.

Food and beverage industry: To stabilize goods like mayonnaise, salad dressings, and drinks, microbial surfactants might be employed as emulsifiers in this sector. Additionally, they can be employed to improve the food items' ability to foam.

Pharmaceuticals: Microbial surfactants are utilized in medicine formulations as stabilizers. They can aid in enhancing the bioavailability and solubility of poorly soluble medications.

Cosmetics and personal care goods: Because microbial surfactants may form stable emulsions, foams, and enhance the texture of these goods, they are employed in cosmetic and personal care products such shampoos, soaps, and lotions.

Enhanced microbial enhanced oil recovery (MEOR): This method of improving oil recovery by using microorganisms can also make use of microbial surfactants. They support the extraction and mobilization of oil from reservoirs.

Green cleaning solutions: Microbial surfactants are an excellent element for green cleaning solutions because of their environmentally beneficial properties. They can retain cleaning effectiveness while assisting in the reduction of harsh chemical use in domestic cleaning.

Reduced surface tension: Microbial surfactants help liquids have less surface tension, which is useful for making foams and emulsions as well as cleaning debris and stains off surfaces.

Biomedical and biotechnological applications: Microbial surfactants show promise in tissue engineering, drug delivery systems, and cell adhesion modification, among other biomedical and biotechnological sectors.

6.1.2 Sources of waste resources for microbial surfactants production

Through fermentation, microorganisms use a variety of complex organic substrates to get carbon and energy by breaking them down into simpler forms. They generate important products such as polysaccharides, vitamins, amino acids, and ethanol. One of the secondary metabolites created during these fermentation processes is biosurfactant see table 6.1.

Table 6.1. Biosurfactant as secondary metabolites created during fermentation processes.

Biosurfactants	Carbon sources	References
Lipopeptide	Mineral oil	Pardhi *et al* (2021a)
	Sucrose	Kanna *et al* (2014)
Rhamnolipid	Chicken feather	Ozdal *et al* (2017)
	Olive oil/*n*-hexadecane	Abouseoud *et al* (2007)
	Waste frying oil	Wadekar *et al* (2012)
	Soybean oil refinery waste	Abalos *et al* (2001)
Polymeric	Diesel	Joshi and Shekhawat (2014)
***Bacillus* sp. Iturin**	Glucose/rapeseed oil, crude oil	Bayoumi *et al* (2010)
Phospholipids	Glucose/diesel/crude oil	Adamu *et al* (2015)
Lipopeptide	Cassava wastewater	Nitschke *et al* (2006)
	Dextrose	López-prieto *et al* (2019)
	Coal tar creosote	Bezza and Chirwa (2015)
	Olive oil/diesel/crude oil/kerosene	Liu *et al* (2016)
	Soybean, sweet potato residues	Wang *et al* (2008)
Surfactin	Crude oil	Pereira *et al* (2013)
Other generaTrehalose-2,3,4,2°-tetraester	Sucrose/molasses, crude oil	Bayoumi *et al* (2010)
Phospholipid	Dextrose	Astuti *et al* (2019)
Glycolipid	Waste frying oil	Ibrahim (2017)
	Heavy oil	Astuti *et al* (2019)
	Crude oil	Vyas and Dave (2011)
Polymeric	Corn waste oil	Araújo *et al* (2019)
	Soybean/corn/diesel	Nogueira *et al* (2020)
Lipopeptide	Waste frying oil	Elazzazy *et al* (2015)
	Crude oil	Gargouri *et al* (2017)
	Gasoline	Kamal *et al* (2015)
Exopolysaccharide	Sugarcane molasses	Fusconi *et al* (2010)
Rhamnolipid	Glycerol	Dubeau *et al* (2009)
	Hexadecane	Nayak *et al* (2009)
	Crude oil	Płaza *et al* (2008)
Bioemulsan	Aliphatic hydrocarbons	Franzetti *et al* (2009)

6.2 Microorganisms for microbial surfactants production from waste resources

Certain microorganisms have the ability to synthesize structurally varied biosurfactants using a variety of substrates that may be detrimental to other germs that do not create biosurfactants. The microorganisms' genetic make-up, physiological circumstances, and the different nutrients they consume all have a role in the chemistry and production of the biosurfactant that is created. Oil-contaminated locations, such as those affected by crude oil pollution, petrochemical industrial waste, tannery wastewater, used food oils, and oil reservoirs, are the main locations where samples are taken in order to identify possible biosurfactant producers. Additionally, it has been observed that extremophiles from marine settings may create biosurfactants that are extremely stable.

6.2.1 Bacteria

6.2.1.1 Pseudomonas aeruginosa

6.2.1.1.1 Media and growth conditions

A 3 l laboratory fermentor (Scigenics India Pvt. Ltd, Chennai) with a working capacity of 2.1 l was used to produce biosurfactants. Mineral media (g l^{-1}, 1.0 K_2HPO_4, 0.2 $MgSO_4{\bullet}7H_2O$, 0.05 $FeSO_4{\bullet}7H_2O$, 0.1 $CaCl_2{\bullet}2H_2O$, 0.001 $Na_2MoO_4{\bullet}2H_2O$, 30NaCl) were used to cultivate the strain. Elsewhere, the circumstances for culture were optimized (Thavasi 2006). The following parameters apply to the culture: pH 8.0, temperature 38 °C, salinity 30% (w/v), substrate concentration of 2.0% (waste motor lubricant oil/peanut oil cake), and dissolved oxygen (DO) of 8.0 mg l^{-1}.

6.2.1.1.2 Estimation of growth and biosurfactant production

Over the course of 168 h, five milliliter samples of culture broth were taken at 12 h intervals. The broth culture was filtered using Millipore filter paper (0.45 ml), dried at 80 °C in a hot air oven, and weighed after the biomass was calculated gravimetrically. The biomass was stated as milligrams per milliliter, or dry weight. Biosurfactant concentration in the culture broth was estimated according to the procedure described by (Li *et al* 1984). The concentration of the biosurfactant was given as mg ml^{-1}. The culture broth was twice extracted using a 2:1 v/v mixture of methanol and chloroform after centrifuging it for 20 min at 4 °C at 6000 rpm. Rotating evaporation was used to remove the solvents, and the residue was partly purified using a silica gel (60–120 mesh) column eluted with methanol and chloroform at gradient ratios of 20:1 to 2:1 (v/v). Following the pooling of the fractions and the solvents' evaporation, the residue was dialyzed against distilled water and lyophilized (Thavasi *et al* 2011).

6.2.1.1.3 Estimation of emulsification activity

30 ml test tubes were filled with 5 ml of Tris buffer (pH 8.0) and partially purified biosurfactant (5 mg). The emulsification activity of hydrocarbons such as waste

motor lubricant oil, crude oil, peanut oil, diesel, kerosene, naphthalene, anthracene, and xylene was examined. The aforementioned biosurfactant solution was mixed with 5 mg of hydrocarbon, thoroughly agitated for 20 min, and then allowed to stand for another 20 min. At 610 nm, the emulsified mixture's optical density was measured, and the findings were reported as D_{610} (Rosenberg *et al* 1979). Triton X-100 (1 mg ml^{-1}) (Hi-Media, Laboratories, Mumbai, India) was used to compare the biosurfactant's emulsification activity. The concentration and emulsification research conditions were kept consistent with the biosurfactant's. The height of the emulsion layer was measured and the emulsion's stability was tracked for a whole day.

6.2.1.2 Bacillus subtilis

6.2.1.2.1 Media and growth conditions

After *Bacillus subtilis* was isolated from wastewater, it was kept frozen in 35% glycerol at a temperature of −80 °C.

6.2.1.2.2 Inoculum and media preparation

After being produced in nutrient broth, the inoculum was incubated at 30 °C for 24 h. The minimal medium used as the basal medium contained the following elements: trace elements (mg l^{-1}): calcium chloride, 1.2; ferrous sulfate heptahydrate, 1.65; manganese sulfate tetrahydrate, 1.5; sodium salt of EDTA, 2.2; ammonium nitrate, 1.0; potassium di-hydrogen phosphate, 6.0; di-hydrogen phosphate, 2.7; magnesium sulfate heptahydrate, 0.1; and magnesium sulfate heptahydrate, 0.1. Separate additions of several carbon sources were made to the autoclaved medium. Using the method outlined by Kulkarni *et al* aqueous extracts of the agrowastes-orange, potato, banana, and bagasse (Kulkarni *et al* 2015). 10% (w/v) suspensions of each form of dry waste were autoclaved to create stock solutions of agricultural waste. Clear filtrates were re-autoclaved after the extracts were filtered using muslin cloth and Whatman filter paper No. 1. At 4% (v/v), these extracts were added to the medium. After being individually autoclaved in distilled water, the desired amounts of molasses were added to the medium (Rane *et al* 2017).

6.2.1.2.3 Biosurfactant extraction and purification

The cultures were nurtured in the above-described conditions at 30 °C for 24 h in order to create biosurfactant. At 4 °C, the culture was centrifuged for 30 min at 12 000 rpm to extract the crude biosurfactant. A Wilhelmy plate-equipped DCAT 11 Tensiometer (DataPhysics Corporation, Germany) was used to measure the surface tension of the culture supernatants. Acid precipitation was used to purify the biosurfactant. After having been acidified with 6 N HCl, the supernatant was stored for precipitation at 4 °C for a whole night. After centrifuging the precipitate for 30 min at 4 °C at 12 000 rpm, it was suspended in the smallest amount of distilled water, which was neutralized using 1 N NaOH, and stirred with a magnetic stirrer for 1 h to dissolve it entirely. This method was described by (Ghribi *et al* 2012), whereas the organic phase was recovered after the aqueous solution was combined with an equal amount of chloroform:methanol (2:1). At least three iterations of the process were carried out to guarantee optimal biosurfactant recovery (Rane *et al* 2017).

6.2.1.3 Acinetobacter calcoaceticus

Acinetobacter calcoaceticus P1–1A, a strain known to generate biosurfactants, was employed in this investigation. From an oily, polluted sample taken from an offshore oil and gas site in Atlantic Canada, it was separated. The isolation was carried out at Memorial University's Northern Region Persistent Organic Pollution Control (NRPOP) laboratory. This strain had significant potential for increasing emulsification index (EI) and lowering ST, and it was verified to be a promising producer of biosurfactants (Cai *et al* 2014).

6.2.1.3.1 Inoculum preparation

The *A. calcoaceticus* P1–1A strain was cultivated in a 250 ml sterilized Erlenmeyer flask using a 50 ml nutritional salt medium and 0.02 v/v (2 vol.%) n-hexadecane as the carbon source. Following that, the medium was filled with a loopful of the bacterial colony from the agar plate. $(NH_4)_2SO_4$, 5 g l^{-1}; NaCl, 15 g l^{-1}; KCl, 1 g l^{-1}; KH_2PO_4, 3 g l^{-1}; Na_2HPO_4, 1.2 g l^{-1}; $FeSO_4 \bullet 7H_2O$, 0.002 g l^{-1}; $MgSO_4$, 0.2 g l^{-1}; and yeast extract, 0.5 g l^{-1} made up the nutritional salt medium. One liter of the nutritional salt medium was mixed with a 0.5 ml element solution comprising $ZnSO_4$, 0.29 g l^{-1}; $CaCl_2$, 0.1 g l^{-1}; $CuSO_4$, 0.25 g l^{-1}; and $MnSO_4$, 0.4 g l^{-1}. For 48 h, the cultures were cultivated in the Innova 43 incubator shaker at 30 °C and 200 revolutions per minute (Moshtagh *et al* 2021).

6.2.1.3.2 Biosurfactant production

Each 250 ml flask held 50 ml of medium containing 0.04 v/v (4 vol.%) of the inoculums at $OD_{660} = 0.5$, such that an equal amount of the bacterial suspensions reached every flask. For six days, the cultivation was done at 200 rpm. A variety of waste carbon sources, such as crude glycerol, refined waste cooking oil, and waste cooking oil, were employed and contrasted with commercial carbon sources, such as n-hexadecane, diesel, pure glycerol, and ethanol. These studies allowed for the selection and addition of the waste carbon source that best supported bacterial growth to the medium.

Central Composite Design (CCD) of Response Surface Methodology (RSM) was used to plan trials for biosurfactant production optimization. The step change was divided by the difference readings among each variable and the center point value to produce the coded dimensionless Xi, which was then utilized in statistical computations. After applying the CCD, the data were fitted using the connection between the variables that were independent and their responses to a predictive the second-order polynomial equation.

The concentration of the carbon source (refined waste from cooking oil), the nitrogen source ($(NH_4)_2SO_4$), pH, temperature, sodium chloride, and five levels of each at six repetitions at the central locations were the chosen independent process variables for this investigation. The EI served as the response variable. One mol l^{-1} of NaOH and one mol l^{-1} of HCl were used to alter the pH. Randomization was used in the experiments to account for any biases. Every experiment was carried out twice. The negative control was a flask that contained no inoculum.

The culture broth was centrifuged at 10 000 rpm for 15 min to extract the cell-free supernatant (crude biosurfactant), and it was then filtered through filter paper with a pore size of 0.45 μm. To assess the productivity and performance of the crude biosurfactant, measurements were made of EI, ST, the biosurfactant concentration, and the critical micelle concentration (CMC). Every measurement was done three times (Moshtagh *et al* 2021).

6.2.2 Yeasts

Because of their pathogenic and/or opportunist qualities, bacterial biosurfactants may have limited use in the food and pharmaceutical sectors. Consequently, yeasts have become more prominent in biosurfactant investigations than bacteria. Furthermore, certain yeasts produce more biosurfactants than bacteria because they convert substrates at a faster pace (Sharma and Saharan 2016, Oyeleke *et al* 2017). The majority of yeast species also have the benefit of being GRAS-approved, meaning that their products may be used in a variety of industrial sectors because they do not pose a danger of toxicity or pathogenicity (Souza *et al* 2017, Nwaguma *et al* 2019). The genus *Candida* stands out in light of the expanding industrial need for yeasts for a variety of applications due to the range of species it produces that have been documented in the literature as creating biosurfactants (Ribeiro, Guerra, *et al* 2020, Camarate *et al* 2022). But subsequent reports have shown that a number of additional yeast genera, including *Rhodotorula* (Derguine-Mecheri *et al* 2021), Saccharomyces (Ribeiro, de Veras, *et al* 2020), and *Wickerhamomyces* (Fernandes *et al* 2020, Souza *et al* 2018), may also be biosurfactant makers that have larvicidal, environmental, antibacterial, antioxidant, and flocculent qualities. Furthermore, their recovery percentages are greater than those of biosurfactants derived from other microbes. Numerous investigations have reported on rising biosurfactant production yields in yeast fermentation, ranging from 10 to 120 g l^{-1} (Deshpande and Daniels 1995, Solaiman *et al* 2004, Ashby *et al* 2005, Morita *et al* 2007). Both water-soluble substances like glycerol and carbohydrates as well as water-immiscible substrates like vegetable oils and hydrocarbons can support the growth of yeast (Souza *et al* 2017, García-Reyes *et al* 2018). Since hydrophobic substrates have a high moisture and protein content that causes their rapid degradation and influences the ability to survive of microbes produced, which utilize the set of energy and carbon sources for growth, the cultivation conditions and medium composition determine the generation and composition of the biosurfactants (Rau *et al* 2005, Günther *et al* 2015). Because it improves the solubility of water-insoluble molecules and aids their transport to the cell, this combination of carbon sources with insoluble substrates promotes intracellular diffusion (Shekhar *et al* 2015).

6.3 Types of waste resources from microbial surfactants production

Agro-industrial wastes are essentially organic wastes that are naturally biodegradable and rich in various nutrients, including proteins, lignin, cellulose, starch, and hemicelluloses, as well as minerals and vitamins. With modernization and civilization, the volume of waste leftovers from various agro-industries is rising at a rapid pace. It is noteworthy to emphasize that Asia produces 4.4 billion tons of waste products annually, with India accounting for about 350 million tons of rubbish from various sources (Madurwar *et al* 2013). These leftovers are frequently disposed of untreated, which has an adverse effect on the environment (Domínguez Rivera *et al* 2019). However, a lot of these agro-industrial waste materials are nutritious, and since they are nutrient-rich, they may foster the development of microorganisms and serve as a substrate for their fermentation by a variety of microbes (Carolin *et al* 2023).

The most prevalent and affordable material with a high carbohydrate content is agro-industrial waste, which is derived from agricultural and processing industries and used to produce biosurfactants on a big scale at the industrial level. Agro-waste products like rice water and wastewater from cereal processing are possible sources of carbohydrates that might be used to make biosurfactants (Cerda *et al* 2019).

6.3.1 Agriculture waste

Tropical agronomic crop leftovers have the potential to be low-cost, high-carbon substrates that aid in fermentation, microbial growth, and even the production of bacteria (Martinez-Burgos *et al* 2021).

6.3.1.1 Oil crops

In the various oil industries, a significant amount of waste is formed during the oil extraction process. These leftover products include oil cakes, oil soap stocks, high-fat by-products, semi-solid and water-soluble wastewater, residues of fatty acids, etc (Dumont and Narine 2007). Numerous plant-based oils, such as castor oil, jojoba oil, mesua oil, and palm oil, are potentially sources of biosurfactant synthesis due to their carbohydrate-enriched bio-composition, but they are not suitable for human consumption due to their disagreeable fragrance, color, and toxicity (Guo-liang *et al* 2005, Zhang *et al* 2009, Fenibo *et al* 2019, Liu *et al* 2020).

Sunflower seed oil is directly hydrolyzed by the bacterial microorganisms' lipase secretion, and therefore serves as a better source of carbon for the synthesis of rhamnolipids. When cultured in olive oil wastewater, *Pseudomonas aeruginosa* 47 T2 has been reported to create rhamnolipids with a yield of up to 8.1 g l^{-1} (Guo-liang *et al* 2005, Zhang *et al* 2009, Fenibo *et al* 2019, Liu *et al* 2020). We may reduce our expensive manufacturing expenses by taking into account these less expensive fuels and oil wastes.

6.3.1.2 Cereal crops

A significant quantity of starchy wastewater and husks are created during the starch extraction process from many corps, including rice, wheat, maize, cassava, and potatoes. These materials are discovered to be excellent carbon sources for the synthesis of biosurfactants.

Potato agriculture is one of the primary sources of inexpensive starchy substrate. Eighty percent of potato industrial waste is water, seventeen percent is carbohydrate, and the other components are small. They are therefore a high source of carbs, which include both sugar and starch. In essence, Fox and Bala created a potato medium and induced the synthesis of surfactin by *B. subtilis* ATCC 21 332 using potato substrate (Fox and Bala 2000). The cassava industry generates a large amount of wastewater from the processing of cassava tubers into starch or flour. This wastewater contains several by-products, such as sugarcane peels, pomace, sievate, and stumps that are high in carbohydrates. Although these products typically pollute the environment, they can be used as feedstock for biosynthesis, which is a promising solution to the problem (Aro *et al* 2010)

Wastewater from cassava plants has the potential to serve as a substrate for the fermentation process that uses *B. subtilis* to make surfactin (Nitschke and Pastore 2006). By altering the microorganism, Siddhartha *et al* reported producing rhamnolipids; another type of biosurfactant *P. aeruginosa* was the fermenting bacterium that his group employed (Costa *et al* 2009).

6.3.1.3 Fruit and vegetable wastes

The cashew nut is found at the end of the cashew apple, which is a portion of the cashew tree. Essentially, this is a faux fruit that is extracted from cashew nuts. The majority of these leftover cashew apples are not wasted; instead, they are used to make juice, jam, and syrup. These are rich in carbohydrates, which allows many microbes to use them as a source of carbon for the manufacture of biosurfactants (Rocha *et al* 2006). The food business produces and extracts juices from a variety of fruits and vegetables, leaving behind a huge amount of waste. This waste includes residues such as banana peels, apple pomace, sugarcane bagasse, various citrus peels, pineapple peels, carrot peels, and more. Potato peels and sugarcane bagasse were used as carbon sources in Das and Kumar's 2018 study on *P. aeruginosa* AJ15's rhamnolipid biosurfactant synthesis during submerged fermentation (Das and Kumar 2018).

6.3.2 Industrial waste

6.3.2.1 Dairy industry waste

Every day, the dairy sector produces a significant quantity of wastewater, derivatives, and by-products (whey, butter, and milk) (Adesra *et al* 2021). Whey is essentially a liquid phase by-product that is produced during the production of casein products. Because it contains a sizable quantity of lactose, it may be fermented. These waste materials generally have a high biological oxygen demand (BOD), which can contaminate water sources if they are disposed of untreated. However, a sizable portion of these waste materials are recycled to create other beneficial products, such as animal feed. Since it is known that these dairy sector effluents encourage the growth of bacteria, they may be utilized to produce biosurfactants (Dubey and Juwarkar 2004, Costa *et al* 2009, Adesra *et al* 2021). Daniel and colleagues used deproteinized whey in a two-stage yeast culture technique to produce a high output of sophorolipids in *Cryptococcus curvatus* ATCC 20509 and *Candida bombicola* ATCC 22214 (Daniel *et al* 1998). Additionally, Daverey *et al* showed that *Candida bombicola* ATCC 22214 produced sophorolipids utilizing dairy effluent from the dairy sector (Daverey *et al* 2011). Many researchers have published studies on the production of rhamnolipids by *P. aeruginosa* strains (Dubey and Juwarkar 2004, Thanomsub *et al* 2006, Colak and Kahraman 2013, Patowary *et al* 2016) and lipopeptide production by *Bacillus* spp. using whey waste as substrate (Joshi *et al* 2008, Gomaa 2013, Decesaro *et al* 2020).

Many carbohydrates, peptides, and amino acids make up a sizable amount of whey waste that is dumped as effluent; as such, these substances are excellent sources of carbon and nitrogen. Whey waste may be used at the industrial level to produce biosurfactants at a low cost and to help control dairy waste in the process.

6.3.2.2 Brewery waste

100 ml of culture media was contained in 250 ml Erlenmeyer flasks, which were used for biosurfactant production. The standardized culture was added to the flasks at a concentration of 5% (v/v). (g l^{-1}) $CaCl_2$ (0.1), KH_2PO_4 (1.0), $MgSO_4{\bullet}7H_2O$ (0.5), NaCl (0.1), and Trub (2%, v/v) made up the medium. Furthermore, in accordance with the complete factorial design (22), which assessed the two components yeast extract (YE) and peptone (PB), these were given to the medium at varying doses (Nazareth *et al* 2021). The variables' values were 2.0 (−1) and 12.0 (+1) for YE and 0.4 (−1) and 1.4 (+1) for PB, expressed in g l^{-1}. Two central points (CP) were also included in the design, and surface tension (ST) was chosen as the response variable. Every experiment was run in duplicate for 72 h at 30 °C while maintaining aseptic conditions. The original pH was adjusted to 7.0 and the stirring rate was kept at 200 rpm. After the experiments were completed, the culture broth was centrifuged for 20 min at 9000 rpm to extract the cells and solid particles from Trub. The final pH was then determined. After adding a 4.0 M HCl solution to bring the supernatant's pH down to 2.0, it was refrigerated at 4 °C for the whole night to allow precipitate formation. After centrifuging the precipitated biosurfactant for 20 min at 9000 rpm, twice through acidified water (pH 2.0), it was resuspended in Milli-Q® water (Millipore, USA) (Nazareth *et al* 2021).

6.3.2.3 Olive oil industry waste

Olive oil mill waste effluent (OMWE), a massive amount of liquid waste from the olive oil extraction process, is produced. With a high organic content (20–60 kg per COD per m^3), the OMWE is a black liquor (Marques 2001). The OMWE is toxic to human health due to the presence of polyphenols (Ancuța and Sonia 2020) Its chemical make-up makes it beneficial because it includes essential organic components including sugars, organic acids, and nitrogenous chemicals that support microbial development. Owing to its high concentration of lipids, sugars, phenols, polysaccharides, and carbohydrates, OMWE is difficult to treat and may be carcinogenic to the environment (Ben Saad *et al* 2020).

6.3.3 Municipal waste

Urbanization, population growth, a booming economy, and a sharp rise in people's living standards have contributed to an acceleration in the production of municipal solid garbage (Udoh and Vinogradov 2019, Koka *et al* 2020, Hosseinalizadeh *et al* 2021, Rajmohan *et al* 2021). This presents a serious challenge to long-term growth and the environment (Ayeleru *et al* 2018, Harris-Lovett *et al* 2019). Therefore, implementing a successful municipal solid waste management (MSWM) strategy, such as bioconversion of waste materials, is considered essential for the upcoming years as it addresses environmental pollution and offers a resource recovery alternative (Udoh and Vinogradov 2019, Heidari *et al* 2019, Khanal *et al* 2020, Mohan *et al* 2020, Sadeghi Ahangar *et al* 2021 Shah *et al* 2021).

About 93.1 crore tons of food waste were created in 2019, according to the United Nations Environment Program. Of this, 61% came from residential settings, 26% from food service, and 13% from retail. This suggests that 17% of the world's food output may have gone to waste. Therefore, preventing food waste is essential for everyone engaged in the delivery, processing, and marketing of food (Varjani *et al* 2021). It is critical to handle these items properly toward the conclusion of their lifespan to minimize the negative effects that untreated, decaying food may have on the environment and society. In addition to losing the energy required to produce food that is not consumed, improperly managed food waste contaminates rivers with nutrients and leachate flow, releases greenhouse gases during decomposition, and can function as a disease vector (Luo *et al* 2021).

One abundant and perhaps valuable source of feedstock is food waste. Numerous studies have demonstrated that food waste makes up the majority of the organic waste composition in India's MSW stream, which has the greatest percentage of recyclable and biodegradable garbage (Udoh and Vinogradov 2019, Singh 2020, Singh *et al* 2020). It is mostly made up of leftover food scraps, vegetable trash, decomposing vegetables, and leaves. Nevertheless, most MSW with a greater organic/biodegradable content is produced globally by low- and middle-income nations (Beyene *et al* 2018). Food security, environmental preservation, and other sustainable development goals (SDGs) depend increasingly on the valuation of food waste (Sindhu *et al* 2020, Manu *et al* 2021). Food waste has a high concentration of organic substance, therefore conventional burning and disposal techniques can release harmful gases that could harm the environment and public health (Dou *et al* 2018, Udoh and Vinogradov 2019, Ites *et al* 2020). Thus, employing these wastes as a biosurfactant synthesis substrate provides a sustainable approach to valuation. Lowering manufacturing costs and emissions, food waste in MSW may be manually sorted and utilized as a substrate for the synthesis of biosurfactants.

6.3.3.1 Sewage sludge

Managing the massive volume of waste activated sludge (WAS) produced by wastewater treatment operations is urgently needed (Mahmoud *et al* 2006). However, because of the high Wc of sludge, disposing of it remains a significant technological difficulty (Zhang *et al* 2016). Dewatering lowers the volume of sludge and water content present in sludge (Wc), which makes it an effective way to lower the cost of downstream treatment (Fu *et al* 2009, Shi *et al* 2015). Reducing bound water is the primary strategy for enhancing sludge dewatering, according to earlier studies (Qi *et al* 2011, Li *et al* 2016). Nevertheless, bound water in sludge flocs cannot be efficiently removed by conventional mechanical dewatering treatment, nor can it lower the Wc of sludge (Novak *et al* 2003, Jin *et al* 2004). Sludge may be made more dewaterable by adding surfactants. Surfactants can be used in sludge dewatering treatments to neutralize the negative charge on the sludge surface once they are added. As a result, there is less binding between the loosely bound EPSs (LB-EPSs) and tightly bound EPSs (TB-EPSs) of the sludge flocs, which greatly facilitates the dissolving of EPSs in the sludge, mostly proteins and polysaccharides. Next, the EPSs that are freed from the sludge floc surface and bound water mix to reach the sludge liquid phase (Hong *et al* 2015). Moreover, due of their great hydration ability, LB-EPSs and TB-EPSs are converted into soluble EPSs (S-EPSs). As a result of the floc structure being destroyed, both the water retained inside the sludge and the interstitial water trapped inside the flocs are freed. Together, these elements aid in the release of bound water, which enhances the dewaterability of sludge (Fu *et al* 2009).

6.3.3.2 Food waste

Large amounts of starchy wastewater and husks are created during the starch extraction process from several crops, including rice, wheat, maize, cassava, and potatoes. These materials have been discovered to be excellent carbon sources for the synthesis of biosurfactants. Potato agriculture is one of the primary sources of inexpensive starchy substrate. Eighty percent of potato industrial waste is water, seventeen percent is carbohydrate, and the other components are small. They are therefore a high source of carbs, which include both sugar and starch. In essence, Fox and Bala created a potato medium and encouraged the synthesis of surfactins by *B. subtilis* ATCC 21332 using potato substrate (Fox and Bala 2000). Additionally, researchers Noah *et al* and Thomson *et al* have documented the formation of surfactin when examining several potato processing effluent types. The bacteria responsible for producing surfactin from potato effluents was *B. subtilis* 21332.

6.4 Production and purification of microbial surfactants from waste resources

6.4.1 Production process

Through fermentation, microorganisms use a variety of complex organic substrates to get carbon and energy by breaking them down into simpler forms. They generate important products such as polysaccharides, vitamins, amino acids, and ethanol. One of the secondary metabolites created during these fermentation processes is biosurfactants. Depending on the kind of microbe, biosurfactants are produced using both submerged and solid-state fermentations.

6.4.1.1 Fermentation process

6.4.1.1.1 Submerged fermentation

Since bacteria and yeasts that produce biosurfactants need water to thrive to their full potential, submerged production procedures are perfect for them. Extracellular substances called biosurfactants are produced by the bacteria in the fermentation broth and are easily purified. Submerged fermentation (SmF) has a drawback in that certain important molecules may have been found to leak out of the liquid component during recovery. Numerous researchers have created the mineral salt media and investigated the shake flask technique of submerged biosurfactant manufacturing (Pardhi *et al* 2020). De Rienzo *et al* (2016) used *P. aeruginosa* ATCC 9027 and *Burkholderia thailandensis* E264 to produce rhamnolipids in a 10 l laboratory-scale bioreactor. *P. aeruginosa* and *Candida bombicola* were shown to have 20 and 34 g l^{-1} sophorolipids in a 50 l bioreactor, respectively (Shah *et al* 2007, Zhu *et al* 2007).

6.4.1.1.2 Solid state fermentation

Solid materials including molasses, wheat bran, cassava dregs, rice husk, cassava bagasse, coffee husk, banana peel, tapioca peel, etc are often inexpensive, renewable waste products high in carbon and protein. These materials are used as a substrate for solid-state fermentation (SSF). *Aspergillus fumigatus*, *Phialemonium* spp., and *Pleurotus ostreatus* have all been reported to produce biosurfactants by successful solid-state fermentations utilizing rice husk with defatted rice bran, soy oil or diesel oil, and sunflower seed oil, respectively (Guimarães Martins *et al* 2006, Velioğlu and Ürek 2015). Furthermore, several bacterial strains, such as *Bacillus pumilus* UFPEDA 448, *Brevibacterium aureum* MSA13, and *Serratia rubidaea* SNAU02, produced more lipopeptides and rhamnolipids utilizing SSF than SmF (Kiran *et al* 2010, Slivinski *et al* 2012, Nalini and Parthasarathi 2014).

6.4.1.2 Extraction process

Even when all necessary parameters have been adjusted to their ideal states, the production process is not complete until the product has undergone appropriate, economically viable recovery and purification—a process known as downstream processing. One of the most crucial factors in the commercialization of biosurfactant production is the product's recovery from the growth medium. Researchers often report on a variety of product recovery techniques, such as solvent extraction, acid precipitation, crystallization, centrifugation, and so on. Furthermore, a few non-conventional techniques such foam fractionation, adsorption–desorption on different media, and ultrafiltration have also been documented. These techniques are all used in accordance with the properties of the biosurfactant, including solubility, surface activity, and capacity to form micelles. Application of these techniques that can recover extremely pure biosurfactant is necessitated by the persistent increase in demand for very pure biosurfactants by certain pharmaceutical, food, and cosmetics sectors (Mukherjee *et al* 2006).

In addition, the solvents commonly used in biosurfactant recovery, such as acetone, methanol, and chloroform, are hazardous to the environment and poisonous by nature. In recent years, researchers have substituted less hazardous and less expensive solvents, such as methyl tertiary-butyl ether (MTBE), for the recovery of biosurfactants (Kuyukina *et al* 2001). These less expensive and hazardous solvents contribute to lowering the recovery costs associated with the synthesis of biosurfactants. Multiple recovery strategies are used for downstream processing when a single recovery operation is insufficient to yield the required output (Reiling *et al* 1986).

6.4.2 Purification process

6.4.2.1 Flocculation process

The traditional cylinder test was used to conduct flocculation investigations. Surfactant, flocculant, or their mixes in the appropriate amounts were introduced while the suspension was being stirred in a beaker with baffles. After adding the reagent, there was one more minute of stirring. After inverting the cylinder five times, the entire suspension was then transferred to it and allowed to settle. There was a discernible liquid slurry contact that descended. This interface's height was measured on a regular basis. The slope of the straight-line segment of the plot of interface height versus settling time was used to estimate settling rates. Following the sedimentation test, the suspension was left undisturbed for a full day. Next, the graduated cylinder was used to directly read the volume that the sediment had taken up. Replicable observations conducted in comparable experimental settings indicate that the amount of sediment and setting rates may be repeated within a 2.5% error. By looking at two different ways to introduce surfactants, the impact of their presence on flocculation and dewatering in the system was examined. The surfactant partly covered the kaolin surface in Mode-I. This sample was subjected to flocculation trials following a 24 h period of equilibration. In every instance, the surfactant coating dose is kept constant at 0.2 mg g^{-1}. This dose is selected in accordance with the extent to which the surfactant reduces surface charge. It has been discovered that, in the absence of any reagents, the amount of negative surface charge of kaolin particles is reduced by almost half by adding 0.2 mg g^{-1} of surfactant (Besra *et al* 2003).

6.4.2.2 Chromatography process

Biosurfactants analysis using liquid chromatography–mass spectroscopy (LC–MS) necessitates an initial purification by eliminating the worst interferences and also concentrating the sample to a noteworthy amount (Biniarz *et al* 2017). To accomplish partitioning between a polar mobile phase and a non-polar stationary phase, the LC–MS makes use of variations in hydrophobicity. Lipopolysaccharide (LP) congeners may be effectively purified and separated using the LC–MS technology. The most appropriate method for characterizing an unidentified lipopolysaccharide is LC–MS.

Biosurfactants are characterized using gas chromatography–mass spectroscopy (GC–MS), where the mass spectroscopy determines the compound's molecular weight. The sample must undergo hydrolytic cleavage between the biosurfactant's peptide/protein or carbohydrate/lipid segments in order for this device to function. Fatty acid derivatization to fatty acid methyl esters (FAME) and subsequent conversion to trimethylsilyl (TMS) derivatives are used to analyze the GC–MS data (Vandana and Singh 2018). Because it can distinguish, measure, and purify individual components of a biosurfactant mixture as well as separate a mixture of surface-active molecules, high-performance liquid chromatography (HPLC) is a unique type of column chromatography used in chemical and biological analysis (Biniarz *et al* 2017). It has been documented that biosurfactants can be purified, quantified, and characterized using HPLC (Janek *et al* 2010). For instance, LP was purified by HPLC utilizing reversed-phase (RP)-HPLC, a semi-preparative C18 column, and a mobile phase of 0.1% trifluoroacetic acid/methanol/H2O (Pathak and Keharia 2014).

6.5 Applications of microbial surfactants from waste resources

6.5.1 Bioremediation

6.5.1.1 Oil spills

Reduced nutritional availability, oxidative stress, and altered soil chemical composition are all results of petroleum contamination. The main detrimental effects of petroleum are that it slows down the absorption of nutrients, reduces photosynthetic pigments, hinders root growth, causes foliar deformation and tissue necrosis, destroys biological membranes, interferes with metabolic pathway signaling, and messes with the architecture of plant roots. Plants can die as a result of LMW hydrocarbons getting into their cells. Furthermore, the development of cancer and other illnesses is facilitated by petroleum and its components. Previous studies discovered that those who were exposed to petroleum pollution had nervous system depression, narcosis, and irritation of the mucous membranes surrounding their eyes. Utilizing naturally occurring or intentionally produced living organisms or their by-products, bioremediation is a technique for reducing (degrading, detoxifying, mineralizing, or converting) pollutants in contaminated environments. Usually, for this aim, living things (plants and microorganisms) that can endure and thrive in contaminated soil are used (Chen and Zhong 2019). The Energy Research Institute (TERI) in India developed the Oilzapper, a consortium of bacteria that breaks down crude oil and oily sludge. By immobilizing five bacterial isolates (obtained from hydrocarbon-contaminated locations) with an appropriate carrier material (powdered corn cob), this microbial consortium was produced. The immobilized culture, also known as an oilzapper, may be aseptically packaged into sterile polythene bags and carried to the polluted location. It has a three-month shelf life at room temperature. It has been effectively applied to remediate greasy sludge and clean up crude oil spills. Over 40 000 metric tons of oily sludge, oil-polluted soil, and drill cuttings have been treated in different places. In different parts of India and the

Middle East, more than 30 000 metric tons of oily sludge and oil-contaminated soil are being treated.

6.5.1.2 Heavy metal removal

In order to increase the bioavailability of organic molecules like hydrocarbons during a cleaning process, biosurfactants are employed to mobilize and eliminate contaminants via pseudo-solubilization and emulsification. Although biosurfactants must be continuously replenished with new amounts of these substances, there are undeniable benefits to using biosurfactants in heavy metals remediation because microorganisms capable of producing surfactant compounds do not need to be able to survive in heavy metal-contaminated soil. Biosurfactants can be applied to a small area of contaminated soil. After being removed from the ground, the biosurfactant–metal complex is put into a large cement mixer. The soil is then redeposited into the Earth after the biosurfactant–metal complex is treated to precipitate the biosurfactant and leave the metal behind. Usually, water flowing through the soil washes away the metal-surfactant complex because of the strong bond formed between the negatively charged tensio-active chemical and the positively charged metal (Sarubbo *et al* 2015). Juwarkar *et al* investigated the removal of lead and cadmium by a biosurfactant generated by *P. aeruginosa* BS2 and used studies in columns to ascertain the removal of the heavy metals by a rhamnolipid. During 36 h, at a concentration of 0.1%, the rhamnolipid eliminated over 92% of the cadmium and 88% of the lead (Juwarkar *et al* 2007b). Lipopolysaccharides (LPSs) are a major class of biosurfactant molecules. The first heavy metal extraction experiments on them were carried out by Langley and Beveridge, who demonstrated that LPSs increased the outer cell walls' hydrophilicity, which made it simpler for bacterial cells to absorb metallic cations (Langley and Beveridge 1999). Other types of biosurfactants produced by species of the genus *Candida*, which are mostly sophorolipids in nature, have also demonstrated the capacity to eliminate over 90% of cations in air-dissolved flotation processes and columns (Albuquerque *et al* 2012).

6.5.2 Agriculture

Because of their many functions and role in biological control as either antifungal agents or inducers of induced systemic resistance, biosurfactants are promising prospects for crop protection applications in the future. Biosurfactants frequently function as biological control agents by rupturing the pathogen's cell wall and opening up channels inside the cell wall (Raaijmakers *et al* 2006). Among the several types of biosurfactants, cyclic lipopeptides like fengycin, iturin, and surfactin, and glycolipids such cellobiose lipids and rhamnolipids shield plants against phytopathogenic fungi by acting as antifungal agents (Banat *et al* 2010).

Stanghellini and Miller discussed the potential application of biosurfactants as biological control agents (Stanghellini and Miller 1997). They describe how rhamnolipids can disrupt oomycete plant disease zoospore membranes and cause lysis of the zoospores. Since then, the important role of rhamnolipids in defense against several phytopathogenic fungi has been discussed in a number of articles. Rhamnolipid biosurfactants were shown to have direct antifungal effects by inhibiting *Botrytis cinerea*'s mycelium development and spore germination (Varnier *et al* 2009). The primary di-rhamnolipid component in this cell-free medium may be the cause of the medium's noticeably higher antifungal efficacy. It is differentiated by improved lysis features over mono-rhamnolipid to tear the spore membranes. This is particularly true for diseases on plants that generate zoospores. Cyclic lipopeptides are a kind of biosurfactant that have antifungal activity against phytopathogenic fungi, much like rhamnolipids. The use of biosurfactants in the hydrophilization process produces uniform fertilizer dispersion, high wettability, and the suppression of pesticide toxicants (Gayathiri *et al* 2022).

6.5.2.1 Soil improvement

Abiotic stress is imposed on cultivated crop plants by the presence of both organic and inorganic contaminants, which reduces the productivity of agricultural land. The process of bioremediation is necessary to improve the quality of such hydrocarbon- and heavy metal-contaminated soil. Microorganisms that produce biosurfactants or biosurfactants themselves can be utilized to remove heavy metals and hydrocarbons successfully (Sun *et al* 2006). Thanks to its capacity to increase bioavailability and facilitate the biodegradation of hydrophobic substances, biosurfactants are used in a variety of technologies, including soil washing and clean-up combination technologies, to effectively remove metal and hydrocarbons, respectively (Maier *et al* 2001, Mulligan 2005, Shin *et al* 2006, Hickey *et al* 2007, Juwarkar *et al* 2007a, Aşçı *et al* 2008, Robles-González *et al* 2008, Santos *et al* 2008, Sheng *et al* 2008, Neto *et al* 2009, Xiaohong *et al* 2009, Zhao and Wong 2009, Coppotelli *et al* 2010, Gottfried *et al* 2010, Kang *et al* 2010, Liu *et al* 2010, Partovinia *et al* 2010). Biosurfactants speed up a very significant process called the desorption of hydrophobic contaminants that are firmly attached to soil particles. This is highly important to the process of bioremediation. Additionally, biosurfactants can speed up the breakdown of certain chemical pesticides that have accumulated in agricultural soil (Nielsen and Sørensen 2003, White *et al* 2006, Wattanaphon *et al* 2008, Sharma *et al* 2009, Singh *et al* 2009, Zhang *et al* 2011). Numerous findings indicate that biosurfactants play a part in soil remediation, which enhances the quality of agricultural soil. There have been studies on the biodegradation of pesticides aided by surfactin and chlorinated hydrocarbons aided by glycolipids (Mata-Sandoval *et al* 2001). The biodegradation accelerator activity of *Lactobacillus pentosus* biosurfactant has been established by reducing octane hydrocarbon from soil from 58.6% to 62.8%. It has been noted that a species of *Burkholderia* that produces biosurfactants and was isolated from soil polluted with oil may be a good fit for bioremediation of various pesticide contaminations (Wattanaphon *et al* 2008, Moldes *et al* 2011).

6.5.2.2 Pesticides delivery

In addition to herbicides, insecticides, and fungicides, surfactants are essential. Currently utilized in the pesticide industry, synthetic surfactants serve as emulsifying, dispersing, spreading, and wetting agents, improving the effectiveness of pesticides. Additionally, because these surfactants have defensive qualities, they are utilized in contemporary agriculture as insecticides (Rostás and Blassmann 2009). Various surfactant types, including cationic, amphoteric, non-ionic, and anionic ones, are now employed in a number of pesticide production sectors (Mulqueen 2003). Surfactants are therefore often utilized in pesticide composition. The surfactant used in pesticide formulations, however, accumulates in soil and has an impact on plant development, texture, and color. Moreover, these dangerous chemicals seep into the groundwater from the soil (Blackwell 2000). It is well known that pesticide residues can linger in soil for years and can also transfer to water and the air. These even stay on the fruit and vegetable surface (Street 1969). Furthermore, synthetic surfactants are thought to be strong organic pollutants in soil (Petrović and Barceló 2004). Given the negative effects of pesticides and the surfactants that are connected with them, it is necessary to replace these hazardous surfactants in the multimillion-dollar pesticide businesses with ecologically friendly biosurfactants in order to prevent contamination (Hopkinson *et al* 1997). Another solution to this environmental issue might involve investigating soil microorganisms that can use the chemical surfactant found in agricultural soil as a source of carbon. A study describes how surfactant degradation is caused by *Pseudomonas* sp. and *Burkholderia* sp. bacteria found in rice fields (Nishio *et al* 2002). Important agricultural goods like insecticides that are created with the help of biosurfactants can be applied extensively in farming environments. In order to do this, the agrochemical industries must develop efficient formulation technology. Many businesses may do this by combining a variety of biosurfactants with polymers in unique ways to create superior formulations for use in agriculture (Sachdev and Cameotra 2013).

6.5.3 Cosmetics and personal care products

One of the amazing uses for multifunctional biosurfactants is in cosmetics. Their superior surface qualities—emulsification, detergency, solubilization, dispersion, wetting, and foaming effects—determine the applications. Additionally, they demonstrated anti-inflammatory benefits, antioxidant activity, and compatibility with skin that had greater moisturizing characteristics (Patel *et al* 2022). Mannosylerythritol lipids (MELs), sophorolipids, and rhamnolipids are a substitute for the petrochemical-based surfactants used in high-end cosmetic preparations such as anti-aging and wrinkle creams because they have moisturizing, low-irritating, and skin-compatible qualities.

MELs are mostly utilized in preparations that prevent skin roughness and were introduced to the cosmetics industry for their unique liquid-crystal-forming and hydrating properties. Diacylated MEL-B penetrated human skin cells injured by sodium dodecyl sulfate (SDS) and resulted in the formation of lyotropic liquid crystals that moisturize the skin, exhibiting 77.1% vitality and self-assembling ability (Morita *et al* 2011). Concaix (2003) reported sophorolipids function as skin fibroblast metabolic stimulators, promoting skin healing, preservation, and restoration. Additionally, they lessen the excess of subcutaneous fat by encouraging adipocytes to produce leptin, which makes cellulite therapy possible (Pellecier and André 2004). In order to cure damaged hair, MEL-A (0.5%) and MEA-B (0.5%) have been shown to increase the tensile strength of damaged hair up to 122 gf/p and 119.4 gf/p (Morita *et al* 2010).

6.5.4 Food industry

Generally speaking, biosurfactants are used as food emulsifiers, fat stabilizers, and anti-adhesive agents in food formulation. They are also used to adjust the rheological characteristics of wheat dough, regulate the aggregation of fat globules, stabilize aerated systems, enhance the consistency and texture of fat-based goods, and extend the shelf life of starch-containing products. Biosurfactants are used to safeguard food goods because they have the ability to reduce the adherence of pathogenic organisms to solid surfaces or infection sites (Zaman and Hamid 2022). When tested against *Lactobacillus reuteri* (77.6%–78.8%), *Lactobacillus casei* (56.5%–63.8%), *Streptomyces sanguis* 12 (72.9%), *Streptococcus mutans* HG985 (31.4%), *Staphylococcus aureus* (76.8%), S. epidermidis (72.9%), *Streptococcus agalactiae* (66.6%), *P. aeruginosa* (21.2%), and *Escherichia coli* (11.8%) were all shown to have anti-adhesive activity when extracted from *Lactobacillus paracasei* spp. paracasei A20 (Gudiña *et al* 2010).

Vegetables polluted with heavy metals can pose a number of health risks to humans if consumed over time. A glycolipid was shown to have antibacterial action against *E. coli* and to have 73% Cd elimination from garlic and 59% biofilm inhibition (Anjum *et al* 2016). The fruit salad dressing's emulsion stability rose from 51.4 to 62.8% thanks to the biosurfactants (Sridhar *et al* 2015). The lipopeptide-treated muffins were found to have better suppleness and less stickiness and hardness (Kiran *et al* 2010). Diacyl mannosylerythritol, a novel glycolipid, demonstrated an ice-packing factor of 35% for eight hours, which helped to increase the storage capacity of ice slurry (Kitamoto *et al* 2001).

6.5.5 Pharmaceuticals

Biosurfactants exert their toxicity on cell membrane permeability in a way akin to that of detergents. Biosurfactants are a useful substance in the medical and cosmetic industries because of their biological characteristics, which include antibacterial, anti-adhesive, anti-cancer, anti-mycoplasma, and hemolytic. *Aspergillus niger*, *Gliocladium virens*, *Chaetomium globosum*, *Penicillium chrysogenum*, *Aureobasidium pullulans*, *B. cinerea*, *Rhizoctonia solani*, *P. chrysogenum*, *Candida albicums*, *B. pumilus*, *Micrococcus luteus*, and *Sarcina lutea* are among the micro-organisms against which the rhamnolipids have demonstrated antimicrobial activity (Abalos *et al* 2001, El-Sheshtawy and Doheim 2014). New biosurfactant Lunasan has shown antimicrobial activity against *C. albicans* (57%) and *Staphylococcus epidermidis* (57.6%), as well as anti-adhesive activity against *P. aeruginosa* (92%), *Streptococcus agalactiae* (100%), and *Streptococcus sanguis* (100%) (Luna *et al* 2011). With a minimum inhibitory dose of 6.25 $\mu g\ ml^{-1}$ for human breast cancer cell line and 50 μg ml^{-1} for insect cell line C6/36, respectively, rhamnolipid A and B exhibit anti-proliferative action (Thanomsub *et al* 2006). *Lactobacillus kefiranofaciens* ATCC 43761 produces the water-soluble polysaccharide kefiran, which has exhibited anti-cancer activity against breast cancer (MCF-7) cells with an IC50 of 193.89 μg ml^{-1} (Dailin *et al* 2020). Biosurfactants are a good choice for biomedical preparations because of their qualities.

Biosurfactants disrupt the pathways that lead to cancer formation, they have been identified as promising anti-cancer medicines. Numerous intracellular molecular recognition processes, such as signal transmission, cell differentiation, and immune response, are aided by these metabolites (Rodrigues *et al* 2006).

6.5.6 Enhanced oil recovery

By means of mobilization, de-emulsification, solubilization, or emulsification, biosurfactants promote the extraction and biodegradation of oil. In comparison to synthetic surfactants, rhamnolipids and surfactins showed a greater ability to remove petroleum from soil. *Pseudoxanthomonas* spp. G3, *Citrobacter freundii* HM-2, and *Ochrobactrum anthropic* HM-1 glycolipids effectively retrieved 70%, 67%, and 20% of the remaining oil from the sand packed column (Ibrahim 2018). Furthermore, Jain *et al* (2012) used 1% (w/v) biosurfactant to recover >90% of the lubricating oil from sandy soil. Similar to bacteria, 46% increased oil recovery was also recorded for *Fusarium* spp. BS-8 (JQ860113) (Qazi *et al* 2013). 90% Mb, 30% Ni, and 70% Vd have been reported to be removed by rhamnolipid (0.4 mg ml^{-1}). By contrast, 44.5% of the carbon in the hazardous waste hydrodesulfurization (HDS) catalyst made by petroleum refineries was eliminated by lipopeptide (17.34 mg ml^{-1}) (ALsaqer *et al* 2018). Because the hazardous chemicals needed to clean them produce a large amount of hazardous waste, maintaining and cleaning oil storage tanks may be difficult. Extremely viscous semi-solid particles known as oil sludge fraction are accumulated on the walls or bottom of storage tanks and are challenging to remove using standard pumping techniques. Oil-contaminated vessels were cleaned in 15 min using *P. aeruginosa* SH 29 biosurfactant (Diab and Din 2013).

6.6 Challenges and future prospects

6.6.1 Challenges

Many businesses now place biosurfactants at the top of their priority list due to the demand for industrial sustainability. Biosurfactants provide a viable alternative to chemical surfactants derived from non-renewable resources, as they may be manufactured from inexpensive renewable feedstocks. The fact that biosurfactants are strong enough for industrial usage yet less harmful to the environment makes them appealing as well. Glycolipids, sophorolipids generated by *Candida* yeasts, MELs produced by *Pseudozyma* yeasts, and rhamnolipids produced by *Pseudomonas* are now the most promising biosurfactants. Even with the present enthusiasm surrounding these substances, a few lingering issues still exist. This assessment identifies issues that still need to be resolved and shows the likelihood that a new generation of microbial biosurfactants will soon be commercially exploited (Marchant and Banat 2012).

6.6.1.1 Cost of production

Whether or not biosurfactants are seen to be effective in small-scale trials and research, manufacturing and cost concerns will ultimately determine whether or not they are used as ingredients in large-scale commercial goods. Safety is the most important factor to take into account. Thus far, no indication has been made of any significant safety or health concerns about any of the biosurfactants that have been studied, and most definitely not the primary ones that are being studied at this time, namely sophorolipids, rhamnolipids, and MELs. Rhamnolipids have been reported to function as immune modulators (McClure and Schiller 1996), furthermore, it has been demonstrated that they function as virulence factors in *P. aeruginosa* infections (Zulianello *et al* 2006). The concentrations of biosurfactants generated, the substrates required to manufacture them, and the necessary downstream processing are other significant manufacturing challenges. Sophorolipids and MELs may already be manufactured at yields more than 100 g l^{-1} (Kitamoto *et al* 2001, Banat *et al* 2010, Makkar *et al* 2011), whereas laboratory strains of *P. aeruginosa* produce only 10–20 g l^{-1} of rhamnolipids.

There have been occasional reports of over-producer strains, however, information regarding whether the strains used to generate rhamnolipids commercially perform considerably better than this is often unavailable (Lang and Wullbrandt 1999).

The fact that glycolipid biosurfactants may be made from a variety of renewable substrates—some of which can even be regarded as waste—is one of its main advantages. Low molecular mass glycolipids are very simple to separate and purify (Smyth *et al* 2010), but the procedure becomes more difficult if an oily substrate is employed and if any of the substrate is left over after fermentation. The use of cost-effective technologies that leverage waste substrates for biosurfactant manufacturing and employ less expensive renewable substrates might greatly lower expenses (Makkar *et al* 2011).

6.6.2 Scale-up

6.6.2.1 Competition with chemical surfactants

Chemicals are a component of daily existence. Chemicals make up all matter, both alive and dead, and they are present in almost every produced good. Many chemicals have the potential to greatly improve health, wellbeing, and quality of life when utilized appropriately (Zhang *et al* 2012, Liang *et al* 2017).

Typically, chemicals utilized as surfactants are used as emulsifiers or surface energy reducers, particularly at the oil–water interface. Based on its solubility, micelle concentration, reduction capacity, and wettability, a surfactant is judged appropriate or inappropriate for a certain application. Large-scale applications have been used for chemical surfactants, which are typically derived from petrochemical or oleochemical sources (Ambaye *et al* 2021).

6.6.3 Future prospects

6.6.3.1 Use of genetic engineering to improve yields

Recombinant strains have various industrial uses, the biotechnological use of genetic engineering recombinant DNA technology to make biosurfactants is now gaining prominence in the scientific community (Hu *et al* 2019, Wu *et al* 2019). It is widely recognized that the microbe has the ability to break down various substrates while producing biosurfactants. To create the biosurfactants, however, little knowledge of the bacteria is needed for their cloning, functional characterization, degradative properties, and molecular features (Satpute *et al* 2016). The synthesis of biosurfactants may be enhanced by introducing certain genes when carbon and hydrocarbons are present as a source. This has created a new field of study for researchers who want to use recombinant DNA technology to create new microorganisms that will increase the production and effectiveness of biosurfactants in a variety of industrial settings (Sekhon 2012). According to Bachmann *et al* (2014) and Soares da Silva *et al* (2019) the modification of the chemical characteristics, microbial strain enhancement by recombinant DNA technology can offer improved biosurfactants manufacturing yield at a reduced cost. Additionally, they discovered that the biosurfactants generated by this process method are also resistant to extremely high pH, salt, and temperature conditions. Therefore, these parameters must be improved if cloned microbial strains using recombinant DNA technology are to reach their full potential for producing biosurfactants on an industrial scale. Additionally, the genetic make-up of the microorganisms limiting the output of the biosurfactant has to be investigated (Satpute *et al* 2010).

Using recombinant *E. coli* pSKA clones containing the BioS gene, srfA (Kandasamy *et al* 2019) compared the production of biosurfactants from olive oil and demonstrated that the cloning and expression of modified genes and enzyme activities in *E. coli* recombinants led to improved production of biosurfactants as compared to biosurfactants produced from its parent strain of *Bacillus* sp. SK320.

Similar results for biosurfactant generation were also reported by (Sekhon 2012) by a comparison of mutant gene expression from microbial cloning. When compared to the original strain, they discovered that the recombinant strain produced twice as much biosurfactant. Additionally, they stated that the recombinant strains might enhance the qualities of the product, resolve issues pertaining to the system's decreased secretion of protein, and extract and purify biosurfactants. These experiments demonstrated how bacteria may use recombinant DNA technology to increase the production of biosurfactants for the environment (Balan *et al* 2019, Kubicki *et al* 2019, Md Badrul Hisham *et al* 2019, Williams *et al* 2019).

6.6.3.2 Development of new application

It is well recognized that biosurfactants offer a major benefit in industrial settings for biotechnological and environmental applications. More research is still needed on the synthesis, large-scale limiting constraints, and the use of biosurfactants in various environmental complexes. Thus, the development of innovative biosurfactants that can quickly recover product and microbially degrade contaminants must be the main focus of study. Through genetic engineering, molecular biology, and surface science, research also concentrated on finding appropriate microbes with high-level metabolic activities. This allowed biosurfactants to be commercially viable for use in a variety of industries, including pharmaceutics, cosmetics, textiles, petroleum, oral hygiene, wastewater treatment, and agriculture.

Future research should also focus on accurately identifying biomolecules and developing precise monitoring and testing protocols to identify the top biosurfactant producers—a process that is currently unclear. More investigation is required to explore the biosurfactant route utilizing the concepts of genomics and proteomics at the gene and species level.

6.7 Conclusion

The broad applications of biosurfactants in nanotechnology product formulation, petroleum, personal healthcare product manufacturing, pharmaceutical, agricultural, agrochemical, and food processing industries, as well as in rehabilitating contaminated environments and their sustainability, were demonstrated by this review. Natural, environmentally benign, and more sustainable than chemical or manufactured surfactants are biosurfactants. By using biotechnological methods and low-cost materials, they may be created using bioresources at a reduced cost of manufacturing. However, because of its high adsorption to soil particles, limited bioavailability and insolubility, and hydrophobicity to pollutants, its biotechnological and environmental applications process may be hindered. Optimizing growth/production conditions with economically viable renewable substrates and effective multi-step downstream processing can help address this problem. Producing a more lucrative biosurfactant would benefit from this. Furthermore, further research is required to determine the ideal preparation conditions for these materials given their societal and economic advantages.

References

Abalos A, Pinazo A, Infante M R, Casals M, Garcia F and Manresa A 2001 Physicochemical and antimicrobial properties of new rhamnolipids produced by *Pseudomonas* a eruginosa AT10 from soybean oil refinery wastes *Langmuir* **17** 1367–71

Abouseoud M, Maachi R and Amrane A 2007 Biosurfactant production from olive oil by *Pseudomonas fluorescens Comm. Curr. Res. Educ. Top. Trends Appl. Microbiol* **1** 340–47

Adamu A, Ijah U J J, Riskuwa M L, Ismail H Y and Ibrahim U B 2015 Study on biosurfactant production by two Bacillus species *Int. J. Sci. Res. Knowledge* **3** 13

Adesra A, Srivastava V K and Varjani S 2021 Valorization of dairy wastes: integrative approaches for value added products *Indian J. Microbiol.* **61** 270–8

Albuquerque C F, Luna-Finkler C L, Rufino R D, Luna J M, de Menezes C T B, Santos V A and Sarubbo L A 2012 Evaluation of biosurfactants for removal of heavy metal ions from aqueous effluent using flotation techniques *Int. Rev. Chem. Eng.* **4** 156–61

ALsaqer S, Marafi M, Banat I M and Ismail W 2018 Biosurfactant-facilitated leaching of metals from spent hydrodesulphurization catalyst *J. Appl. Microbiol.* **125** 1358–69

Ambaye T G, Vaccari M, Prasad S and Rtimi S 2021 Preparation, characterization and application of biosurfactant in various industries: a critical review on progress, challenges and perspectives *Environ. Technol. Innov.* **24** 102090

Ancuța P and Sonia A 2020 Oil press-cakes and meals valorization through circular economy approaches: a review *Appl. Sci.* **10** 7432

Anjum F, Gautam G, Edgard G and Negi S 2016 Biosurfactant production through Bacillus sp. MTCC 5877 and its multifarious applications in food industry *Bioresour. Technol.* **213** 262–9

Araújo H W C, Andrade R F S, Montero-Rodríguez D, Rubio-Ribeaux D, Alves da Silva C A and Campos-Takaki G M 2019 Sustainable biosurfactant produced by *Serratia marcescens* UCP 1549 and its suitability for agricultural and marine bioremediation applications *Microb. Cell Fact.* **18** 1–13

Aro S O, Aletor V A, Tewe O O and Agbede J O 2010 Nutritional potentials of cassava tuber wastes: a case study of a cassava starch processing factory in south-western Nigeria *Livest. Res. Rural Dev.* **22** 42–7

Aşçı Y, Nurbaş M and Sağ Açıkel Y 2008 A comparative study for the sorption of Cd(II) by K-feldspar and sepiolite as soil components, and the recovery of Cd(II) using rhamnolipid biosurfactant *J. Environ. Manage.* **88** 383–92

Ashby R D, Nuñez A, Solaiman D K Y and Foglia T A 2005 Sophorolipid biosynthesis from a biodiesel co-product stream *J. Am. Oil Chem. Soc.* **82** 625–30

Astuti D I, Purwasena I A, Putri R E, Amaniyah M and Sugai Y 2019 Screening and characterization of biosurfactant produced by *Pseudoxanthomonas sp.* G3 and its applicability for enhanced oil recovery *J. Petrol. Explor. Prod. Technol.* **9** 2279–89

Ayeleru O O, Okonta F N and Ntuli F 2018 Municipal solid waste generation and characterization in the City of Johannesburg: a pathway for the implementation of zero waste *Waste Manage. (Oxford)* **79** 87–97

Bachmann R T, Johnson A C and Edyvean R G J 2014 Biotechnology in the petroleum industry: an overview *Int. Biodeterior. Biodegradation* **86** 225–37

Balan S S, Kumar C G and Jayalakshmi S 2019 Physicochemical, structural and biological evaluation of Cybersan (trigalactomargarate), a new glycolipid biosurfactant produced by a marine yeast, *Cyberlindnera saturnus* strain SBPN-27 *Process Biochem.* **80** 171–80

Banat I M, Franzetti A, Gandolfi I, Bestetti G, Martinotti M G, Fracchia L, Smyth T J and Marchant R 2010 Microbial biosurfactants production, applications and future potential *Appl. Microbiol. Biotechnol.* **87** 427–44

Bayoumi R A, Haroun B M, Ghazal E A and Maher Y A 2010 Structural analysis and characteristics of biosurfactants produced by some crude oil utilizing bacterial strains *Aust. J. Basic Appl. Sci.* **4** 3484–98

Ben Saad A, Jerbi A, Khlif I, Ayedi M and Allouche N 2020 Stabilization of refined olive oil with phenolic monomers fraction and purified hydroxytyrosol from olive mill wastewater *Chem. Afr.* **3** 657–65

Besra L, Sengupta D K, Roy S K and Ay P 2003 Influence of surfactants on flocculation and dewatering of kaolin suspensions by cationic polyacrylamide (PAM-C) flocculant *Sep. Purif. Technol.* **30** 251–64

Beyene H D, Werkneh A A and Ambaye T G 2018 Current updates on waste to energy (WtE) technologies: a review *Renew. Energy Focus* **24** 1–11

Bezza F A and Chirwa E M N 2015 Production and applications of lipopeptide biosurfactant for bioremediation and oil recovery by *Bacillus subtilis* CN2 *Biochem. Eng. J.* **101** 168–78

Bhardwaj G 2013 Biosurfactants from fungi: a review *J. Pet. Environ. Biotechnol.* **4** 1000160

Biniarz P, Łukaszewicz M and Janek T 2017 Screening concepts, characterization and structural analysis of microbial-derived bioactive lipopeptides: a review *Crit. Rev. Biotechnol.* **37** 393–410

Blackwell P 2000 Management of water repellency in Australia, and risks associated with preferential flow, pesticide concentration and leaching *J. Hydrol.* **231–2** 384–95

Cai Q, Zhang B, Chen B, Zhu Z, Lin W and Cao T 2014 Screening of biosurfactant producers from petroleum hydrocarbon contaminated sources in cold marine environments *Mar. Pollut. Bull.* **86** 402–10

Camarate M C, Merma A G, Hacha R R and Torem M L 2022 Selective bioflocculation of ultrafine hematite particles from quartz using a biosurfactant extracted from candida stellata yeast *Sep. Sci. Technol.* **57** 36–47

Carolin C F, Senthil Kumar P, Mohanakrishna G, Hemavathy R V, Rangasamy G and M Aminabhavi T 2023 Sustainable production of biosurfactants via valorisation of industrial wastes as alternate feedstocks *Chemosphere* **312** 137326

Cerda A, Mejias L, Rodríguez P, Rodríguez A, Artola A, Font X, Gea T and Sánchez A 2019 Valorisation of digestate from biowaste through solid-state fermentation to obtain value added bioproducts: a first approach *Bioresour. Technol.* **271** 409–16

Chen S and Zhong M 2019 Bioremediation of petroleum-contaminated soil *Environ. Chem. Recent Pollut. Control Approach.* **34** 1–12

Colak A K and Kahraman H 2013 The use of raw cheese whey and olive oil mill wastewater for rhamnolipid production by recombinant *Pseudomonas aeruginosa Environ. Exp. Biol.* **11** 125–30

Concaix F B 2003 Use of sophorolipids comprising diacetyl lactones as agent for stimulating skin fibroblast metabolism *Google Patents* US09/701,414

Coppotelli B M, Ibarrolaza A, Dias R L, Del Panno M T, Berthe-Corti L and Morelli I S 2010 Study of the degradation activity and the strategies to promote the bioavailability of phenanthrene by *Sphingomonas paucimobilis* strain 20006FA *Microb. Ecol.* **59** 266–76

Costa S G, Lépine F, Milot S, Déziel E, Nitschke M and Contiero J 2009 Cassava wastewater as a substrate for the simultaneous production of rhamnolipids and polyhydroxyalkanoates by *Pseudomonas aeruginosa J. Ind. Microbiol. Biotechnol.* **36** 1063–72

Dailin D J, Elsayed E A, Malek R A, Hanapi S Z, Selvamani S, Ramli S, Sukmawati D, Sayyed R Z and El Enshasy H A 2020 Efficient kefiran production by *Lactobacillus kefiranofaciens* ATCC 43761 in submerged cultivation: influence of osmotic stress and nonionic surfactants, and potential bioactivities *Arabian J. Chem.* **13** 8513–23

Daniel H-J, Reuss M and Syldatk C 1998 Production of sophorolipids in high concentration from deproteinized whey and rapeseed oil in a two stage fed batch process using *Candida bombicola* ATCC 22214 and *Cryptococcus curvatus* ATCC 20509 *Biotechnol. Lett.* **20** 1153–6

Das A J and Kumar R 2018 Utilization of agro-industrial waste for biosurfactant production under submerged fermentation and its application in oil recovery from sand matrix *Bioresour. Technol.* **260** 233–40

Daverey A, Pakshirajan K and Sumalatha S 2011 Sophorolipids production by Candida bombicola using dairy industry wastewater *Clean Technol. Environ. Policy* **13** 481–8

De Rienzo M A D, Kamalanathan I D and Martin P J 2016 Comparative study of the production of rhamnolipid biosurfactants by *B. thailandensis* E264 and *P. aeruginosa* ATCC 9027 using foam fractionation *Process Biochem.* **51** 820–7

Decesaro A, Machado T S, Cappellaro Â C, Rempel A, Margarites A C, Reinehr C O, Eberlin M N, Zampieri D, Thomé A and Colla L M 2020 Biosurfactants production using permeate from whey ultrafiltration and bioproduct recovery by membrane separation process *J. Surfactants Deterg.* **23** 539–51

Derguine-Mecheri L, Kebbouche-Gana S and Djenane D 2021 Biosurfactant production from newly isolated *Rhodotorula* sp. YBR and its great potential in enhanced removal of hydrocarbons from contaminated soils *World J. Microbiol. Biotechnol.* **37** 1–18

Deshpande M and Daniels L 1995 Evaluation of sophorolipid biosurfactant production by candida bombicola using animal fat *Bioresour. Technol.* **54** 143–50

Diab A and Din G E 2013 Application of the biosurfactants produced by Bacillus spp.(SH 20 and SH 26) and *P. aeruginosa* SH 29 isolated from the rhizosphere soil of an Egyptian salt marsh plant for the cleaning of oil-contaminataed vessels and enhancing the biodegradat *Afr. J. Environ. Sci. Technol.* **7** 671–9

Domínguez Rivera Á, Martínez Urbina M Á and López y López V E 2019 Advances on research in the use of agro-industrial waste in biosurfactant production *World J. Microbiol. Biotechnol.* **35** 155

Dou Z, Toth J D and Westendorf M L 2018 Food waste for livestock feeding: feasibility, safety, and sustainability implications *Glob. Food Secur.* **17** 154–61

Dubey K and Juwarkar A 2004 Determination of genetic basis for biosurfactant production in distillery and curd whey wastes utilizing Pseudomonas aeruginosa strain BS2 *Ind. J. Biotechnol.* **3** 74–81

Dubeau D, Déziel E, Woods D E and Lépine F 2009 Burkholderia thailandensis harbors two identical rhl gene clusters responsible for the biosynthesis of rhamnolipids *BMC Microbiol.* **9** 1–12

Dumont M-J and Narine S S 2007 Soapstock and deodorizer distillates from North American vegetable oils: review on their characterization, extraction and utilization *Food Res. Int.* **40** 957–74

Elazzazy A M, Abdelmoneim T S and Almaghrabi O A 2015 Isolation and characterization of biosurfactant production under extreme environmental conditions by alkali-halo-thermophilic bacteria from Saudi Arabia *Saudi J. Biol. Sci.* **22** 466–75

El-Sheshtawy H S and Doheim M M 2014 Selection of *Pseudomonas aeruginosa* for biosurfactant production and studies of its antimicrobial activity *Egyp. J. Pet.* **23** 1–6

Fenibo E O, Douglas S I and Stanley H O 2019 A Review on microbial surfactants: production, classifications, properties and characterization *J. Adv. Microbiol.* **18** 1–22

Fernandes N, de A T, de Souza A C, Simoes L A, Dos Reis G M F, Souza K T, Schwan R F and Dias D R 2020 Eco-friendly biosurfactant from *Wickerhamomyces anomalus* CCMA 0358 as larvicidal and antimicrobial *Microbiol. Res.* **241** 126571

Fox S L and Bala G A 2000 Production of surfactant from *Bacillus subtilis* ATCC 21332 using potato substrates *Bioresour. Technol.* **75** 235–40

Franzetti A, Caredda P, Ruggeri C, La Colla P, Tamburini E, Papacchini M and Bestetti G 2009 Potential applications of surface active compounds by Gordonia sp. strain BS29 in soil remediation technologies *Chemosphere* **75** 801–7

Fu J, Jiang B and Cai W 2009 Effect of synthetic cationic surfactants on dewaterability and settleability of activated sludge *Int. J. Environ. Pollut.* **37** 113–31

Fusconi R, Assunção R M N, de Moura Guimarães R, Rodrigues Filho G and da Hora Machado A E 2010 Exopolysaccharide produced by Gordonia polyisoprenivorans CCT 7137 in GYM commercial medium and sugarcane molasses alternative medium: FT-IR study and emulsifying activity *Carbohydr. Polym.* **79** 403–8

Gargouri B, Contreras M, del M, Ammar S, Segura-Carretero A and Bouaziz M 2017 Biosurfactant production by the crude oil degrading Stenotrophomonas sp. B-2: chemical characterization, biological activities and environmental applications *ESPR* **24** 3769–79

García-Reyes S, Yáñez-Ocampo G, Wong-Villarreal A, Rajaretinam R K, Thavasimuthu C, Patiño R and Ortiz-Hernández M L 2018 Partial characterization of a biosurfactant extracted from *Pseudomonas sp.* B0406 that enhances the solubility of pesticides *Environ. Technol.* **39** 2622–31

Gayathiri E, Prakash P, Karmegam N, Varjani S, Awasthi M K and Ravindran B 2022 Biosurfactants: potential and eco-friendly material for sustainable agriculture and environmental safety—a review *Agronomy* **12** 662

Ghribi D, Abdelkefi-Mesrati L, Mnif I, Kammoun R, Ayadi I, Saadaoui I, Maktouf S and Chaabouni-Ellouze S 2012 Investigation of antimicrobial activity and statistical optimization of *Bacillus subtilis* SPB1 biosurfactant production in solid-state fermentation *Bio. Med. Res. Int.* **2012** 373682

Gomaa E Z 2013 Antimicrobial activity of a biosurfactant produced by *Bacillus licheniformis* strain M104 grown on whey *Braz. Arch. Biol. Technology* **56** 259–68

Gottfried A, Singhal N, Elliot R and Swift S 2010 The role of salicylate and biosurfactant in inducing phenanthrene degradation in batch soil slurries *Appl. Microbiol. Biotechnol.* **86** 1563–71

Gudiña E J, Rocha V, Teixeira J A and Rodrigues L R 2010 Antimicrobial and antiadhesive properties of a biosurfactant isolated from *Lactobacillus paracasei* ssp. paracasei A20 *Lett. Appl. Microbiol.* **50** 419–24

Guimarães Martins V, Kalil S J, Elit T E and Vieira Costa J A 2006 Solid state biosurfactant production in a fixed-bed column bioreactor *Z. Naturforsch.* C **61** 721–6

Günther M, Grumaz C, Lorenz S, Stevens P, Lindemann E, Hirth T, Sohn K, Zibek S and Rupp S 2015 The transcriptomic profile of *Pseudozyma aphidis* during production of mannosylerythritol lipids *Appl. Microbiol. Biotechnol.* **99** 1375–88

Guo-liang Z, Yue-ting W, Xin-ping Q and Qin M 2005 Biodegradation of crude oil by *Pseudomonas aeruginosa* in the presence of rhamnolipids *J. Zhejiang Univ. Sci.* B **6** 725–30

Harris-Lovett S, Lienert J and Sedlak D 2019 A mixed-methods approach to strategic planning for multi-benefit regional water infrastructure *J. Environ. Manage.* **233** 218–37

Heidari R, Yazdanparast R and Jabbarzadeh A 2019 Sustainable design of a municipal solid waste management system considering waste separators: a real-world application *Sustain. Cities Soc.* **47** 101457

Hickey A M, Gordon L, Dobson A D W, Kelly C T and Doyle E M 2007 Effect of surfactants on fluoranthene degradation by *Pseudomonas alcaligenes* PA-10 *Appl. Microbiol. Biotechnol.* **74** 851–6

Hong C, Si Y, Xing Y, Wang Z, Qiao Q and Liu M 2015 Effect of surfactant on bound water content and extracellular polymers substances distribution in sludge *RSC Adv.* **5** 23383–90

Hopkinson M J, Collins H M, Goss G R and Pesticides A C E−35 on 1997 *Pesticide Formulations and Application Systems: 17th Volume* **v. 17** (ASTM)

Hosseinalizadeh R, Izadbakhsh H and Shakouri H 2021 A planning model for using municipal solid waste management technologies-considering energy, economic, and environmental impacts in Tehran-Iran *Sustain. Cities Soc.* **65** 102566

Hu F, Liu Y and Li S 2019 Rational strain improvement for surfactin production: enhancing the yield and generating novel structures *Microb. Cell Fact.* **18** 1–13

Ibrahim H M M 2018 Characterization of biosurfactants produced by novel strains of Ochrobactrum anthropi HM-1 and Citrobacter freundii HM-2 from used engine oil-contaminated soil *Egypt. J. Petrol.* **27** 21–9

Ites S, Smetana S, Toepfl S and Heinz V 2020 Modularity of insect production and processing as a path to efficient and sustainable food waste treatment *J. Clean. Prod.* **248** 119248

Jain R M, Mody K, Mishra A and Jha B 2012 Physicochemical characterization of biosurfactant and its potential to remove oil from soil and cotton cloth *Carbohydr. Polym.* **89** 1110–6

Janek T, Łukaszewicz M, Rezanka T and Krasowska A 2010 Isolation and characterization of two new lipopeptide biosurfactants produced by *Pseudomonas fluorescens* BD5 isolated from water from the Arctic Archipelago of Svalbard *Bioresour. Technol.* **101** 6118–23

Jin B, Wilén B-M and Lant P 2004 Impacts of morphological, physical and chemical properties of sludge flocs on dewaterability of activated sludge *Chem. Eng. J.* **98** 115–26

Joshi S, Bharucha C, Jha S, Yadav S, Nerurkar A and Desai A J 2008 Biosurfactant production using molasses and whey under thermophilic conditions *Bioresour. Technol.* **99** 195–9

Joshi P A and Shekhawat D B 2014 Screening and isolation of biosurfactant producing bacteria from petroleum contaminated soil *Eur. J. Exp. Biol* **4** 164–9

Juwarkar A A, Nair A, Dubey K V, Singh S K and Devotta S 2007a Biosurfactant technology for remediation of cadmium and lead contaminated soils *Chemosphere* **68** 1996–2002

Juwarkar A A, Nair A, Dubey K V, Singh S K and Devotta S 2007b Biosurfactant technology for remediation of cadmium and lead contaminated soils *Chemosphere* **68** 1996–2002

Kamal I, Blaghen M, Lahlou F Z and Hammoumi A 2015 Evaluation of biosurfactant production by Aeromonas salmonicida sp. degrading gasoline *Int. J. Appl. Microbiol. Biotechnol. Res* **3** 389–95

Kanna R, Gummadi S N and Kumar G S 2014 Production and characterization of biosurfactant by Pseudomonas putida MTCC 2467 *J. Biol. Sci.* **14** 436

Kandasamy R, Rajasekaran M, Venkatesan S K and Uddin M 2019 New trends in the biomanufacturing of green surfactants: biobased surfactants and biosurfactants *Next Generation Biomanufacturing Technologies* (ACS Publications) pp 243–60

Kang S-W, Kim Y-B, Shin J-D and Kim E-K 2010 Enhanced biodegradation of hydrocarbons in soil by microbial biosurfactant, sophorolipid *Appl. Biochem. Biotechnol.* **160** 780–90

Karanth N G K, Deo P G and Veenanadig N K 1999 Microbial production of biosurfactants and their importance *Curr. Sci.* **77** 116–26

Khanal S K, Varjani S, Lin C S K and Awasthi M K 2020 Waste-to-resources: opportunities and challenges *Bioresour. Technol.* **317** 123987

Kiran G S, Thomas T A and Selvin J 2010 Production of a new glycolipid biosurfactant from marine nocardiopsis lucentensis MSA04 in solid-state cultivation *Colloids Surf.,* B **78** 8–16

Kitamoto D, Ikegami T, Suzuki G T, Sasaki A, Takeyama Y, Idemoto Y, Koura N and Yanagishita H 2001 Microbial conversion of n-alkanes into glycolipid biosurfactants, mannosylerythritol lipids, by *Pseudozyma* (*Candida antarctica*) *Biotechnol. Lett.* **23** 1709–14

Kitamoto D, Yanagishita H, Endo A, Nakaiwa M, Nakane T and Akiya T 2001 Remarkable antiagglomeration effect of a yeast biosurfactant, diacylmannosylerythritol, on ice-water slurry for cold thermal storage *Biotechnol. Progr.* **17** 362–5

Koka R K, Pilli K, Das S and Meghana K T 2020 Municipal solid waste management and MSW compost impact on soil fertility *Current Research in Soil Fertility* (Delhi: AkiNik Publications) ch 2 p 33

Kubicki S, Bollinger A, Katzke N, Jaeger K-E, Loeschcke A and Thies S 2019 Marine biosurfactants: biosynthesis, structural diversity and biotechnological applications *Mar. Drugs* **17** 408

Kulkarni S O, Kanekar P P, Jog J P, Sarnaik S S and Nilegaonkar S S 2015 Production of copolymer, poly (hydroxybutyrate-co-hydroxyvalerate) by *Halomonas campisalis* MCM B-1027 using agro-wastes *Int. J. Biol. Macromol.* **72** 784–9

Kuyukina M S, Ivshina I B, Philp J C, Christofi N, Dunbar S A and Ritchkova M 2001 Recovery of Rhodococcus biosurfactants using methyl tertiary-butyl ether extraction *J. Microbiol. Methods* **46** 149–56

Lang S and Wullbrandt D 1999 Rhamnose lipids–biosynthesis, microbial production and application potential *Appl. Microbiol. Biotechnol.* **51** 22–32

Langley S and Beveridge T J 1999 Effect of O-side-chain-lipopolysaccharide chemistry on metal binding *Appl. Environ. Microbiol.* **65** 489–98

Liu B, Liu J, Ju M, Li X and Yu Q 2016 Purification and characterization of biosurfactant produced by Bacillus licheniformis Y-1 and its application in remediation of petroleum contaminated soil *Mar. Pollut. Bull.* **107** 46–51

Li Y, Yuan X, Wu Z, Wang H, Xiao Z, Wu Y, Chen X and Zeng G 2016 Enhancing the sludge dewaterability by electrolysis/electrocoagulation combined with zero-valent iron activated persulfate process *Chem. Eng. J.* **303** 636–45

Li Z-Y, Lang S, Wagner F, Witte L and Wray V 1984 Formation and identification of interfacial-active glycolipids from resting microbial cells *Appl. Environ. Microbiol.* **48** 610–7

Liang X F, Wang H D, Yi H and Li D 2017 Warship reliability evaluation based on dynamic bayesian networks and numerical simulation *Ocean Eng.* **136** 129–40

Liu K, Sun Y, Cao M, Wang J, Lu J R and Xu H 2020 Rational design, properties, and applications of biosurfactants: a short review of recent advances *Curr. Opin. Colloid Interface Sci.* **45** 57–67

Liu J-F, Mbadinga S, Yang S-Z, Gu J-D and Mu B-Z 2015 Chemical structure, property and potential applications of biosurfactants produced by bacillus subtilis in petroleum recovery and spill mitigation *Int. J. Mol. Sci.* **16** 4814–37

Liu W W, Yin R, Lin X G, Zhang J, Chen X M, Li X Z and Yang T 2010 Interaction of biosurfactant-microorganism to enhance phytoremediation of aged polycyclic aromatic hydrocarbons (PAHS) contaminated soils with alfalfa (*Medicago sativa* L.) *Huan Jing Ke Xue* **31** 1079–84

López-Prieto A, Martínez-Padrón H, Rodríguez-López L, Moldes A B and Cruz J M 2019 Isolation and characterization of a microorganism that produces biosurfactants in corn steep water *CyTA-J. Food* **17** 509–16

Luna J M, Rufino R D, Sarubbo L A, Rodrigues L R M, Teixeira J A C and de Campos-Takaki G M 2011 Evaluation antimicrobial and antiadhesive properties of the biosurfactant Lunasan produced by *Candida sphaerica* UCP 0995 *Curr. Microbiol.* **62** 1527–34

Luo L, Kaur G, Zhao J, Zhou J, Xu S, Varjani S and Wong J W C 2021 Optimization of water replacement during leachate recirculation for two-phase food waste anaerobic digestion system with off-gas diversion *Bioresour. Technol.* **335** 125234

Madurwar M V, Ralegaonkar R V and Mandavgane S A 2013 Application of agro-waste for sustainable construction materials: a review *Constr. Build. Mater.* **38** 872–8

Mahmoud N, Zeeman G, Gijzen H and Lettinga G 2006 Interaction between digestion conditions and sludge physical characteristics and behaviour for anaerobically digested primary sludge *Biochem. Eng. J.* **28** 196–200

Maier R M, Neilson J W, Artiola J F, Jordan F L, Glenn E P and Descher S M 2001 Remediation of metal-contaminated soil and sludge using biosurfactant technology *Int. J. Occup. Med. Environ. Health* **14** 241–8

Makkar R S, Cameotra S S and Banat I M 2011 Advances in utilization of renewable substrates for biosurfactant production *AMB Express* **1** 1–19

Manu M K, Li D, Liwen L, Jun Z, Varjani S and Wong J W C 2021 A review on nitrogen dynamics and mitigation strategies of food waste digestate composting *Bioresour. Technol.* **334** 125032

Marchant R and Banat I M 2012 Microbial biosurfactants: challenges and opportunities for future exploitation *Trends Biotechnol.* **30** 558–65

Marques I P 2001 Anaerobic digestion treatment of olive mill wastewater for effluent re-use in irrigation *Desalination* **137** 233–9

Martinez-Burgos W J *et al* 2021 Agro-industrial wastewater in a circular economy: characteristics, impacts and applications for bioenergy and biochemicals *Bioresour. Technol.* **341** 125795

Mata-Sandoval J C, Karns J and Torrents A 2001 Influence of rhamnolipids and triton X-100 on the biodegradation of three pesticides in aqueous phase and soil slurries *J. Agric. Food Chem.* **49** 3296–303

McClure C D and Schiller N L 1996 Inhibition of macrophage phagocytosis by *Pseudomonas aeruginosa* rhamnolipids *in vitro* and *in vivo Curr. Microbiol.* **33** 109–17

Md Badrul Hisham N H, Ibrahim M F, Ramli N and Abd-Aziz S 2019 Production of biosurfactant produced from used cooking oil by *Bacillus* sp. HIP3 for heavy metals removal *Molecules* **24** 2617

Md F 2012 Biosurfactant: production and application *J. Pet. Environ. Biotechnol.* **3** 1000124

Mohan S, Varjani S, Pant D, Sauer M and Chang J-S 2020 Circular bioeconomy approaches for sustainability *Bioresour. Technol.* **318** 124084

Moldes A B, Paradelo R, Rubinos D, Devesa-Rey R, Cruz J M and Barral M T 2011 Ex Situ treatment of hydrocarbon-contaminated soil using biosurfactants from lactobacillus pentosus *J. Agric. Food Chem.* **59** 9443–7

Morita T, Ishibashi Y, Hirose N, Wada K, Takahashi M, Fukuoka T, Imura T, Sakai H, Abe M and Kitamoto D 2011 Production and characterization of a glycolipid biosurfactant, mannosylerythritol lipid B, from sugarcane juice by ustilago scitaminea NBRC 32730 *Biosci. Biotechnol. Biochem.* **75** 1371–6

Morita T, Kitagawa M, Yamamoto S, Sogabe A, Imura T, Fukuoka T and Kitamoto D 2010 Glycolipid biosurfactants, mannosylerythritol lipids, repair the damaged hair *J. Oleo Sci.* **59** 267–72

Morita T, Konishi M, Fukuoka T, Imura T and Kitamoto D 2007 Microbial conversion of glycerol into glycolipid biosurfactants, mannosylerythritol lipids, by a basidiomycete yeast, *Pseudozyma antarctica* JCM 10317T *J. Biosci. Bioeng.* **104** 78–81

Moshtagh B, Hawboldt K and Zhang B 2021 Biosurfactant production by native marine bacteria (*Acinetobacter calcoaceticus* P1-1A) using waste carbon sources: impact of process conditions *Can. J. Chem. Eng.* **99** 2386–97

Mukherjee S, Das P and Sen R 2006 Towards commercial production of microbial surfactants *Trends Biotechnol.* **24** 509–15

Mulligan C N 2005 Environmental applications for biosurfactants *Environ. Pollut.* **133** 183–98

Mulqueen P 2003 Recent advances in agrochemical formulation *Adv. Colloid Interface Sci.* **106** 83–107

Muthusamy K, Gopalakrishnan S, Ravi K T and Sivachidambaram P 2008 Biosurfactants: properties, commercial production and application *Curr. Sci.* **94** 736–47

Nalini S and Parthasarathi R 2014 Production and characterization of rhamnolipids produced by *Serratia rubidaea* SNAU02 under solid-state fermentation and its application as biocontrol agent *Bioresour. Technol.* **173** 231–8

Nayak A S, Vijaykumar M H and Karegoudar T B 2009 *Characterization of biosurfactant produced by Pseudoxanthomonas sp. PNK-04 and its application in bioremediation. Int. Biodeterioration & Biodegradation* **63** 73–9

Nazareth T C, Zanutto C P, Maass D, de Souza A A U, Guelli Ulson de Souza S M and de A 2021 Bioconversion of low-cost brewery waste to biosurfactant: an improvement of surfactin production by culture medium optimization *Biochem. Eng. J.* **172** 108058

Neto D C, Meira J A, Tiburtius E, Zamora P P, Bugay C, Mitchell D A and Krieger N 2009 Production of rhamnolipids in solid-state cultivation: characterization, downstream processing and application in the cleaning of contaminated soils *Biotechnol. J.* **4** 748–55

Nielsen T H and Sørensen J 2003 Production of cyclic lipopeptides by *Pseudomonas fluorescens* strains in bulk soil and in the sugar beet rhizosphere *Appl. Environ. Microbiol.* **69** 861–8

Nishio E, Ichiki Y, Tamura H, Morita S, Watanabe K and Yoshikawa H 2002 Isolation of bacterial strains that produce the endocrine disruptor, octylphenol diethoxylates, in paddy fields *Biosci. Biotechnol. Biochem.* **66** 1792–8

Nitschke M and Pastore G M 2006 Production and properties of a surfactant obtained from Bacillus subtilis grown on cassava wastewater *Bioresource Technol.* **97** 336–41

Nogueira I B, Rodríguez D M, da Silva Andradade R F, Lins A B, Bione A P, da Silva I G S, Franco L, de O and de Campos-Takaki G M 2020 Bioconversion of agroindustrial waste in the production of bioemulsifier by Stenotrophomonas maltophilia UCP 1601 and application in bioremediation process *Int. J. Chem. Eng.* **2020** 9434059

Novak J T, Sadler M E and Murthy S N 2003 Mechanisms of floc destruction during anaerobic and aerobic digestion and the effect on conditioning and dewatering of biosolids *Water Res.* **37** 3136–44

Nwaguma I V, Chikere C B and Okpokwasili G C 2019 Effect of cultural conditions on biosurfactant production by *Candida sp.* isolated from the sap of Elaeis guineensis *Biotechnol. J. Int.* **23** 1–14

Oyeleke S B, Oyewole O A, Aliyu G M, Shaba A M, Ikekwem C C and Ayisa T T 2017 Production and characterization of biosurfactants by Candida boleticola H09 and Rhodotorula bogoriensis H15 for crude oil recovery and cleaning of oil contaminated fabrics *Int. J. Life Sci. Technol.* **10** 109–123

Pacwa-Płociniczak M, Płaza G A, Piotrowska-Seget Z and Cameotra S S 2011 Environmental applications of biosurfactants: recent advances *Int. J. Mol. Sci.* **12** 633–54

Pardhi D, Panchal R and Rajput K 2020 Screening of biosurfactant producing bacteria and optimization of production conditions for *Pseudomonas guguanensis* D30 *Biosci. Biotechnol. Res. Commun.* **13** 170–9

Pardhi D S, Panchal R R, Raval V H, Joshi R G, Poczai P, Almalki W H and Rajput K N 2022 Microbial surfactants: a journey from fundamentals to recent advances *Front. Microbiol.* **13** 1–23

Partovinia A, Naeimpoor F and Hejazi P 2010 Carbon content reduction in a model reluctant clayey soil: slurry phase n-hexadecane bioremediation *J. Hazard. Mater.* **181** 133–9

Patel P, Bhatt S, Patel H, Sayyed R Z and Aguilar-Marcelino D 2022 Biosurfactant a biomolecule and its potential applications *Microbial Surfactants* (Boca Raton, FL: CRC Press) pp 63–81

Pathak K V and Keharia H 2014 Application of extracellular lipopeptide biosurfactant produced by endophytic *Bacillus subtilis* K1 isolated from aerial roots of banyan (*Ficus benghalensis*) in microbially enhanced oil recovery (MEOR) *3 Biotech.* **4** 41–8

Patowary R, Patowary K, Kalita M C and Deka S 2016 Utilization of paneer whey waste for cost-effective production of rhamnolipid biosurfactant *Appl. Biochem. Biotechnol.* **180** 383–99

Pellecier F and André P 2004 Cosmetic use of sophorolipids as subcutaneous adipose cushion regulation agents and slimming application *World Pat.* **108063** 208

Petrović M and Barceló D 2004 Analysis and fate of surfactants in sludge and sludge-amended soils *TrAC—Trends Anal. Chem.* **23** 762–71

Pradhan A and Bhattacharyya A 2018 An alternative approach for determining critical micelle concentration: dispersion of ink in foam *J. Surfactants Deterg.* **21** 745–50

Qazi M A, Subhan M, Fatima N, Ali M and Ahmed S 2013 Role of biosurfactant produced by *Fusarium sp.* BS-8 in enhanced oil recovery (EOR) through sand pack column *Int. J. Biosci. Biochem. Bioinform.* **3** 598

Qi Y, Thapa K B and Hoadley A F A 2011 Application of filtration aids for improving sludge dewatering properties—a review *Chem. Eng. J.* **171** 373–84

Raaijmakers J M, De Bruijn I and De Kock M J D 2006 Cyclic lipopeptide production by plant-associated *Pseudomonas spp.*: diversity, activity, biosynthesis, and regulation *Mol. Plant-Microbe Interact.* **19** 699–710

Rajmohan K S, Ramya C and Varjani S 2021 Trends and advances in bioenergy production and sustainable solid waste management *Energy Environ.* **32** 1059–85

Rane A N, Baikar V V, Ravi Kumar D V and Deopurkar R L 2017 Agro-industrial wastes for production of biosurfactant by *Bacillus subtilis* ANR 88 and its application in synthesis of silver and gold nanoparticles *Front. Microbiol.* **8** 1–12

Rau U, Nguyen L A, Roeper H, Koch H and Lang S 2005 Fed-batch bioreactor production of mannosylerythritol lipids secreted by *Pseudozyma aphidis Appl. Microbiol. Biotechnol.* **68** 607–13

Reiling H E, Thanei-Wyss U, Guerra-Santos L H, Hirt R, Käppeli O and Fiechter A 1986 Pilot plant production of rhamnolipid biosurfactant by *Pseudomonas aeruginosa Appl. Environ. Microbiol.* **51** 985–9

Ribeiro B G, de Veras B O, dos Santos Aguiar J, Guerra J M C and Sarubbo L A 2020 Biosurfactant produced by *Candida utilis* UFPEDA1009 with potential application in cookie formulation *Electron. J. Biotechnol.* **46** 14–21

Ribeiro B G, Guerra J M C and Sarubbo L A 2020 Biosurfactants: production and application prospects in the food industry *Biotechnol. Progr.* **36** e3030

Robles-González I V, Fava F and Poggi-Varaldo H M 2008 A review on slurry bioreactors for bioremediation of soils and sediments *Microb. Cell Fact.* **7** 1–16

Rocha e Silva N M P, Rufino R D, Luna J M, Santos V A and Sarubbo L A 2014 Screening of *Pseudomonas species* for biosurfactant production using low-cost substrates *Biocatal. Agric. Biotechnol.* **3** 132–9

Rocha M P, Oliveira A H S, Souza M C M and Gonçalves L R B 2006 Natural cashew apple juice as fermentation medium for biosurfactant production by *Acinetobacter calcoaceticus World J. Microbiol. Biotechnol.* **22** 1295–9

Rodrigues L, Banat I M, Teixeira J and Oliveira R 2006 Biosurfactants: potential applications in medicine *J. Antimicrob. Chemother.* **57** 609–18

Ron E Z and Rosenberg E 2001 Natural roles of biosurfactants *Environ. Microbiol.* **3** 229–36

Rosenberg E, Zuckerberg A, Rubinovitz C and Gutnick D L 1979 Emulsifier of arthrobacter RAG-1: isolation and emulsifying properties *Appl. Environ. Microbiol.* **37** 402–8

Rostás M and Blassmann K 2009 Insects had it first: surfactants as a defence against predators *Proc. R. Soc. B: Biol. Sci.* **276** 633–8

Sachdev D P and Cameotra S S 2013 Biosurfactants in agriculture *Appl. Microbiol. Biotechnol.* **97** 1005–16

Sadeghi Ahangar S, Sadati A and Rabbani M 2021 Sustainable design of a municipal solid waste management system in an integrated closed-loop supply chain network using a fuzzy approach: a case study *J. Ind. Prod. Eng.* **38** 323–40

Santos D, Rufino R, Luna J, Santos V and Sarubbo L 2016 Biosurfactants: multifunctional biomolecules of the 21st century *Int. J. Mol. Sci.* **17** 401

Santos E C, Jacques R J S, Bento F M, Peralba M do C R, Selbach P A, Sá E L S and Camargo F A O 2008 Anthracene biodegradation and surface activity by an iron-stimulated *Pseudomonas* sp *Bioresour. Technol.* **99** 2644–9

Sarubbo L A, Rocha R B, Luna J M, Rufino R D, Santos V A and Banat I M 2015 Some aspects of heavy metals contamination remediation and role of biosurfactants *Chem. Ecol.* **31** 707–23

Sarubbo L, Sobrinho H, Luna J and Rufino R 2014 *Biosurfactants: Classification, Properties and Environmental Applications* (Houston, TX: Studium Press) pp 303–30

Satpute S K, Bhuyan S S, Pardesi K R, Mujumdar S S, Dhakephalkar P K, Shete A M and Chopade B A 2010 Molecular genetics of biosurfactant synthesis in microorganisms *Biosurfactants* (Springer) pp 14–41

Satpute S K, Kulkarni G R, Banpurkar A G, Banat I M, Mone N S, Patil R H and Cameotra S S 2016 Biosurfactant/s from Lactobacilli species: properties, challenges and potential biomedical applications *J. Basic Microbiol.* **56** 1140–58

Sekhon K 2012 Biosurfactant production and potential correlation with esterase activity *J. Pet. Environ. Biotechnol.* **3** 1000133

Shah V, Jurjevic M and Badia D 2007 Utilization of restaurant waste oil as a precursor for sophorolipid production *Biotechnol. Progr.* **23** 512–5

Shah A V, Srivastava V K, Mohanty S S and Varjani S 2021 Municipal solid waste as a sustainable resource for energy production: state-of-the-art review *J. Environ. Chem. Eng.* **9** 105717

Sharma D and Saharan B S 2016 Functional characterization of biomedical potential of biosurfactant produced by *Lactobacillus helveticus Biotechnol. Rep.* **11** 27–35

Sharma S, Singh P, Raj M, Chadha B S and Saini H S 2009 Aqueous phase partitioning of hexachlorocyclohexane (HCH) isomers by biosurfactant produced by *Pseudomonas aeruginosa* WH-2 *J. Hazard. Mater.* **171** 1178–82

Shekhar S, Sundaramanickam A and Balasubramanian T 2015 Biosurfactant producing microbes and their potential applications: a review *Crit. Rev. Environ. Sci. Technol.* **45** 1522–54

Sheng X, He L, Wang Q, Ye H and Jiang C 2008 Effects of inoculation of biosurfactant-producing *Bacillus sp.* J119 on plant growth and cadmium uptake in a cadmium-amended soil *J. Hazard. Mater.* **155** 17–22

Shi Y *et al* 2015 Synergetic conditioning of sewage sludge via Fe2+/persulfate and skeleton builder: effect on sludge characteristics and dewaterability *Chem. Eng. J.* **270** 572–81

Shin K-H, Kim K-W and Ahn Y 2006 Use of biosurfactant to remediate phenanthrene-contaminated soil by the combined solubilization–biodegradation process *J. Hazard. Mater.* **137** 1831–7

Sindhu R, Binod P, Nair R B, Varjani S, Pandey A and Gnansounou E 2020 Waste to wealth: valorization of food waste for the production of fuels and chemicals *Current Developments in Biotechnology and Bioengineering* (Amsterdam: Elsevier) pp 181–97

Singh P B, Sharma S, Saini H S and Chadha B S 2009 Biosurfactant production by *Pseudomonas sp*. and its role in aqueous phase partitioning and biodegradation of chlorpyrifos *Lett. Appl. Microbiol.* **49** 378–83

Singh S 2020 Solid waste management in urban India: imperatives for improvement *J. Contemp. Issues Bus. Gov.* **25** 87–92

Singh S, Kumar V, Singh S, Dhanjal D S, Datta S, Sharma D, Singh N K and Singh J 2020 Biosurfactant-based bioremediation *Bioremediation of Pollutants* (Amsterdam: Elsevier) pp 333–58

Slivinski C T, Mallmann E, de Araújo J M, Mitchell D A and Krieger N 2012 Production of surfactin by *Bacillus pumilus* UFPEDA 448 in solid-state fermentation using a medium based on okara with sugarcane bagasse as a bulking agent *Process Biochem.* **47** 1848–55

Smyth T, Perfumo A, Marchant R and Banat I 2010 Isolation and analysis of low molecular weight microbial glycolipids *Handbook of Hydrocarbon and Lipid Microbiology* (Berlin: Springer) pp 3705–23

Soares da Silva R de C F, de Almeida D G, Brasileiro P P F, Rufino R D, de Luna J M and Sarubbo L A 2019 Production, formulation and cost estimation of a commercial biosurfactant *Biodegradation* **30** 191–201

Solaiman D K Y, Ashby R D, Nuñez A and Foglia T A 2004 Production of sophorolipids by candida bombicola grown on soy molasses as substrate *Biotechnol. Lett.* **26** 1241–5

Souza K S T, Gudiña E J, Azevedo Z, de Freitas V, Schwan R F, Rodrigues L R, Dias D R and Teixeira J A 2017 New glycolipid biosurfactants produced by the yeast strain wickerhamomyces anomalus CCMA 0358 *Colloids Surf.* B **154** 373–82

Souza K S T, Gudiña E J, Schwan R F, Rodrigues L R, Dias D R and Teixeira J A 2018 Improvement of biosurfactant production by *Wickerhamomyces anomalus* CCMA 0358 and its potential application in bioremediation *J. Hazard. Mater.* **346** 152–8

Sridhar B, Karthik R, Pushpam A C, Vanitha M C, Nadu T and Nadu T 2015 Production and purification of biosurfactants from *Saccharomyces cerevisiae* and *Pseudomonas aeruginosa* and its application on fruit salads *Int. J. Adv. Res. Eng. Technol.* **6** 97–104

Stancu M M 2015 Response of rhodococcus erythropolis strain IBBPo1 to toxic organic solvents *Braz. J. Microbiol.* **46** 1009–18

Stanghellini M E and Miller R M 1997 Biosurfactants: their identity and potential efficacy in the biological control of zoosporic plant pathogens *Plant Dis.* **81** 4–12

Street J C 1969 Methods of removal of pesticide residues *Can. Med. Assoc. J.* **100** 154–60

Sun X, Wu L and Luo Y 2006 Application of organic agents in remediation of heavy metals-contaminated soil *Ying Yong Sheng Tai Xue Bao= J. Appl. Ecol.* **17** 1123–8

Thanomsub B, Pumeechockchai W, Limtrakul A, Arunrattiyakorn P, Petchleelaha W, Nitoda T and Kanzaki H 2006 Chemical structures and biological activities of rhamnolipids produced by *Pseudomonas aeruginosa* B189 isolated from milk factory waste *Bioresour. Technol.* **97** 2457–61

Thavasi R 2006 *Biosurfactants from marine hydrocarbonoclastic bacteria and their application in marine oil pollution abatement PhD Thesis* Annamalai University, Chidambaram

Thavasi R, Nambaru V R M S, Jayalakshmi S, Balasubramanian T and Banat I M 2011 Biosurfactant production by *Pseudomonas aeruginosa* from renewable resources *Indian J. Microbiol.* **51** 30–6

Tripathy D B and Mishra A 2016 Sustainable biosurfactants *Sustain. Inorg. Chem.* **1** 175–92

Udoh T and Vinogradov J 2019 Experimental Investigations of behaviour of biosurfactants in brine solutions relevant to hydrocarbon reservoirs *Colloids Interf.* **3** 24

Vandana P and Singh D 2018 Review on biosurfactant production and its application *Int. J. Curr. Microbiol. Appl. Sci.* **7** 4228–41

Varjani S, Shah A V, Vyas S and Srivastava V K 2021 Processes and prospects on valorizing solid waste for the production of valuable products employing bio-routes: a systematic review *Chemosphere* **282** 130954

Varnier A, Sanchez L, Vatsa P, Boudesocque L, Garcia-Brugger A, Rabenoelina F, Sorokin A, RENAULT J, Kauffmann S and Pugin A 2009 Bacterial rhamnolipids are novel MAMPs conferring resistance to *Botrytis* cinerea in grapevine *Plant, Cell Environ.* **32** 178–93

Velioğlu Z and Ürek R Ö 2015 Biosurfactant production by pleurotus ostreatus in submerged and solid-state fermentation systems *Turk. J. Biol.* **39** 160–6

Wattanaphon H T, Kerdsin A, Thammacharoen C, Sangvanich P and Vangnai A S 2008 A biosurfactant from *Burkholderia cenocepacia* BSP3 and its enhancement of pesticide solubilization *J. Appl. Microbiol.* **105** 416–23

White J C, Parrish Z D, Gent M P N, Iannucci-Berger W, Eitzer B D, Isleyen M and Incorvia Mattina M 2006 Soil amendments, plant age, and intercropping impact p, p ′-DDE bioavailability to cucurbita pepo *J. Environ. Qual.* **35** 992–1000

Williams W, Kunorozva L, Klaiber I, Henkel M, Pfannstiel J, Van Zyl L J, Hausmann R, Burger A and Trindade M 2019 Novel metagenome-derived ornithine lipids identified by functional screening for biosurfactants *Appl. Microbiol. Biotechnol.* **103** 4429–41

Wu Q, Zhi Y and Xu Y 2019 Systematically engineering the biosynthesis of a green biosurfactant surfactin by *Bacillus subtilis* 168 *Metab. Eng.* **52** 87–97

Xiaohong P E, Xinhua Z and Lixiang Z 2009 Effect of biosurfactant on the sorption of phenanthrene onto original and H_2O_2-treated soils *J. Environ. Sci.* **21** 1378–85

Zaman M and Hamid B 2022 Biosurfactants production and applications in food *Microbial Surfactants: Volume 2: Applications in Food and Agriculture* (CRC Press) ch 7 pp 133–49

Zhang C, Wang S and Yan Y 2011 Isomerization and biodegradation of beta-cypermethrin by *Pseudomonas aeruginosa* CH7 with biosurfactant production *Bioresour. Technol.* **102** 7139–46

Zhang H, Xiang H, Zhang G, Cao X and Meng Q 2009 Enhanced treatment of waste frying oil in an activated sludge system by addition of crude rhamnolipid solution *J. Hazard. Mater.* **167** 217–23

Zhang W, Cao B, Wang D, Ma T, Xia H and Yu D 2016 Influence of wastewater sludge treatment using combined peroxyacetic acid oxidation and inorganic coagulants re-flocculation on characteristics of extracellular polymeric substances (EPS) *Water Res.* **88** 728–39

Zhang W, Xu D, Niu Z, Yin K, Liu P and Chen L 2012 Isolation and characterization of *Pseudomonas* sp. DX7 capable of degrading sulfadoxine *Biodegradation* **23** 431–9

Zhao Z and Wong J W C 2009 Biosurfactants from acinetobacter calcoaceticus BU03 enhance the solubility and biodegradation of phenanthrene *Environ. Technol.* **30** 291–9

Zhu Y, Gan J, Zhang G, Yao B, Zhu W and Meng Q 2007 Reuse of waste frying oil for production of rhamnolipids using *Pseudomonas aeruginosa* zju. u1M *J. Zhejiang Univ. -Sci.* A **8** 1514–20

Zulianello L, Canard C, Köhler T, Caille D, Lacroix J-S and Meda P 2006 Rhamnolipids are virulence factors that promote early infiltration of primary human airway epithelia by *Pseudomonas aeruginosa Infect. Immun.* **74** 3134–47

Chapter 7

Microbial surfactants in soil remediation

Mohd Yusuf and Shafat Ahmad Khan

From bioremediation to phytoremediation and electrokinetic remediation, microbial surfactants have the potential to revolutionize the field of soil remediation, offering sustainable and eco-friendly solutions to the challenges posed by soil pollution. In recent years, biosurfactants have gained significant attention as a multifunctional technology with the potential to revolutionize the reclamation of heavy metal-contaminated sites. This chapter provides a comprehensive overview of biosurfactants, exploring their structural classification and key properties. It highlights their pivotal role as bioremediation agents for heavy metals in contaminated soil, emphasizing their biodegradability, low toxicity, and efficiency in sequestering heavy metals. The chapter also discusses the mechanisms by which biosurfactants enhance heavy metal availability for microbial remediation and their potential synergies with metal-resistant microorganisms. The chapter highlights biosurfactant insights towards sustainable and environmentally friendly solutions for modern heavy metal pollution challenges in soil reclamation.

7.1 Introduction

Soil is a vital component of our ecosystem, playing a pivotal role in supporting plant growth, filtering water, and serving as a habitat for numerous microorganisms. However, due to various human activities and industrial processes, soil contamination has become a pressing global concern. Soil remediation, the process of restoring contaminated soil to its natural or usable state, has emerged as a critical field of environmental science and engineering. In this chapter, we delve into the innovative and sustainable approach of using microbial surfactants in soil remediation. Soil remediation encompasses a series of techniques and strategies aimed at mitigating the adverse effects of soil pollution [1–3]. Soil pollution occurs when contaminants such as heavy metals, pesticides, hydrocarbons, or hazardous chemicals infiltrate the soil, rendering it unsuitable for agricultural, industrial, or residential purposes. Soil remediation seeks to eliminate or reduce these

doi:10.1088/978-0-7503-5989-4ch7

contaminants to safe levels, thereby restoring the soil's functionality and minimizing environmental risks. Choosing the most appropriate remediation technology for a contaminated site is a critical decision that must align with the site's environmental characteristics. The foremost priority is to reduce risks to both the environment and human health without simply relocating the contamination elsewhere. In this regard, biosurfactants stand out as promising remediation agents due to their unique qualities, including biodegradability, low toxicity, and proven effectiveness [4–10].

The selection of biosurfactants should be a well-informed process, considering various factors such as the characteristics and properties of the pollutants, treatment capacity requirements, financial considerations, regulatory obligations, and project timelines [5–7]. The type, concentration, and chemical nature of contaminants should be carefully assessed to ensure that the chosen biosurfactants are well-suited for addressing specific pollutant classes. Additionally, the scale and severity of contamination should align with the treatment capacity offered by the selected biosurfactants. Financial aspects, including production and application costs, must be evaluated to determine the cost-effectiveness of biosurfactant-based remediation compared to alternative methods. Compliance with environmental regulations is non-negotiable, making it essential to choose biosurfactants that meet all relevant regulatory standards. Moreover, project timelines and any time constraints should influence the selection process, as some biosurfactants may expedite remediation more effectively than others [6]. Furthermore, site-specific factors such as soil type, hydrogeology, and proximity to sensitive ecosystems should be taken into account to ensure compatibility. Finally, an environmental impact assessment should be conducted to prevent unintended adverse consequences and to monitor and adapt the remediation effort in real-time based on evolving site conditions. In conclusion, selecting biosurfactants as remediation agents requires a comprehensive evaluation of various factors to make informed choices that effectively address contamination while safeguarding the environment and human health [2, 8].

The importance of soil remediation cannot be overstated. Contaminated soil poses a myriad of threats to the environment, human health, and biodiversity. Polluted soil can lead to the contamination of groundwater, the release of harmful gases into the atmosphere, and the disruption of ecosystems. Additionally, when agricultural soils are contaminated, it can jeopardize food safety and security, impacting human nutrition and livelihoods. Soil remediation, therefore, represents a crucial step in safeguarding our environment and ensuring sustainable land use.

In recent years, there has been a growing interest in harnessing the potential of microbial surfactants as a powerful tool in soil remediation. Microbial surfactants are natural compounds produced by microorganisms, such as bacteria and fungi, and they possess unique properties that enable them to enhance the mobilization and solubilization of hydrophobic contaminants in soil [9–11]. Their ability to reduce the surface tension of water and interact with hydrophobic molecules makes them valuable agents in the process of soil remediation. This chapter explores the latest advancements in microbial surfactant research, shedding light on their mechanisms of action, environmental benefits, and applications in various remediation strategies.

7.2 Mechanisms of soil remediation using microbial surfactants

Microbial surfactants are versatile compounds produced by microorganisms, and they play a crucial role in soil remediation by facilitating the removal of hydrophobic contaminants through a series of key mechanisms: emulsification, dispersion, solubilization, and mobilization [2, 12–14].

7.2.1 Emulsification

Definition: Emulsification involves the creation of stable emulsions, where water-immiscible hydrophobic contaminants become dispersed in water as tiny droplets.

Mechanism: Microbial surfactants possess amphiphilic properties, with one end of the molecule being hydrophilic (water-attracting) and the other hydrophobic (water-repellent). When added to contaminated soil, microbial surfactants adsorb to the surface of hydrophobic contaminants, with their hydrophobic tails anchoring into the contaminant molecules and their hydrophilic heads facing outwards towards the surrounding water. This arrangement reduces the interfacial tension between the hydrophobic contaminant and water, allowing for the formation of emulsions.

Significance: Emulsification increases the surface area of hydrophobic contaminants exposed to water, making it easier for other remediation processes to take place. It enhances the bioavailability of contaminants to microorganisms, promoting biodegradation, and making them more amenable to removal.

7.2.2 Dispersion

Definition: Dispersion refers to the even distribution of contaminants throughout the soil matrix.

Mechanism: Microbial surfactants act as dispersants by reducing the cohesive forces between soil particles and contaminants. Their hydrophilic ends interact with water, while their hydrophobic tails interact with the contaminants, effectively preventing the contaminants from clumping together.

Significance: Dispersion ensures that contaminants are uniformly distributed in the soil, preventing their localized accumulation and facilitating their interaction with remediation agents such as microbes or chemical reagents.

7.2.3 Solubilization

Definition: Solubilization involves transforming poorly water-soluble contaminants into more soluble forms.

Mechanism: Microbial surfactants can solubilize hydrophobic contaminants by incorporating them into micelles. Micelles are small structures formed by surfactant molecules in which the hydrophobic tails cluster together, creating a hydrophobic core while the hydrophilic heads face outward and interact with water molecules. Contaminants are encapsulated within these micelles, effectively increasing their solubility in water.

Significance: Solubilization enhances the transport of contaminants through the soil matrix and into the aqueous phase. This process is especially important for contaminants that are otherwise sparingly soluble in water.

7.2.4 Mobilization

Definition: Mobilization involves the physical movement of contaminants within the soil, making them more accessible for removal or treatment.

Mechanism: Microbial surfactants, by virtue of emulsification, dispersion, and solubilization, reduce the adhesion forces between contaminants and soil particles. This reduction in adhesion allows contaminants to move more freely in the soil matrix when subjected to hydraulic forces, such as rainfall, irrigation, or groundwater flow.

Significance: Mobilization is essential for transporting contaminants to collection points where they can be captured, treated, or further remediated. It aids in the overall removal and remediation of contaminants from the soil.

Therefore, microbial surfactants employ a combination of emulsification, dispersion, solubilization, and mobilization mechanisms to enhance the removal and treatment of hydrophobic contaminants in soil remediation. These mechanisms not only improve the bioavailability of contaminants but also contribute to the overall effectiveness of remediation strategies, promoting the restoration of contaminated soils to a more pristine state.

7.3 Factors affecting the effectiveness of microbial surfactants in soil remediation

The effectiveness of microbial surfactants in soil remediation can be influenced by various factors, which need to be carefully considered when designing and implementing remediation strategies. Understanding these factors is crucial to optimize the use of microbial surfactants for efficient and sustainable soil clean-up. Here are some key factors that can affect the effectiveness of microbial surfactants in soil remediation.

7.3.1 Contaminant type and concentration

The type and concentration of contaminants present in the soil significantly impact the effectiveness of microbial surfactants. Different surfactants may be more suitable for specific types of contaminants, and higher contaminant concentrations may require higher surfactant doses.

7.3.2 Surfactant type and concentration

The choice of microbial surfactant is critical. Different surfactant molecules have varying properties and affinities for specific contaminants. Selecting the right surfactant for the job is essential. Additionally, the concentration of the surfactant applied to the soil should be optimized to ensure effective remediation without excessive usage.

7.3.3 Soil characteristics

Soil properties, such as texture, organic matter content, pH, and mineral composition, can influence the performance of microbial surfactants. Highly organic soils may require different surfactants than clayey soils, and the pH of the soil can affect the stability of surfactant molecules.

7.3.4 Microbial activity

The presence and activity of indigenous microorganisms in the soil can impact the degradation and utilization of surfactants. Competing microbial populations may affect the overall effectiveness of bioremediation processes.

7.3.5 Environmental conditions

Factors like temperature, moisture content, and oxygen availability in the soil environment can influence the activity of both surfactant-producing microorganisms and contaminant-degrading microorganisms. Optimal conditions should be maintained for effective remediation.

7.3.6 Surfactant persistence

The persistence of microbial surfactants in the soil environment is a critical factor. Some surfactants degrade rapidly, while others may persist for extended periods. The choice of surfactant should align with the remediation timeline and objectives.

7.3.7 Surfactant toxicity

Surfactants, if present in high concentrations, can be toxic to soil organisms, including beneficial microbes. Balancing surfactant concentrations to enhance contaminant removal without causing harm to the soil ecosystem is essential.

7.3.8 Hydraulic conditions

The movement of surfactants and contaminants in the soil can be affected by hydraulic conditions such as rainfall, irrigation, or groundwater flow. Proper understanding and management of these conditions are necessary for effective surfactant-based remediation.

7.3.9 Regulatory and safety considerations

Compliance with regulatory standards and safety guidelines is crucial when using microbial surfactants in soil remediation. Understanding and meeting legal requirements is essential to ensure the safe and responsible use of surfactants.

In some cases, surfactant-based remediation may be used in conjunction with other remediation techniques, such as bioremediation or chemical treatments. Compatibility and synergistic effects between these methods should be considered. Continuous monitoring and evaluation of remediation progress and the performance of microbial surfactants are essential. Adjustments to surfactant application rates or strategies may be needed based on monitoring results. Nevertheless, the effectiveness of microbial surfactants in soil remediation is influenced by a complex interplay of factors related to contaminants, surfactants, soil, microorganisms, and environmental conditions. A holistic approach that considers these factors and their interactions is necessary to design and implement successful surfactant-based soil remediation strategies.

7.4 Applications of microbial surfactants in soil remediation

Microbial surfactants, also known as biosurfactants, have shown great promise in various applications within the field of soil remediation. These natural compounds, produced by microorganisms, possess unique properties that make them valuable tools for addressing soil contamination [12–15]. Below, we describe in detail some of the key applications of microbial surfactants in soil remediation:

- *Enhanced bioremediation:* Microbial surfactants are used to boost bioremediation processes. They achieve this by increasing the bioavailability of hydrophobic contaminants in the soil. These surfactants emulsify, solubilize, and disperse contaminants, making them more accessible to naturally occurring or introduced microorganisms. This facilitates the biodegradation of contaminants in the soil.
- *Phytoremediation support:* Phytoremediation involves using plants to remove or neutralize contaminants from soil. Microbial surfactants aid in phytoremediation by improving the uptake of contaminants by plant roots. They enhance the solubility of hydrophobic contaminants, making it easier for the contaminants to move from the soil into the plant roots. Once inside the plant, contaminants can be accumulated or degraded.
- *Soil washing:* Soil washing is a physical remediation technique that relies on washing contaminants out of the soil using water-based solutions. Microbial surfactants are employed as washing agents to enhance the solubility and mobility of contaminants. They help in releasing contaminants from soil particles, making it easier to separate and treat them in the wash solution.
- *Chemical oxidation and reduction: In situ* chemical oxidation and reduction are remediation methods that involve introducing chemicals into the contaminated soil to transform or degrade contaminants. Microbial surfactants are added to these chemical solutions to ensure better distribution within the soil. This improved distribution increases the contact between the chemicals and contaminants, enhancing the overall efficiency of the chemical treatment.
- *Soil flushing and pump-and-treat systems:* Soil flushing and pump-and-treat systems are groundwater remediation techniques that involve the extraction and treatment of contaminated groundwater and associated soils. Microbial surfactants can be introduced into these systems to mobilize and solubilize hydrophobic contaminants in the subsurface. This aids in the removal and treatment of contaminants.
- *Aiding soil erosion control:* Microbial surfactants also play a role in preventing soil erosion. By reducing the surface tension of water, they improve water

infiltration into the soil, which helps prevent soil erosion and promotes vegetation growth. This application is particularly valuable in areas affected by soil erosion due to contamination or other factors.

- *Enhanced oil recovery (EOR)/petroleum products:* In the context of oil spills, microbial surfactants assist in enhancing oil recovery efforts. By emulsifying oil, they improve the mobilization and recovery of spilled oil from contaminated soil. This is particularly crucial in the remediation of oil-contaminated coastal or terrestrial environments.
- *Mitigation of soil wettability issues:* Some contaminated soils exhibit hydrophobicity, which hinders water and nutrient penetration. Microbial surfactants counteract this problem by wetting the soil surface and enhancing water infiltration. This is vital for both bioremediation and phytoremediation efforts.
- *Green and sustainable remediation:* Microbial surfactants are eco-friendly and sustainable alternatives to synthetic surfactants, which may have adverse ecological impacts. Their use aligns with green and sustainable remediation principles, minimizing harm to the environment while effectively treating soil contaminants.
- *Custom formulations for specific contaminants:* Researchers and practitioners can customize microbial surfactant formulations to suit specific soil contamination scenarios. This flexibility allows for the optimization of remediation strategies based on the type and extent of contamination.

However, microbial surfactants offer a wide array of applications in soil remediation, ranging from enhancing bioremediation to supporting various physical and chemical treatment methods. Their versatility, biodegradability, and environmentally friendly nature make them valuable tools for addressing soil contamination while minimizing the environmental impact of remediation efforts. Other specific applications are described herein.

7.4.1 Bioremediation of hydrocarbons

Hydrocarbons, such as petroleum hydrocarbons and polycyclic aromatic hydrocarbons (PAHs), are common contaminants in soil and water. Bioremediation, a sustainable approach, can be enhanced by microbial surfactants in the following ways:

Emulsification: Microbial surfactants can emulsify hydrophobic hydrocarbons, breaking them down into smaller, more bioavailable droplets. This increases the surface area for microbial degradation.

Solubilization: Surfactants can solubilize hydrocarbons in water by forming micelles. This enhances the dissolution of hydrocarbons, making them more accessible to hydrocarbon-degrading microorganisms.

Enhanced biodegradation: Microbial surfactants facilitate the attachment of hydrocarbon-degrading bacteria to the hydrophobic hydrocarbon molecules, promoting their biodegradation. They also reduce the inhibitory effects of hydrocarbons on microbial growth.

In recent years, the utilization of surfactants has gained considerable attention as a technology for soil washing. This approach is particularly valuable because hydrophobic pollutants tend to strongly adhere to soil particles, which are generally poorly soluble in water. Surfactants, which can be either synthetic or naturally derived, play a pivotal role in detergent formulations due to their amphipathic nature [7]. Their molecular structure, characterized by both polar and apolar components, enables surfactants to solubilize soil contaminants, thus facilitating their removal. By promoting emulsification, surfactant molecules enhance the solubility of hydrophobic pollutants, making them more readily available to oil-degrading microorganisms during bioremediation processes. Surfactants can be categorized into two primary groups based on their origin and composition: synthetic surfactants and green surfactants. Synthetic surfactants, the more wide-spread and cost-effective of the two, are manufactured synthetically from non-renewable sources, resulting in structures distinct from natural cellular components. On the other hand, green surfactants encompass both biobased surfactants and biosurfactants. Biobased surfactants, while typically produced through chemical synthesis, contain fats, sugars, or amino acids from renewable sources in their composition, rendering them moderately biocompatible [9–12]. In contrast, bio-surfactants, also known as microbial surfactants, are considered highly biocompatible and ecologically safe. These surfactants are produced by living cells, primarily bacteria and yeasts, without the need for organic synthesis intermediaries.

When microorganisms possessing the capability to degrade hydrocarbons also produce biosurfactants, the combined effect can significantly accelerate the bio-remediation of hydrocarbon-contaminated environments. The capacity of certain bacterial genera, including *Pseudomonas*, *Bacillus*, *Acenetobacter*, *Alcaligenes*, *Rhodococcus*, *Corynebacterium*, and others, to enhance the degradation of petro-leum oil through biosurfactant production has been extensively investigated by various scientists [13–16]. Biosurfactants tend to amass at the juncture of two immiscible fluids or at the boundary of a fluid and a solid surface. Pacwa-Płociniczak *et al* described the mechanism for biosurfactant action on a polluted entity [16]. Their primary function involves diminishing both surface tension, which occurs at the interface between a liquid and air, and interfacial tension, which arises between two liquid phases. Through this reduction in tension, biosurfactants play a crucial role in mitigating the repulsive forces that typically exist between disparate phases. This, in turn, facilitates the seamless blending and enhanced interaction of these distinct phases, promoting a more efficient and harmonious integration of otherwise incompatible components. In addition, the absorption capabilities of organo-layered double hydroxides (LDHs) exhibited a pronounced linear enhance-ment corresponding to the escalating quantity of loaded sodium dodecyl sulfate (SDS) on the LDH matrices [17]. Notably, all absorption isotherms demonstrated a linear relationship. Furthermore, an intriguing observation emerged as the SDS loading increased—the surface areas of the organo-LDHs underwent a substantial reduction. This decline in surface area was attributed to the SDS molecules, which acted as a preventative barrier, hindering the LDHs from presenting their exposed surfaces effectively (figure 7.1).

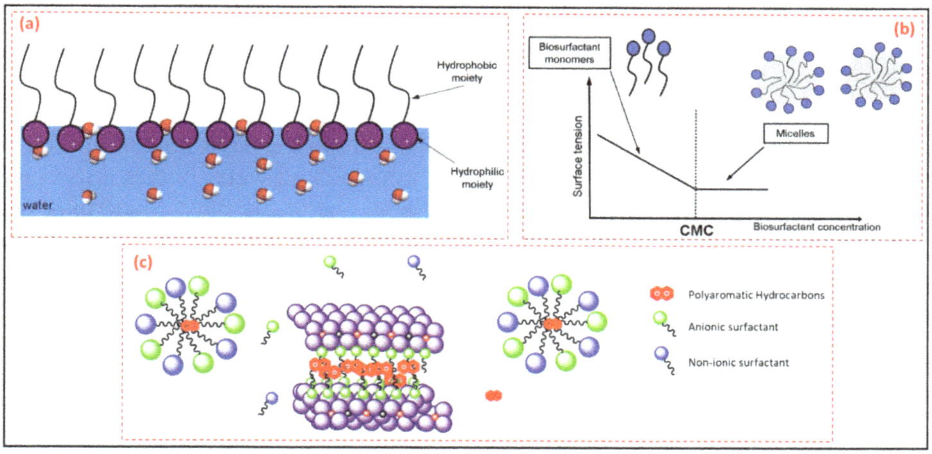

Figure 7.1. (a) Accumulation of biosurfactants at the interface between liquid and air. (b) The relationship between biosurfactant concentration, surface tension and formation of micelles. (c) The mechanism of sorption of polyaromatic hydrocarbons by organo-LDH in an aqueous solution with a high surfactant concentration ((a), (b) and (c) adapted from references [16, 17] respectively under CC BY MDPI, 2011, 2023).

Biosurfactants exhibit remarkable capabilities for enhancing the solubility and mobility of hydrocarbons, making them particularly appealing for bioremediation efforts targeting crude oil contamination [18, 19]. Numerous studies have highlighted the effectiveness of biosurfactants in remediating soils contaminated with various organic and inorganic pollutants. For instance, a biosurfactant produced by *Pseudomonas aeruginosa* was found to significantly increase the rate of oil degradation by up to 90%, underscoring its potential as a powerful tool to stimulate petroleum degradation. Additionally, crude biosurfactants derived from *Pseudomonas cepacia* CCT6659 facilitated the recovery of nearly three-quarters of oil from contaminated sand. The observed 10%–20% increase in the rate of oil biodegradation by *Bacillus* sp. and *Candida sphaerica* in the presence of their respective biosurfactants suggests that these natural surfactants could significantly enhance hydrocarbon degradation in soil [20–24]. Furthermore, biosurfactants have been explored in combination with cost-effective synthetic surfactants or in novel formulations to create more stable products, potentially opening new avenues in the petroleum market.

7.4.2 Bioremediation of heavy metals

Metals serve as crucial components in enzymatic and metabolic processes across all life forms, acting as catalysts or cofactors. Nevertheless, their necessity is confined to small quantities, and when present in elevated concentrations, these metals become highly toxic to various life forms. The release of heavy metals into the environment is predominantly attributed to a wide array of human activities, with the majority originating from various anthropogenic sources. These sources encompass a spectrum of industrial and commercial processes, such as mining, ore processing, leather tanning, fossil fuel combustion, and electroplating, as well as the utilization of various products, including paints, pigments, preservatives, fertilizers, and chemicals. These activities and products are characterized by their significant utilization of toxic heavy metals, making them substantial contributors to environmental heavy metal contamination. The unregulated discharge of these heavy metals into the soil and aquatic ecosystems has far-reaching consequences, posing serious threats to both the environment and human health. The unchecked release of heavy metals into soil and water bodies is associated with a cascade of detrimental effects. Among these, one of the most concerning is the bioaccumulation of heavy metals within the food chain. As heavy metals leach into the environment, they are absorbed by plants and aquatic organisms, entering the very foundation of the food web [24]. Over time, these contaminants become increasingly concentrated as they move up the trophic levels. Ultimately, this bioaccumulation culminates in a heightened risk to human health, as these heavy metals find their way into the food we consume. Thus, understanding and mitigating the sources and consequences of heavy metal contamination represent critical endeavors in safeguarding both the environment and the well-being of human populations.

The accumulation of excessive levels of these noxious heavy metals in water bodies and soil poses a significant threat to both aquatic and terrestrial ecosystems, endangering the well-being of plant and animal life [25–27]. Furthermore, such toxic metals can detrimentally impact soil fertility and reduce crop productivity. Heavy metals, such as lead, cadmium, and mercury, are toxic contaminants that pose serious risks to ecosystems and human health. Microbial surfactants aid in the bioremediation of heavy metals by:

Complexation and precipitation: Surfactants can complex with heavy metals, forming insoluble complexes or precipitates. This reduces the bioavailability and mobility of the heavy metals, preventing their further spread.

Enhancing phytoextraction: In phytoremediation, plants can accumulate heavy metals in their tissues. Microbial surfactants can improve the uptake of heavy metals by plant roots, increasing the efficiency of this remediation method.

7.4.3 Bioremediation of pesticides

Pesticides, including herbicides, insecticides, and fungicides, can contaminate soil and water, posing environmental and health risks. Microbial surfactants contribute to the bioremediation of pesticides by:

Emulsifying pesticides: Surfactants emulsify pesticide compounds, dispersing them into the aqueous phase. This increases their accessibility to pesticide-degrading microorganisms.

Stabilizing pesticide-degrading enzymes: Microbial surfactants can stabilize enzymes produced by pesticide-degrading microorganisms, enhancing their activity and longevity in the soil.

7.4.4 Bioremediation of explosives

Explosives, such as nitroaromatic compounds (e.g., TNT) and nitramines (e.g., RDX), are persistent organic pollutants found in military and industrial sites. Microbial surfactants aid in their bioremediation by:

Emulsification and dispersion: Surfactants emulsify and disperse explosive compounds, increasing their accessibility to microbial degradation. This is especially important for TNT, which is highly insoluble in water.

Biodegradation enhancement: Microbial surfactants promote the growth and activity of explosive-degrading microorganisms. They enhance the breakdown of complex explosive molecules into less toxic products.

In particular, microbial surfactants play a vital role in the bioremediation of hydrocarbons, heavy metals, pesticides, and explosives. They improve the bioavailability and degradation of these contaminants, making bioremediation processes more effective and environmentally friendly. These applications showcase the potential of microbial surfactants in addressing diverse types of soil and water pollution.

7.5 Limitations of using microbial surfactants in soil remediation

The utilization of microbial surfactants in soil remediation brings numerous advantages, but it also presents certain limitations that need to be considered in the context of their application. One of the primary limitations is the specificity of microbial strains. Microbial surfactants are produced by specific strains of microorganisms, and their effectiveness may be limited to certain types of contaminants. As a result, choosing the appropriate microbial strain for a particular contamination scenario is critical, and this specificity can restrict their broader applicability. Another significant constraint is the production costs associated with microbial surfactants. Producing these surfactants can be more expensive than their synthetic counterparts. The fermentation processes required to cultivate the microorganisms and extract the surfactants can be resource-intensive, potentially impacting the economic feasibility of large-scale soil remediation projects. Furthermore, the production of microbial surfactants is time-consuming. It often necessitates several days or even weeks for the cultivation of microorganisms and the extraction of surfactants. In situations where rapid remediation is imperative, the lengthy production process may not align with project timelines.

Environmental conditions also play a pivotal role in the effectiveness of microbial surfactants. Since these surfactants are produced by living microorganisms, factors such as temperature, pH, and nutrient availability can influence their production. Suboptimal conditions may reduce surfactant production and, consequently, the efficiency of the remediation process. Additionally, microbial surfactants typically have a shorter shelf life compared to synthetic alternatives. This limitation can necessitate on-site production or frequent reapplication in long-term remediation projects. While the biodegradability of microbial surfactants is generally advantageous, it can also be a limitation in certain cases. In some scenarios, the surfactants may degrade too rapidly, resulting in a shorter effective remediation window. Regulatory approval for the use of microbial surfactants in soil remediation may vary by region. Acquiring the necessary permits and approvals can be a bureaucratic hurdle that delays or complicates their implementation. Moreover, the effectiveness of microbial surfactants in treating persistent contaminants, such as certain chlorinated compounds and heavy metals, may be limited. Complementary remediation methods may be required to address these stubborn pollutants adequately. Lastly, soil remediation sites can be complex, with varying soil types, contamination levels, and geological factors. Microbial surfactants may not perform consistently in all environments, necessitating tailored approaches for each site.

Therefore, while microbial surfactants offer environmentally friendly and biocompatible options for soil remediation, their use is not without limitations. Site-specific factors, cost considerations, and the nature of contaminants should all be carefully evaluated when deciding whether to incorporate microbial surfactants into a remediation strategy.

7.6 Conclusion

The microbial surfactants offer a promising significant environmentally friendly solution for soil remediation, characterized by their biodegradability, low toxicity, and contaminant solubilization capabilities. However, their effective application requires careful consideration of limitations, including specificity to certain contaminants, production costs, time-intensive processes, environmental dependencies, and regulatory hurdles. To harness their potential fully, a site-specific approach is essential, considering contaminant profiles, environmental conditions, project timelines, and regulatory constraints. Often, microbial surfactants are best utilized in conjunction with other remediation methods to address complex contamination scenarios comprehensively. While challenges exist, ongoing research and innovation in this field continue to expand the role of microbial surfactants in soil remediation, offering a promising avenue for sustainable environmental clean-up and human health protection.

References

[1] Tripathy D B and Mishra A 2016 Sustainable biosurfactants *Sustain. Inorg. Chem.* **1** 175–92

[2] Liu J F, Mbadinga S M, Yang S Z, Gu J D and Mu B Z 2015 Chemical structure, property and potential applications of biosurfactants produced by *Bacillus subtilis* in petroleum recovery and spill mitigation *Int. J. Mol. Sci.* **16** 4814–37

[3] Almakki A, Jumas-Bilak E, Marchandin H and Licznar-Fajardo P 2019 Antibiotic resistance in urban runoff *Sci. Total Environ.* **667** 64–76

[4] Yusuf M 2019 Synthetic dyes: a threat to the environment and water ecosystem ed M Shabbir *Textiles and Clothing: Environmental Concerns and Solutions* (Beverly: Scrivener Publishing) pp 11–26

[5] Abbasi H, Sharafi H, Alidost L, Bodagh A, Zahiri H S and Noghabi K A 2013 Response surface optimization of biosurfactant produced by *Pseudomonas aeruginosa* MA01 isolated from spoiled apples *Prep. Biochem. Biotech.* **43** 398–414

[6] Yusuf M, Shabbir M and Mohammad F 2017 Natural colorants: historical, processing and sustainable prospects *Nat. Prod. Bioprospect.* **7** 123–45

[7] Yusuf M, Ahmad A, Shahid M, Khan M I, Khan S A, Manzoor N and Mohammad F 2012 Assessment of colorimetric, antibacterial and antifungal properties of woollen yarn dyed with the extract of the leaves of henna (*Lawsonia inermis*) *J. Clean. Prod.* **27** 42–50

[8] Yusuf M, Khan S A, Shabbir M and Mohammad F 2017 Developing a shade range on wool by madder (*Rubia cordifolia*) root extract with gallnut (*Quercus infectoria*) as biomordant *J. Nat. Fibers* **14** 597–607

[9] Malvar J L, Santos J L, Martín J, Aparicio I and Alonso E 2021 Occurrence of the main metabolites of the most recurrent pharmaceuticals and personal care products in Mediterranean soils *J. Environ. Manage.* **278** 111584

[10] Yusuf M, Khan M A and Mohammad F 2016 Investigations of the colourimetric and fastness properties of wool dyed with colorants extracted from Indian madder using reflectance spectroscopy *Optik* **127** 6087–93

[11] Deleu M and Paquot M 2004 From renewable vegetables resources to microorganisms: new trends in surfactants *C.R. Chim.* **7** 641–6

[12] Bhardwaj G, Cameotra S S and Chopra H K 2013 Biosurfactants from fungi: a review *J. Pet. Environ. Biotechnol.* **4** 1–6

[13] Jagtap S, Yavankar S, Pardesi K and Chopade B 2010 Production of bioemulsifier by acinetobacter species isolated from healthy human skin. Indian *J. Exp. Biol.* **48** 70–6

[14] Stancu M M 2015 Response of *Rhodococcus* erythropolis strain IBBPo1 to toxic organic solvents *Braz. J. Microbiol.* **46** 1009–18

[15] Morita T, Konishi M, Fukuoka T, Imura T and Kitamoto D 2007 Physiological differences in the formation of the glycolipid biosurfactants, mannosylerythritol lipids, between *Pseudozyma antarctica* and *Pseudozyma aphidis Appl. Microbiol. Biotechnol.* **74** 307–15

[16] Pacwa-Płociniczak M, Płaza G A, Piotrowska-Seget Z and Cameotra S S 2011 Environmental applications of biosurfactants: recent advances *Int. J. Mol. Sci.* **12** pp. 633–54

[17] Tiwari M and Tripathy D B 2023 Soil contaminants and their removal through surfactant-enhanced soil remediation: a comprehensive review *Sustainability* **15** 13161
Abbasian F, Lockington R, Megharaj M and Naidu R 2016 A review on the genetics of aliphatic and aromatic hydrocarbon degradation *Appl. Biochem. Biotechnol.* **178** 224–50

[18] Abdel-Mawgoud A M, Lépine F and Déziel E 2010 Rhamnolipids: diversity of structures, microbial origins and roles *Appl. Microbiol. Biotechnol.* **86** 1323–36

[19] Abouseud M, Yataghene A, Amrane A and Maachi R 2008 Biosurfactant production by free and alginate entrapped cells of *Pseudomonas fluorescens J. Ind. Microbiol. Biotechnol.* **35** 1303–8

[20] Al-Wasify R S and Hamed S R 2014 Bacterial biodegradation of crude oil using local isolates *Int. J. Bacteriol.* **2014** 863272

[21] Santos D K F, Rufino R D, Luna J M, Santos V A and Sarubbo L A 2016 Biosurfactants: multifunctional biomolecules of the 21st century *Int. J. Mol. Sci.* **17** 401

[22] Almeida D G, Soares da Silva Rdcf, Luna J M, Rufino R D, Santos V A and Sarubbo L A 2017 Response surface methodology for optimizing the production of biosurfactant by *Candida tropicalis* on industrial waste substrates *Front. Microbiol.* **8** 157

[23] Gysel N, Dixit P, Schmitz D A, Engling G, Cho A K, Cocker D R and Karavalakis G 2018 Chemical speciation, including polycyclic aromatic hydrocarbons (PAHs), and toxicity of particles emitted from meat cooking operations *Sci. Total Environ.* **633** 1429–36

[24] da Silva S, Gonçalves I, Gomes de Almeida F C, Padilha da Rocha e Silva N M, Casazza A A, Converti A and Asfora Sarubbo L 2020 Soil Bioremediation: overview of technologies and trends *Energies* **13** 4664

[25] Chakraborty J and Das S 2014 Biosurfactant-based bioremediation of toxic metals *Microbial Biodegradation and Bioremediation* (Amsterdam: Elsevier) pp 167–201
[26] Mishra S, Lin Z, Pang S, Zhang Y, Bhatt P and Chen S 2021 Biosurfactant is a powerful tool for the bioremediation of heavy metals from contaminated soils *J. Hazard. Mater.* **418** 126253
[27] Rahmat M, Kiran S, Gulzar T, Yusuf M, Nawaz R, Khalid J, Fatima N, Ullah A and Azam M 2023 Plant-assisted synthesis and characterization of MnO_2 nanoparticles for removal of crystal violet dye: an environmental remedial approach *Environ. Sci. Pollut. Res.* **30** 57587–98

Chapter 8

Microbial surfactants in water remediation

Deepjyoti Paul, Ashish Kumar Agrahari, Pratichi Singh and Lekshmi Narendrakumar

Water is an indispensable need of all living beings. However, due to both natural and anthropogenic activities, the available water sources globally are shrinking. Additionally, there is also increasing pollution of water bodies that has put almost half the population of the world in water-stressed conditions. The major pollutants of the water bodies include hydrocarbons, heavy metals, pesticides, plastic and radioactive waste apart from the biological contaminants. These pollutants pose a threat to humans, animals and plants alike and hence effective removal of these contaminants is a growing and critical need. In recent years, bioremediation, which uses microorganisms or microorganism derived products, has gained in popularity due to its many advantages like low-toxicity, biodegradability and stability over a wide range of conditions. Biosurfactants, that are surface-active agents produced by certain microorganisms, are considered as one of the efficient 'green technologies' by which bioremediation of water bodies can be achieved. This chapter expounds the different properties of an ideal biosurfactant that could be used for bioremediation, its mechanism of action, factors affecting biosurfactant production and its various potential applications in the biodegradation of water pollutants. The authors also recapitulate some eminent studies on screening of efficient microbial biosurfactant producers and its production for wider usage.

8.1 Introduction

Water is one amongst the most important requirements of any life form. It was identified that life originated in water, and all organisms apart from few extremophiles perish in parched conditions. Although 71% of the Earth is made up of water, a vast majority of it is in the oceans and seas and not potable. Only 3% of the total water is freshwater of which only 1.2% can be used for drinking ('Earth's Fresh Water')[1]. For humans, water is important for consumption, agricultural use,

[1] https://education.nationalgeographic.org/resource/earths-fresh-water

doi:10.1088/978-0-7503-5989-4ch8

industrial use and many others. However, billions of people around the globe are facing immense water scarcity. It is estimated that by 2025, half of the world will be living in water-stressed conditions (Klobucista 2021). Broadly, water scarcity is of two types (i) physical scarcity and (ii) economic scarcity. Physical scarcity occurs when the population of the region exceeds the water availability in the region while economic scarcity occurs when there is no proper water infrastructure[2]. The increased population has in turn led to increased food demand. The expansion of industries and residential areas led to landfilling of natural water sources and parallel expansion of agricultural lands led to demand for more green water availability. The demand of water globally has increased tremendously and Mr Rajendra Singh, known as the 'water man of India' said that the third world war would be for water. One major problem that aggravates water scarcity is water pollution. Discharge of industrial effluents to lakes and rivers is a main source of water pollution globally. The strong chemicals in these industrial effluents have been identified to not only contaminate surface water, but seep down and contaminate the groundwater systems. The emerging organic contaminants (EOCs) in these water bodies significantly reduce their quality for drinking (Bhattacharya *et al* 2018). Hence, it is very important to preserve the existing and dying water resources.

Water which is contaminated can be treated and remediated to some extent for its use. Water remediation can be defined as the several processes that improve the quality of contaminated water. The process of water remediation involves several steps and also there are different types of water remediation. The first step of water remediation is to understand the type and concentration of contaminants in the water. The second step involves choosing the right decontamination method and also the management and disposal of the contaminant. The third phase involves the checking and monitoring of the treated water to ensure compliance with the regulations. One of the emerging areas of water remediation is the use of microbial surfactants to treat contaminated water. Microbial surfactants are tensio-active biomolecules produced by certain microorganisms like *Pseudomonas aeruginosa*, *Bacillus subtilis*, *Acinetobacter calcoaceticus*, *Candida* spp., *Aspergillus* spp., *Pichia* spp. and others. The microbial surfactants or biosurfactants are today widely used for biotechnological and industrial applications (Tripathy and Mishra, 2016; Inamuddin and Prasad 2021). The increasing popularity of biosurfactants in industries over chemical or synthetic surfactants is because of its several advantages like low toxicity, biodegradability and stability making them more eco-friendly (Kumar and Dubey 2023).

Water pollution by organic and inorganic contaminants is a major cause of concern around the globe. The present chapter on microbial surfactants in water remediation recapitulates the different mechanisms of water remediation using microbial surfactants, factors affecting the biosurfactants used in water remediation, their applications, necessity and limitations. The chapter would serve as a preliminary yet primary source of easy understanding of the emerging topic. The chapter would be useful for students, researchers and industrialists interested in the topic.

[2] https://www.cfr.org/backgrounder/water-stress-global-problem-thats-getting-worse

8.2 Mechanisms of water remediation using microbial surfactants

Microbial surfactants are amphiphilic surface-active molecules that have the property to reduce the surface and interfacial tensions. They are broadly classified based upon their molecular weight and chemical structure (Pardhi *et al* 2022). Further, based on the molecular weight of the biosurfactant, they can be either low molecular weight or high molecular weight. The high molecular weight biosurfactants are also known as bioemulsans. The property of biosurfactants depends largely on their microbial origin and culturing conditions of the source microbe. The different microbial surfactants based on their chemical composition are lipopeptides, glycolipids, and lipoprotein, fatty acids and neutral lipids, phospholipids, polymeric microbial surfactants and particulate biosurfactants. Of these, glycolipids, lipopeptides and lipoprotein, fatty acids and neutral lipids, phospholipids are of low molecular weight while polymeric microbial surfactants and particulate biosurfactants are of higher molecular weight (Pardhi *et al* 2022). Some researchers studied glycopeptides include rhamnolipids produced by *P. aeruginosa* (Gaur *et al* 2023), trehalolipids produced by *Rhodococcus erytropolis* (Ciapina *et al* 2006) and sophorolipids produced by several yeast strains.

Any biosurfactant's general mode of action is to lower the surface to interfacial tension at the interface between water and air. They are also capable of stabilising oil in water emulsions quite well. For the same reason, biosurfactants are used as both emulsifiers and de-emulsifiers. The biosurfactants reduce the interfacial tension to form micelles, vesicles or bilayers which are the active forms (Shiomi 2015). The crucial micelle concentration (CMC) is the point at which micelle production takes place. The general mechanism of action of surfactants in water remediation is schematically represented in figure 8.1.

Biosurfactants are produced by bacteria in both planktonic as well as biofilm stages. Surfactants produced by bacteria in the biofilm stage have been identified to be more effective in bioremediation than those in the planktonic stage (Peele *et al* 2016). The majority of the time, biosurfactant-producing microorganisms are kept apart from environments where pollutants have caused contamination. Though the function of biosurfactants produced by the microbes is not fully understood, it is believed that the biosurfactant helps the bacteria/fungi to survive in an unfavorable environment (Desai and Banat 1997). The majority of the time, biosurfactant-producing microorganisms are kept apart from environments where pollutants have caused contamination. Most of the environmental pollutants particularly in the soil and water are hydrocarbons. These hydrocarbons emulsified and solubilized by microbial surfactants can act as the carbon source for the microbial growth.

8.2.1 Emulsification

Emulsification is the property of surfactants to enhance the bioavailability of the hydrophobic contaminants into the aqueous phase by reducing the surface tension forming small droplets. Biosurfactants with higher molecular weight are better emulsifiers than those with lower molecular weight. The minimum benchmark of surface tension reduction of any surfactant is below 35 mN m^{-1} (Bhakta and Nath 2017).

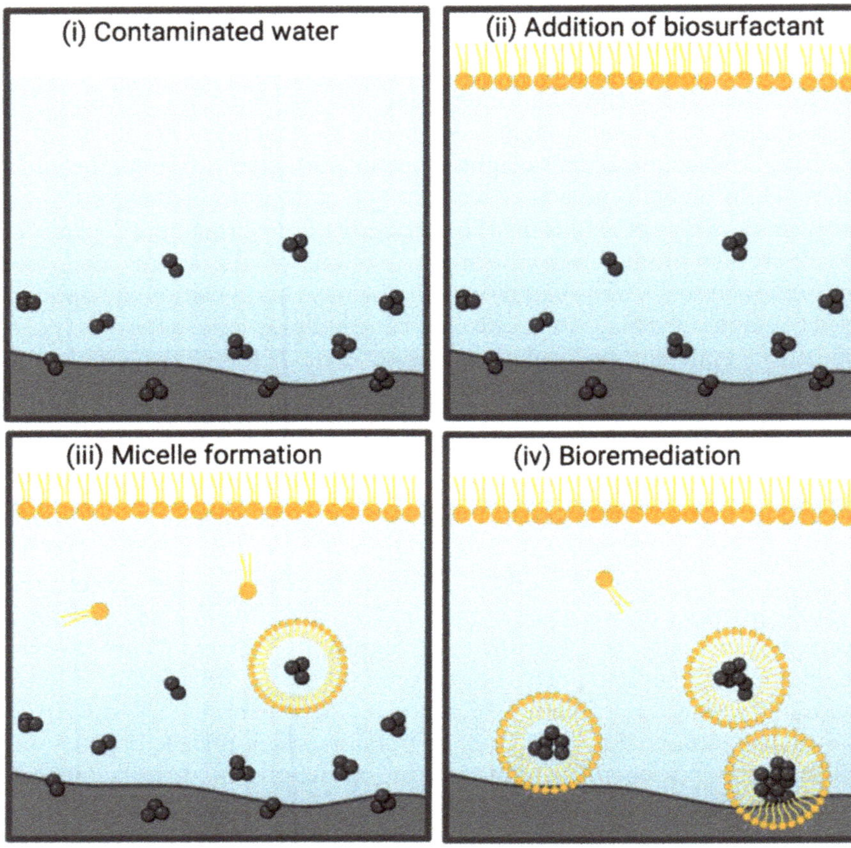

Figure 8.1. Mechanism of action of microbial surfactants. (i) Water contaminated by pollutants. (ii) Surface saturation of the added biosurfactant by lowering the surface tension. (iii) Biosurfactant attaining the critical micelle concentration (CMC) Forming micelle. (iv) Pollutants trapped within the biosurfactant micelle and removal.

But there are reports of many biosurfactants having the ability to emulsify hydrocarbons even when the medium surface tension reduction is above 35 mN m^{-1} (Ponte Rocha *et al* 2009, de Oliveira *et al* 2013). The capability of forming emulsion by the biosurfactant is an important consideration in the applicability of the biosurfactant for water remediation. The emulsification index of a biosurfactant is used to determine its capacity to create emulsions (E24). The emulsification index is calculated by dividing the total height of the emulsion produced by the biosurfactant in a tube by the total height of the aqueous phase plus the height of the emulsion multiplied by hundred.

Further, the emulsification index can vary with physical parameters such as temperature, salinity, pH and presence of particulate materials in the aqueous phase (Lai *et al* 2009). For instance, change in pH of the solution was identified to increase the formation and rigidity of interfacial films, causing the emulsion to coalesce in hydrocarbon remediation (Tadros 2013). Also, salinity plays an important role in emulsion formation and stability as the salt ions cause ionization effect. Additionally, emulsion stability is identified to be greatly influenced by the increasing temperature. At higher temperatures, emulsification of hydrocarbons was identified to cease and enter the de-emulsification phase. A list of some microorganisms producing commercially important bio-emulsifiers and their type is noted in table 8.1.

Table 8.1. List of some microorganisms producing commercially important bio-emulsifiers and their type.

S. No.	Bacteria	Bio-emulsifier type	References
1.	*Candida lipolytica*	Liposan bio-emulsifier	Cirigliano and Carman (1984)
2.	*Saccharomyces cerevisiae*	Mannoprotein bio-emulsifier	Cameron *et al* (1988)
3.	*Acinetobacter radioresistens* KA53	Bio-emulsifier alasan	Barkay *et al* (1999)
4.	*Pseudomonas aeruginosa* 47T2 NCIB40044	Rhamnolipids	Haba *et al* (2000)
5.	*Halomonas eurihalina* F2–7	Exopolysaccharide	Martínez-Checa *et al* (2007)
6.	*Pseudomonas oleovorans*	Extracellular polysaccharide	Freitas *et al* (2009)
7.	*Brevibacterium* sp. PDM-3 strain	Bio-emulsifier degrading polyaromatic hydrocarbons	Reddy *et al* (2010)
8.	*Gordonia* sp. strain BS 29	Cell-bound glycolipid biosurfactant and extracellular bio-emulsifier	Franzetti *et al* (2010)
9.	*Enterobacter cloacae* strain TU	Exopolysaccharide acts as bio-emulsifier	Hua *et al* (2010)
10.	*Brevibacterium aureum* MSA 13	Lipopeptide bio-emulsifier	Kiran *et al* (2010)
11.	*Microbacterium* sp. MC3B-10	Microbactan emulsified	Camacho-Chab *et al* (2013)
12.	*Variovorax paradoxus* 7bCT5	High molecular weight polysaccharides	Franzetti *et al* (2012)
13.	*Solibacillus silvestris* AM	Glycoprotein bio-emulsifier	Markande *et al* (2013)
14.	*Ochrobactrum pseudintermedium* strain C1	Exopolysaccharide bio-emulsifier	Bhattacharya *et al* (2014)
15.	Paenibacillus sp. strain	Low molecular weight oligosaccharide-lipid complex	Gudiña *et al* (2015)
16.	*Acinetobacter* sp. Ab9-ES, Ab33-ES	Glycoprotein bio-emulsifier	Adetunji and Olaniran (2019)
17	*Meyerozyma caribbica*	Proteoglycan bio-emulsifier	Bhaumik *et al* (2020)
18.	*Bacillus safensis YKS2*	Lipopeptide bio-emulsifier	Kalaimurugan *et al* (2022)

8.2.2 Dispersion

Dispersion is the property of surfactants to prevent aggregation of pollutants or particles with each other in a suspension phase. It is attained by reducing the cohesive attraction between the particles. The dispersion ability of biosurfactants has high utility and effectiveness in bioremediation of open water. The dispersion efficiency in percentage is calculated by dividing the value of total dispersed oil by the mass of oil added multiplied by one hundred. When there is an oil spill in sea water, a dispersant is used to induce oil dispersion. Further, the surfactant breaks the oil slick into microdroplets that enhance its biodegradation preventing sediment on the seashore (Marti *et al* 2014). *Rhodococcus erythropolis*, a biosurfactant hyper-producing mutant, has been widely studied for its oil dispersing efficiency (Cai *et al* 2016). *R. erythropolis* SB 1A was efficient in dispersion of crude oil from contaminated water. The study proposed *R. erythropolis* SB 1A to be an efficient oil spill response agent due to its high pollutant dispersion ability. Another study has reported the octadecane dispersion efficacy of *P. aeruginosa* rhamnolipids. The rhamnolipids enhanced the octadecane dispersion by a factor of more than 4 orders of magnitude, from 0.009 to >250 mg l^{-1} of contaminated water. However, the dispersion efficacy of rhamnolipid was identified to be dependent on pH and shaking conditions (Zhang and Miller 1992). Considering the high toxicity of chemical surfactants, development of an efficient biosurfactant that can disperse oils is of high relevance.

8.2.3 Solubilization

The solubilization efficiency is yet another important property of surfactants apart from emulsification, mobilization and dispersion. It is the property of any surface-active agents to increase the solubility of any water immiscible substance. Biosurfactant has the ability to solubilize hydrophobic organic compounds through micelle formation wherein a hydrophobic core is presented by the biosurfactant to the non-soluble compounds. Therefore, the organic molecules have a significantly stronger affinity for the surfactant-enhanced aqueous phase. The molar solubilization ratio (MSR), which is defined as the increase in solubilized HOC concentration (mol l^{-1}) per unit increase in surfactant concentration (mol l^{-1}) in the solution, indicates the solubility ability of any surfactant (Zhong *et al* 2016). The weight or molar solubilization ratio, which is the ratio of the solubilizate to the surfactant utilized, is the most widely used format for presenting solubilization efficiency. Like the other properties of biosurfactants, the solubilization ability also depends on various physicochemical parameters of both the surfactant and the environment. Some studies have also shown that the substrate in which the biosurfactant producer is grown is known to affect the solubilizing ability. Zhang and Miller showed that the structure of biosurfactant produced by the microorganism has an important role in the solubilization of hydrophobic substances (Zhang and Miller 1992). Rhamnolipid biosurfactants produced by *Rhodococcus fascians* have been identified to be efficient in solubilizing polycyclic aromatic hydrocarbons (PAHs), particularly anthracene (Kim *et al*

2019). Also, solubilization is an important factor in the bioremediation of nonaqueous phase liquids (NAPLs) (McCray *et al* 2001). This si similar for solubilization of hydrophobic organic compounds (HOCs). For the bioremediation of HOC, biosurfactants above the CMC are used. There are a few studies which have also reported solubilization of *n*-alkanes by rhamnolipids at concentrations near the CMC or sub-CMC (Zhong *et al* 2016). Biosurfactants produced by *Nocardia erythropolis* (ATTC 4277), have also shown micellar solubilization of hydrophobic substances. It was also identified that more hydrophobic biosurfactants have more solubilization ability. Biosurfactant alasan produced by *A. radioresistens* KA53 increased the solubility of PAH by 10 times. Further, some of the biosurfactants are also known to solubilize heavy metals by producing oxidation changes. Analyzing the solubilization ability of the biosurfactant is crucial for optimizing the same for wider applications.

8.2.4 Mobilization

Mobilization property of the biosurfactant is its ability to displace and disperse the contaminant. Mobilization ability of biosurfactants has been widely used for the extraction of crude oil from its reservoir. Within oil wells, crude oil is usually present in a trapped form which is hard to be recovered. For the recovery of such trapped crude oil, biosurfactant-producing microorganisms along with their growth factors/substrates or only essential elements that enhance the growth of indigenous biosurfactant-producing microorganisms are injected into the oil vents. Under suitable conditions, the microorganisms produce biosurfactants that enhance the solubility of crude oil and also produce gases that create the pressure to mobilize the crude oil from oil wells (Singh *et al* 2012, Fenibo *et al* 2019). This technique is collectively termed as microbial enhanced oil recovery (MOER). The most predominantly used microbes for MOER are *X. campestri*, *B. licheniformis*, *P. aeruginosa* (Singh *et al* 2008). The most acclaimed advantage of the MOER technology is the biodegradability of microbial surfactants used for oil mobilization from oil fields and oil spillages. This avoids further processing and harsh chemical treatments of the water from which the oil is mobilized. Using a combination of biosurfactant-producing bacteria along with oil degrading bacteria like *P. putida* has been identified to be efficient in fast and efficient removal of oil spills (Kumara *et al* 2006). High molecular weight biosurfactants that can create stable oil in water emulsions are also widely used in the petroleum industry. Microbial surfactant's efficiency to mobilize crude oil is also greatly dependent on parameters like temperature, pH, substrate and biosurfactant type.

8.3 Factors affecting the effectiveness of microbial surfactants in water remediation

As discussed in the above sections, the different properties of biosurfactants and their efficacy greatly depend on various factors like pH, temperature, shaking

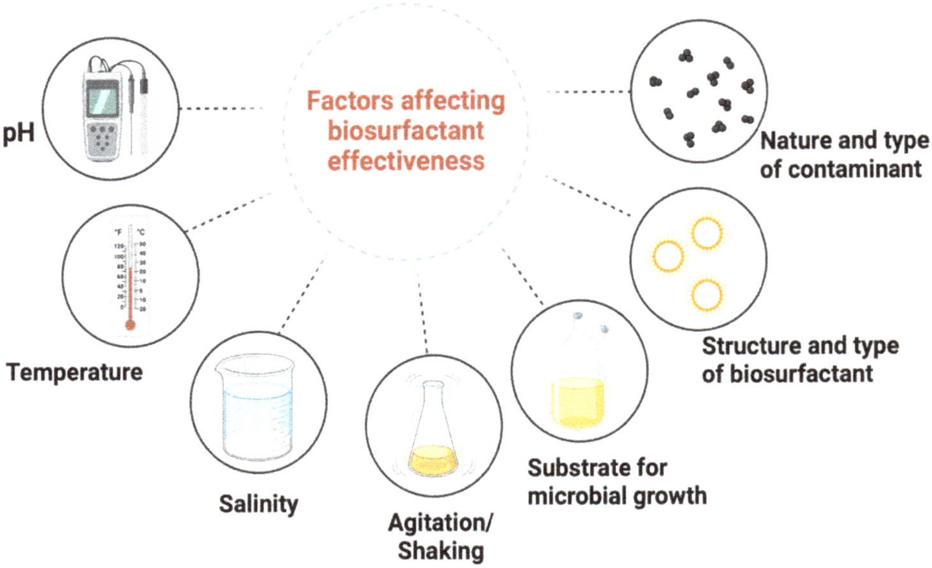

Figure 8.2. Factors affecting the efficacy of biosurfactants in water remediation.

conditions, microbe or microbial consortium used and the substrate in which they are grown. Factors that determine the effectiveness of microbial surfactants in water remediation are schematically represented in figure 8.2. Some important factors are highlighted and explained in the following sections.

8.3.1 pH

There have been various studies that have enumerated the effect of pH on microbial growth and biosurfactants. Abouseoud *et al* reported that *P. fluorescens* strain produced a biosurfactant with a low CMC efficient in the bioremediation of PAH, naphthalene from polluted water. The biosurfactant production and activity was identified to be optimum at pH 7. The pH affected the CMC of the biosurfactant and was important in enhancing the solubilization of contaminants (Abouseoud *et al* 2010). Further, Sakiyo and Németh reported the importance of pH in biosurfactant production by *B. subtilis* DSM10. The maximum amount of surfactant (5 g l^{-1}) was produced when there was no pH control in the fermentor (Sakiyo and Németh 2023). In yet another experiment by Elazzazy *et al*, the biosurfactant production and its efficacy in water remediation increased with increase in pH. At alkaline pH of 11, maximum biosurfactant production was observed after which the production ceased (Elazzazy *et al* 2015). Another bacterium, *Virgibacillus salarius*, was also identified to produce biosurfactants efficiently at higher pH of 9 (Rahimi 2022). The efficiency and stability of these biosurfactants in water bioremediation was identified to be compromised when the pH was lowered to 5. A few other strains of *Bacillus*-produced biosurfactants were capable of reducing the surface tension over a wide range of pH 2–12. *Bacillus* is itself known to thrive in harsh environmental conditions and the surfactants produced by these bacteria were stable over a wide range of temperature and pH and had high utility in water bioremediation (Bezza 2016). Hence, the water pH also greatly affects the efficiency of biosurfactant used for remediation. During emulsion formation, the interfacial film stability is highly dependent on the pH of water. The ionization of polar groups of biosurfactants is greatly influenced by the pH. Additionally, pH also affects the growth, biosurfactant production and efficiency of fungal cultures. Fungi are generally more resistant to acidic pH. Hence, fungal cultures have more utility in the bioremediation of polluted water whose pH is low (Khoramfar *et al* 2020).

8.3.2 Temperature

Another important environmental factor that has a pronounced effect on biosurfactant efficacy in water remediation is the temperature. There has been some research that has emanated from the effect of temperature on surfactant performance in both microcosm and macrocosm environments. Temperature has been found to be an important component since it influences the growth of the microbe that produces the biosurfactant as well as the aspects of the biosurfactant that is created. Wang *et al* had demonstrated that temperature affects the foaming ability and foam stability of biosurfactants (Wang *et al* 2017). In yet another study by Karnanda, the effect of temperature on the bioremediation of Saudi Arabian crude oil was elucidated (Karnanda *et al* 2013). Also, Yuan *et al* had studied the effect of temperature on emulsification by molecular dynamic simulation studies (Yuan *et al* 2003). An ideal biosurfactant should be stable at different temperatures. *Bacillus thuringiensis* and *Bacillus toyonensis* have been shown to produce biosurfactants that are stable at temperatures as high as 120 °C (Darwesh *et al* 2021). In the study, the authors had also demonstrated the ability of biosurfactants (decanoic acid and oleamide) produced from these isolates in remediating crude oil spills surrounding oil recovery sites located in water bodies (Darwesh *et al* 2021). However, the optimum temperature range for any biosurfactant activity for the degradation of hydrocarbons in marine water and freshwater environments have been identified to be 15 °C–20 °C and 20 °C–30 °C (Das and Chandran 2011). Today, increasing research has been focusing on screening biosurfactant-producing thermophiles as they have increasing applicability in mobilizing and extraction of oil from deep hydrothermal oil vents in ocean floors.

8.3.3 Salinity

Yet another indispensable factor that determines the efficacy of biosurfactant activity in open water environments is the salinity. The salinity varies between freshwater and marine water systems and hence the applicability of biosurfactants in these environments also depends on the biosurfactant stability in these varying salinity ranges. Generally, higher salinity suppresses the growth of any microorganisms indirectly, also reducing their efficacy for biosurfactant production. High salinity is identified to adversely affect the production of some key enzymes important for hydrocarbon degradation. Darvishi *et al* had demonstrated the negative impact of salinity on bioremediation of oil spills (Darvishi *et al* 2011). It is necessary to use saline tolerant microorganisms for bioremediation of such saline rich environments. In marine environments, where the salinity ranges from 33 to 37 ppt, a microbial consortium that can tolerate salinity will have higher bioremediation efficacy than halo sensitive microorganisms (Ebadi *et al* 2017). The optimum salinity for hydrocarbon bioremediation has been reported to be 5%–35% and the efficiency ceased when the salinity increased to 50% (Chicca *et al* 2022). Studies of Qin *et al* had identified a halotolerant microbial consortium effective in degrading petroleum contamination from saline rich environments (Qin *et al* 2012). Interestingly, some strains of *Achromobacter* sp, produced biosurfactants suitable for the degradation of crude oil at high saline conditions. The efficacy of biosurfactant production, emulsification and stability was identified to be more at higher salt concentrations (Deng *et al* 2020). Inherent halotolerant bacteria like *Vibrio* spp, have also been exploited to degrade diesel pollution in marine environments. *V. alginolyticus* was identified to be an efficient diesel degrader in saline conditions of 20% (Khalid *et al* 2021). Likewise, several halophilic bacteria, like *Cellulomonas* spp., *Dietzia maris, B. marisflavi*, and *Halomonas eurihalina* have been identified to efficiently degrade organic contaminants at high saline conditions.

8.3.4 Types of contaminants and their concentration

Contamination of water bodies with hydrocarbons and petroleum is common these days. However, there are several other pollutants that decrease the quality of water. Oil spills from ships transporting oil, leakage from underwater oil pipes, drilling sites and storage tanks are major sources of hydrocarbon contamination in water. As of 2022, there have been estimated oil spills of more than 700 tonnes recorded worldwide (Basile *et al* 2023). The oil spilled on water either remains on the surface as a film or sinks to the deeper waters if they are of the higher molecular fractions. Diesel spills are considered to be more toxic than oil spills as diesel comprises aliphatic, aromatic and olefinic compounds (Lominchar *et al* 2018). Another major contaminant of water bodies is the heavy metals. Industrial waste, fertilizer and animal manures, E-waste, sewage sludges, mine tailings amongst others are the major sources of heavy metal pollutants in water bodies. Heavy metal toxicity is a major problem that affects humans, animals, fish and other aquatic plants and animals alike. The concentration of heavy metal pollutants varies from place to place. Antimony, arsenic, chromium, cadmium, beryllium, copper, lead, zinc, selenium, thallium, mercury, nickel, and silver are the most common heavy metal pollutants identified in water bodies. Other major pollutants of water bodies include pesticides, plastics, radioactive waste and explosives.

8.4 Applications of microbial surfactants in water remediation

The increasing population and the emerging urbanization have become a serious environmental concern especially its impact on the water ecosystem. Various studies have been made to understand and provide novel insights into the significant and crucial roles of microbial activities in the cleaning and removal of various contaminants and one such promising method is called bioremediation, i.e. 'biological response to environmental abuse' (Sharma *et al* 2022, Kapahi *et al* 2019). The meaning of 'bio' in bioremediation refers to biological and remediation refers to removal. Hence bioremediation can clearly be defined as a process used for the elimination of environmental contaminants or organic waste or pollutants including hydrocarbons, heavy metals, pesticides, explosives, toxic materials and their derivatives into much fewer toxic substrates using different microbial flora such as bacteria, fungi, algae, actinomycetes and others.

8.4.1 Bioremediation of hydrocarbons

Bioremediation is a process and an efficient method to remove or remediate various hydrocarbon pollutants and this can be done through various microbial activities which directly or indirectly lead to the degradation of complex hydrocarbon pollutants to simpler molecules. Petroleum hydrocarbons, which is one of the major problems in aquatic ecosystems leading to several harmful consequences towards aquatic life. However, this bioremediation process transforms this toxic product into non-toxic substances such as fatty acids, carbon-dioxide, and water without creating any adverse effects on the surrounding systems, unlike different chemical treatments (Sayed *et al* 2021).

These petroleum hydrocarbons are mainly used by bacteria, yeasts, fungi and algae, and the growth rate of these microbial cells depends on the substrate availability which are mainly of three three different types, viz. (i) primary organic, (ii) secondary organic, and (iii) co-metabolism. When it comes to primary substrate, the contamination is seen as the primary source of energy that microorganisms use to replicate further; in contrast, when it comes to secondary substrate, which is broken down by enzymes and provides energy to cells, the contaminant is seen as the secondary substrate (Varjani 2017, Bhakta and Nath 2017). These enzymes produced by a variety of microbes associated with the bioremediation process in the presence of carbon facilitate the breakdown of complex hydrocarbon bonds. The third type is called co-metabolism, in which energy from the transformable molecules is used to sustain microbial growth through the oxidation of other substances. This mostly happens when the enzymes that these bacteria linked with bioremediation make can accelerate the breakdown of the growth substrate to produce energy, and the carbon that is produced from it can also degrade other substances.

The removal of non-volatile oil components from freshwater or marine environments by specific groups of bacteria, fungi, or algae present in the water gradually breaks down certain petroleum hydrocarbons through natural attenuation, which can take several days to months (Dell' Anno *et al* 2021). This bioremediation process is relatively slow. As a result, the procedure has been improved by combining various methods, including bioaugmentation and biostimulation. While the second one stimulates the growth of native bacteria by providing nutrition, electrons, and other growth co-factors, the former one increases the population of native microorganisms with the ability to degrade oil.

8.4.2 Bioremediation of heavy metals

Heavy-metals are non-biodegradable materials emerging in Nature due to rapid industrialization, urbanization, population growth and several anthropogenic activities like mining and smelting operations, use of agricultural chemicals, dumping of sewage sludge and others. The heavy metals once known to be Nature's treasure eventually became highly toxic at higher concentration in water. The major sources of heavy metal contamination in water above the permissible limits are mainly from textile mill effluent (cu), wastewater samples from the electroplating industry (Cr, Cu, Ni), irrigation water (Cd, Cr, As, Cu, Fe, Pb, Zn) (Kapahi and Sachdeva 2019). This heavy metal pollution may cause hazardous impact even at low concentration; hence it became very important to relinquish the deposition of these kinds of metals in the environment. The existing conventional techniques for alleviating heavy metals have several limitations like slow and insufficient removal of metals, high cost and low specificity, hence as an alternative method the biological reduction of heavy metals, i.e., bioremediation has gained popularity in recent times due to its cost-effective and sustainable nature.

Microbes such as bacteria, and algae play an important role in the bioremediation of heavy metals present in water, apart from other microbes and plants. These microbes convert the harmful heavy metals into less toxic products as well as aiding the mineralization of organic substances into natural gases and water. The process of microorganism-mediated remediation mainly involves in *in situ* and *ex situ* techniques, where the former involves the stimulation of indigenous microflora of the ecosystem or the introduction of bio-engineered microorganisms. However, *ex situ* techniques involve transport of the contaminated water from the site to a new place for further treatment and purification. Bioremediation of water depends on various factors like pH, temperature, microbial competition for nutrients and availability of required amounts of gases for cell growth and metabolism (Gadd 2010, Thompson *et al* 2005, Puri and Gupta 2023). Bacteria is one such important microbe to defend against metal toxicity by various mechanisms including bio-adsorption, bioaccumulation, bioprecipitation and bioleaching. Apart from bacteria, microalgae which are eukaryotic, phototrophic microorganisms that exist in aquatic environments are also an important group of microorganisms that can remediate heavy metals from wastewater. There are many reports around the world suggesting that microalgae have a potential role to treat wastewater (Agrawal *et al* 2020a, Goswami *et al* 2021a), and among them an important one is *Scenedesmus incrassatulus*, which is known to remove 25%–78% of several heavy metals like chromium, cadmium, and copper (Goswami *et al* 2022). Apart from, *S. incrassatulus*, other important microalgae including *Chlorella* spp., *Spirulina* spp., and *Nanochloropsis* are known to be involved in the heavy metal bioremediation process from different wastewater plants (Richards and Mullins 2013).

8.4.3 Bioremediation of pesticides

In the modern agricultural system, pesticides occupy a crucial position and are widely used in crop fields, veterinary farms, and other industries to enhance the yield of the products. The various classes of pesticides used in the markets include insecticides, herbicides, fungicides, nematicides, rodenticides etc, however, the inappropriate and unnecessary use significantly enhances the accumulation of these pesticides in the environment and thereafter in the water bodies. The presence of pesticides in water bodies is a major concern as it is one of the potential routes to human exposure, mainly through drinking water. The traditional ways of removal of pesticides during drinking water treatment are not very effective and the advanced techniques are expensive. Therefore, bioremediation has been proposed as a safer alternative for the remediation of pesticide-contaminated water and there are several mechanisms through which microorganisms remove pesticides, such as bio-adsorption, bioaccumulation and biodegradation (Nie *et al* 2020). Biofiltration is one such important bioremediation strategy considered for the removal of pesticides from contaminated water (Helbling 2015). This is due to the fact that the majority of pesticides found in water are mobile, and biofiltration systems enable the construction of a small bioremediation compartment through which water contaminated by pesticides can pass. Additionally, some biofiltration devices' hydrodynamics can be adjusted to allow for high pesticide and other carbon substrate loading rates.

8.4.3.1 Bacterial bioremediation of pesticides

(a) Cypermethrin, a synthetic insecticide used in a large-scale commercial agricultural application is degraded by several bacterial species like *P. aeruginosa*, *Klebsiella* spp., *Escherichia coli*, *Bacillus* spp. and *Corynebacterium* (Galadima *et al* 2021).
(b) Carbofuran, a carbamate group of pesticide widely used around the world to prevent insects in a variety of crop fields like potatoes, corn, and soyabeans is mainly degraded by *Sphingomonas* spp. and *Arthrobacter* spp. Carbaryl pesticide is degraded by many bacterial species such as *Corynebacterium* spp., *S. aureus*, *E. coli*, *Sphingomonas* and others (Mishaqa 2017).
(c) Another critical pesticide Lindane, is banned in many countries, but India is a major producer of the pesticide known to be degraded by some bacterial species of *Pseudomonas* spp., *Mycobacterium* spp., and several microbial consortia (Nie *et al* 2020, González *et al* 2012).

8.4.3.2 Role of fungi in the bioremediation of pesticides

(a) Pesticides such as triazine, phenylurea and chlorinated organophosphorous compounds are degraded by a special kind of fungi named *Auricularia auricula*. Also, white rot fungi have the ability to degrade several other kinds of pesticides such as Heptachlor atrazine, lindane, chlordane mirex, aldrin and DDT (Hai *et al* 2012).
(b) Mancozeb, a significant fungicide, has severe effects on various organ systems in humans and also causes high environmental risk as it has its own ecotoxicity and endocrine action (Axelstad *et al* 2011). This potential fungicide is remediated by two fungal species viz. *Aspergillus niger* and *A. flavus* (Vaksmaa *et al* 2023).

8.4.3.3 Potential role of actinomycetes in pesticide remediation
Several genera of Actinomycetes are known to have great bioremediating properties and have the high capability to degrade pesticides by producing extracellular enzymes and transform these xenobiotics into soluble or less harmful components. These genera include *Arthrobacter*, *Corynecaterium*, *Mycobacterium*, *Nocardia*, *Streptomyces*, *Brevibacterium* etc (Cruz Silva *et al* 2022, Mawang *et al* 2021).

8.4.4 Bioremediation of explosives

Explosives are substances or combinations of substances with a high potential energy that, when released suddenly, could produce enormous amounts of energy along with pressure, heat, light, and sound. Significant explosives, primarily common military explosives, are classified as nitro-aromatic and nitramine compounds. Examples of these compounds include TNT, RDX, and HMX. These compounds can be found in a variety of locations, including firing ranges, industrial production sites, explosive dumping grounds, and soil at destruction ranges (Weiss *et al* 2004, Alothman *et al* 2020). These explosives' transformation products are cytotoxic, have the potential to alter cells, and contaminate surrounding waterways, groundwater, and soil (EPA 2014, Mercimek *et al* 2015). Therefore, in order to prevent risks to the environment and public health, appropriate handling and disposal procedures must be implemented.

Numerous fungal species can neutralize and continue the breakdown of nitro-aromatic pollutants in addition to using cellular enzymes to break down these nitrogenous explosives. (Lewis and Aurand 1997, Spain *et al* 1995). One such group of fungi is Basidiomycetes such as *Phanerochaete chrysosporium* which are found to have the ability to mineralize TNT. Recent studies have also revealed that Trichoderma species, especially *Trichoderma viride*, have a better ability to degrade TNT and its products (Alothman *et al* 2020). Other than fungi, photosynthetic bacteria viz. *Rhodobacter sphaeroides* are also potentially known to degrade explosives *ex situ*, and a *Pseudomonas* spp JLR11 obtained from a wastewater treatment plant was found to have the ability to utilize TNT as its nitrogen source under anoxic conditions. This physiological role of bacteria raises potential interest in using them as an agent to reduce environmental pollution (Esteve-Nuñez *et al* 2000).

8.5 Advantages and limitations of using microbial surfactants in water remediation.

- There have been several advantages of using microbial surfactants in bioremediation of various contaminants from environmental sources, especially water sources. Their eco-friendly nature, biodegradability, non-toxicity, structural diversity, efficiency at lower CMC, stability over a wide range of environmental conditions such as pH, temperature and salinity have attracted the use of biosurfactants in contaminant bioremediation and oil extraction.

- Biosurfactants have high specificity for contaminants due to their molecular variability and structural diversity. This allows the selection of suitable biosurfactants for the remediation of particular contaminants.
- Microbes that produce biosurfactants are being increasingly exploited to mobilize crude oil from deep vents. Unlike the chemicals that were previously used to solubilize and displace oil from trapped rocks, biosurfactants were identified to be milder to the surrounding environment and the flora and fauna. Also, the oil did not require any post recovery treatments as the biosurfactants were self-degradable.
- Further, unlike the chemical surfactants that require intermittent application for consistent efficacy, the self-generation ability of microorganisms when suitable growth conditions and substrates are provided helps in the continuous production of biosurfactants, which aids in the uniform biodegradation of the pollutants.
- One of the major limitations of using biosurfactants for biodegradation of environmental pollutants is their limited production. Because of the high cost of manufacturing and lack of appropriate substrates for the mass production of microorganisms that produce biosurfactants, the environmental applicability of biosurfactants has not yet been accomplished. Identification of a cheap and efficient raw material for the growth of biosurfactant-producing microorganisms could help scale up industrial production of biosurfactants that could in turn advance environmental trials of bioremediation.
- Yet another limitation of using biosurfactants is that some biosurfactants have antimicrobial properties that could inhibit the growth of the natural microbial flora of the environment. This antimicrobial activity could also have a negative impact when a microbial consortium is being applied for bioremediation.
- Soil sorption of the biosurfactants used in marine environments is yet another disadvantage of using them in the field.

8.6 Conclusion

Biosurfactants are undoubtedly one of the most suitable agents for bioremediation of both organic and inorganic pollutants from water bodies. Despite the many advantages of biosurfactants over the chemical surfactants, the use of them is limited to laboratory experiments primarily due to their high production cost. Very few companies are producing biosurfactants for use. The main reason for this is the unavailability of a cheap but rich substrate for the microbial culturing. In this context, it is highly warranted that further research has to be done to develop an efficient culturing medium that supports optimum growth and biosurfactant production of diverse microorganisms. Agro-industrial waste could be utilized as a cheap and efficient source of carbon source for biosurfactant production. Further, the use of advanced technologies such as metagenomics and whole genome sequencing could be deployed for understanding the gene diversity of both culturable and non-culturable microbial communities active in the biodegradation process. Omics approaches like proteomics, metabolomics and lipidomics could also play an instrumental role in understanding the enzymes involved in hydrocarbon and contaminant degradation. By understanding the genes and their functional activity involved in biosurfactant production and bioremediation of slow growing microorganisms, recombinant DNA technology can be used to extrapolate fast growing and tolerant microbes for enhanced production of the surface-active agents. Though there are several literatures that have demonstrated screening of potential biosurfactant producers, their production efficiency, stability and utility, studies on their application in a wider environment are sparse. Careful and controlled use of biosurfactants is definitely a potential direction for enhanced and efficient cleaning of water bodies. Hence, further research on their applicability in open water bodies has to be undertaken.

Conflict of interest

The authors declare that there is no conflict of interest.

Acknowledgments

Dr Deepjyoti Paul and Dr Lekshmi Narendrakumar are thankful to the Department of Biotechnology (DBT), Government of India for the MK Bhan Fellowship. Images are created using BioRender.

References

Abouseoud M, Yataghene A, Amrane A and Maachi R 2010 Effect of pH and salinity on the emulsifying capacity and naphthalene solubility of a biosurfactant produced by pseudomonas fluorescens *J. Hazard. Mater.* **180** 131–6

Adetunji A I and Olaniran A O 2019 Production and characterization of bioemulsifiers from strains isolated from lipid-rich wastewater *3 Biotech* **9** 151

Agrawal K, Bhatt A, Bhardwaj N, Kumar B and Verma P 2020a Integrated approach for the treatment of industrial effluent by physico-chemical and microbiological process for sustainable environment *Combined Application of Physico-Chemical and Microbiological Processes for Industrial Effluent Treatment Plant* ed M P Shah, S Rodriguez-Couto and S S Sengor (Singapore: Springer) pp 119–43

Alothman Z A, Bahkali A H, Elgorban A M, Al-Otaibi M S, Ghfar A A, Gabr S A, Wabaidur S M, Habila M A and Badjah Hadj Ahmed A Y 2020 Bioremediation of explosive TNT by *Trichoderma viride Molecules* **25** 1393

Axelstad M, Boberg J, Nellemann C, Kiersgaard M, Jacobsen P R, Christiansen S *et al* 2011 Exposure to the widely used fungicide mancozeb causes thyroid hormone disruption in rat dams but no behavioral effects in the offspring *Toxicol. Sci.* **120** 439–46

Barkay T, Navon-Venezia S, Ron E Z and Rosenberg E 1999 Enhancement of solubilization and biodegradation of polyaromatic hydrocarbons by the bioemulsifier alasan *Appl. Environ. Microbiol.* **65** 2697–702

Basile A, Cassano A, Rahimpour M R and Makarem M A 2023 *Advanced Technologies in Wastewater Treatment: Oily Wastewaters* (Amsterdam: Elsevier)

Bezza F A 2016 Biosurfactant assisted bioremediation of petroleum and polycyclic aromatic hydrocarbons in aquatic and soil media *PhD Thesis* University of Pretoria

Bhakta and Nath J 2017 *Handbook of Research on Inventive Bioremediation Techniques* (IGI Global)

Bhattacharya M, Biswas D, Sana S and Datta S 2014 Utilization of waste engine oil by Ochrobactrum pseudintermediumstrain C1 that secretes an exopolysaccharide as a bio-emulsifier *Biocatal. Agric. Biotechnol.* **3** 167–76

Bhattacharya S, Gupta A B, Gupta A and Pandey A (ed) 2018 Water remediation *Energy, Environment, and Sustainability* 1st edn (Singapore: Springer)

Bhaumik M, Dhanarajan G, Chopra J, Kumar R, Hazra C and Sen R 2020 Production, partial purification and characterization of a proteoglycan bioemulsifier from an oleaginous yeast *Bioprocess. Biosyst. Eng.* **43** 1747–59

Cai Q, Zhang B, Chen B, Cao T and Lv Z 2016 Biosurfactant produced by a rhodococcus erythropolis mutant as an oil spill response agent *Water Qual. Res. J. Can.* **51** 97–105

Camacho-Chab J C, Guézennec J, Chan-Bacab M J, Ríos-Leal E, Sinquin C, Muñiz-Salazar R, De la Rosa-García S del C, Reyes-Estebanez M and Ortega-Morales B O 2013 Emulsifying activity and stability of a non-toxic bioemulsifier synthesized by Microbacterium sp. MC3B-10 *Int. J. Mol. Sci.* **14** 18959–72

Cameron D R, Cooper D G and Neufeld R J 1988 The mannoprotein of accharomyces cerevisiae is an effective bioemulsifier *Appl. Environ. Microbiol.* **54** 1420–5

Chicca I, Becarelli S and Di Gregorio S 2022 Microbial involvement in the bioremediation of total petroleum hydrocarbon polluted soils: challenges and perspectives *Environments* **9** 52

Ciapina E M P, Melo W C, Santa Anna L M M, Santos A S, Freire D M G and Pereira Jr N 2006 Biosurfactant production by rhodococcus erythropolis grown on glycerol as sole carbon source *Appl. Biochem. Biotechnol.* **131** 880–86

Cirigliano M C and Carman G M 1984 Purification and characterization of liposan, a bioemulsifier from candida lipolytica *Appl. Environ. Microbiol.* **50** 846–50

Darvishi P, Mowla D, Ayatollahi S and Niazi A 2011 Biodegradation of heavy crude oil in wastewater by an efficient strain, ERCPPI-1 *Desalin. Water Treat.* **28** 46–54

Darwesh O M, Mahmoud M S, Barakat K M, Abuellil A and Ahmad M E 2021 Improving the bioremediation technology of contaminated wastewater using biosurfactants produced by novel bacillus isolates *Heliyon* **7** e08616

Das N and Chandran P 2011 Microbial degradation of petroleum hydrocarbon contaminants: an overview *Biotechnol. Res. Int.* **2011** 941810

de Oliveira D W F, Lima França I W, Félix A K N, Martins J J L, Giro M E A, Melo V M M and Gonçalves L R B 2013 Kinetic study of biosurfactant production by Bacillus subtilis LAMI005 grown in clarified cashew apple juice *Colloids Surf.* B **101** 34–43

Dell' Anno F, Rastelli E, Sansone C, Brunet C, Ianora A and Dell' Anno A 2021 Bacteria, fungi and microalgae for the bioremediation of marine sediments contaminated by petroleum hydrocarbons in the Omics Era *Microorganisms* **9** 1695

Deng Z, Jiang Y, Chen K, Li J, Zheng C, Gao F and Liu X 2020 One biosurfactant-producing bacteria sp. a-8 and its potential use in microbial enhanced oil recovery and bioremediation *Front. Microbiol.* **11** 247

Desai J D and Banat I M 1997 Microbial production of surfactants and their commercial potential *Microbiol. Mol. Biol. Rev.: MMBR* **61** 47–64

Dolan F, Lamontagne J, Link R *et al* 2021 Evaluating the economic impact of water scarcity in a changing world *Nat. Commun.* **12** 1915

Ebadi A, Sima N A K, Olamaee M, Hashemi M and Nasrabadi R G 2017 Effective bioremediation of a petroleum-polluted saline soil by a surfactant-producing consortium *J. Advert. Res.* **8** 627–33

Elazzazy A M, Abdelmoneim T S and Almaghrabi O A 2015 Isolation and characterization of biosurfactant production under extreme environmental conditions by alkali-halo-thermophilic bacteria from Saudi Arabia *Saudi J. Biol. Sci.* **22** 466–75

EPA 2014 Technical fact sheet-2,4,6-trinitrotoluene (TNT) *Office of Solid Waste and Emergency Response (5106P)* (Washington, DC: United States Environmental Protection Agency)

Esteve-Nuñez A, Lucchesi G, Philipp B, Schink B and Ramos J L 2000 Respiration of 2,4,6-trinitrotoluene by Pseudomonas sp. strain JLR11 *J. Bacteriol.* **182** 1352–5

Fenibo E O, Ijoma G N, Selvarajan R and Chikere C B 2019 Microbial surfactants: the next generation multifunctional biomolecules for applications in the petroleum industry and its associated environmental remediation *Microorganisms* **7** 581

Franzetti A, Gandolfi I, Bestetti G, Smyth T J and Banat I M 2010 Production and applications of trehalose lipid biosurfactants *Eur. J. Lipid. Sci. Tech.* **112** 617–27

Franzetti A, Gandolfi I, Raimondi C, Bestetti G, Banat I M, Smyth T J, Papacchini M, Cavallo M and Fracchia L 2012 Environmental fate, toxicity, characteristics and potential applications of novel bioemulsifiers produced by variovorax paradoxus 7bCT5 *Bioresour. Technol.* **108** 245–51

Freitas F, Alves V, Carvalheira M, Costa N, Oliveira R and Reis M 2009 Emulsifying behavior and rheological properties of the extracellular polysaccharide produced by Pseudomonas oleovorans grown on glycerol byproduct *Carbohydr. Polym.* **78** 549–56

Gadd G M 2010 Metals, minerals and microbes: geomicrobiology and bioremediation *Microbiology* **156** 609–43

Galadima M, Singh S, Pawar A, Khasnabis S, Dhanjal D S, Anil A G *et al* 2021 Toxicity, microbial degradation and analytical detection of pyrethroids: a review *Environ. Adv.* **5** 100105

Gaur S, Gupta S and Jain A 2023 Production, characterization, and kinetic modeling of biosurfactant synthesis by Pseudomonas aeruginosa Gi |KP 163922|: a mechanism perspective *World J. Microbiol. Biotechnol.* **39** 178

González R, García-Balboa C, Rouco M, Lopez-Rodas V and Costas E 2012 Adaptation of microalgae to lindane: a new approach for bioremediation *Aquat. Toxicol.* **109** 25–32

Goswami R K, Agrawal K, Shah M P and Verma P 2022 Bioremediation of heavy metals from wastewater: a current perspective on microalgae-based future *Lett. Appl. Microbiol.* **75** 701–17

Goswami R K, Agrawal K and Verma P 2021a An overview of microalgal carotenoids: advances in the production and its impact on sustainable development *Bioenergy Research: Evaluating Strategies for Commercialization and Sustainability* ed N Srivastava and M Srivastava (Hoboken, NJ: Wiley) pp 105–28

Gudiña E J, Pereira J F B, Costa R, Evtuguin D V, Coutinho J A P, Teixeira J A and Rodrigues L R 2015 Novel bioemulsifier produced by a paenibacillus strain isolated from crude oil *Microbial Cell Fact.* **14** 14

Haba E, Espuny M J, Busquets M and Manresa A 2000 Screening and production of rhamnolipids by Pseudomonas aeruginosa 47T2 NCIB 40044 from waste frying oils *J. Appl. Microbiol.* **88** 379–87

Hai F I, Modin O, Yamamoto K, Fukushi K, Nakajima F and Nghiem L D 2012 Pesticide removal by a mixed culture of bacteria and white-rot fungi *J. Taiwan Inst. Chem. Eng.* **43** 459–62

Helbling D E 2015 Bioremediation of pesticide-contaminated water resources: the challenge of low concentrations *Curr. Opin. Biotechnol.* **33** 142–48

Hua X, Wu Z, Zhang H, Lu D, Wang M, Liu Y and Liu Z 2010 Degradation of hexadecane by enterobacter cloacae strain tu that secretes an exopolysaccharide as a bioemulsifier *Chemosphere* **80** 951–56

Inamuddin M I A and Prasad R 2021 *Microbial Biosurfactants: Preparation, Properties and Applications* (Springer Nature)

Kalaimurugan D, Balamuralikrishnan B, Govindarajan R K, Al-Dhabi N A, Valan Arasu M, Vadivalagan C, Venkatesan S, Kamyab H, Chelliapan S and Khanongnuch C 2022 Production and characterization of a novel biosurfactant molecule from bacillus safensis YKS2 and assessment of its efficiencies in wastewater treatment by a directed metagenomic approach *Sustainability* **14** 2142

Kapahi M and Sachdeva S 2019 Bioremediation options for heavy metal pollution *J. Health Pollut.* 27**9** 191203

Karnanda W, Benzagouta M S, AlQuraishi A and Amro M M 2013 Effect of temperature, pressure, salinity, and surfactant concentration on ift for surfactant flooding optimization *Arabian J. Geosci.* **6** 3535–44

Khalid F E, Lim Z S, Sabri S, Gomez-Fuentes C, Zulkharnain A and Ahmad S A 2021 Bioremediation of diesel contaminated marine water by bacteria: a review and bibliometric analysis *J. Marine Sci. Eng.* **9** 155

Khoramfar S, Jones K D, Ghobadi J and Taheri P 2020 Effect of surfactants at natural and acidic pH on microbial activity and biodegradation of mixture of benzene and O-xylene *Chemosphere* **260** 127471

Kim C-H, Lee D W, Heo Y M, Lee H, Yoo Y, Kim G-H and Kim J-J 2019 Desorption and solubilization of anthracene by a rhamnolipid biosurfactant from rhodococcus fascians *Water Environ. Res.: Res. Publ. Water Environ. Fed.* **91** 739–47

Kiran G S, Thomas T A, Selvin J, Sabarathnam B and Lipton A P 2010 Optimization and characterization of a new lipopeptide biosurfactant produced by marine Brevibacterium aureum MSA 13 in solid state culture *Bioresour. Technol.* **101** 2389–96

Klobucista C 2021 Water stress: a global problem that's getting worse *Council on Foreign Relations.* 21 https://cfr.org/backgrounder/water-stress-global-problem-thats-getting-worse

Kumar P and Dubey R C 2023 *Multifunctional Microbial Biosurfactants* (Springer Nature)

Kumara M, Leon V, Materano A D S, Ilzins O A, Galindo-Castro I and Fuenmayor S L 2006 Polycyclic aromatic hydrocarbon degradation by biosurfactant-producing Pseudomonas sp. IR1 *' Z. Naturforsch. C J. Biosci.* **61** 203–12

Lai C-C, Huang Y-C, Wei Y-H and Chang J-S 2009 Biosurfactant-enhanced removal of total petroleum hydrocarbons from contaminated soil *J. Hazard. Mater.* **167** 609–14

Lewis A and Aurand D 1997 Putting dispersants to work: Overcoming obstacles *Proc. 1997 Oil Spill Conference* (Washington DC: American Petroleum Institute) pp 157–164

Lominchar M A, Santos A, de Miguel E and Romero A 2018 Remediation of aged diesel contaminated soil by alkaline activated persulfate *Sci. Total Environ.* **622–3** 41–8

Markande A R, Acharya S R and Nerurkar A S 2013 Physicochemical characterization of a thermostable glycoprotein bioemulsifier from Solibacillus silvestris AM1 *Proc. Biochem.* **48** 1800–8

Marti M E, Colonna W J, Patra P, Zhang H, Green C, Reznik G, Pynn M *et al* 2014 Production and characterization of microbial biosurfactants for potential use in oil-spill remediation *Enzyme Microb. Technol.* **55** 31–9

Martínez-Checa F, Toledo F L, El Mabrouki K, Quesada E and Calvo C 2007 Characteristics of bioemulsifier V2-7 synthesized in culture media added of hydrocarbons: chemical composition, emulsifying activity and rheological properties *Bioresour. Technol.* **98** 3130–35

Mawang C I, Azman A S, Fuad A M and Ahamad M 2021 Actinobacteria: an eco-friendly and promising technology for the bioaugmentation of contaminants *Biotechnol. Rep. (Amst)* **32** e00679

McCray J E, Bai G, Maier R M and Brusseau M L 2001 Biosurfactant-enhanced solubilization of NAPL mixtures *J. Contam. Hydrol.* **48** 45–68

Mercimek H A, Dincer S, Guzeldag G, Ozsavli A, Matyar F, Arkut A, Kayis F and Ozdenefe M S 2015 Degradation of 2,4,6-trinitrotoluene by P. aeruginosa and characterization of some metabolites *Braz. J. Microbiol.* **46** 103–11

Mishaqa E S I 2017 Biosorption potential of the microchlorophyte Chlorella vulgaris for some pesticides *J. Fertilizers Pesticides* **8** 5

Nie J, Sun Y, Zhou Y, Kumar M, Usman M, Li J, Shao J, Wang L and Tsang. D C W 2020 Bioremediation of water containing pesticides by microalgae: mechanisms, methods, and prospects for future research *Sci. Total Environ.* **707** 136080

Pardhi D S, Panchal R R, Raval V H, Joshi R G, Poczai P, Almalki W H and Rajput K N 2022 Microbial surfactants: a journey from fundamentals to recent advances *Front. Microbiol.* **13** 982603

Peele K A, Ch V R T and Kodali V P 2016 Emulsifying activity of a biosurfactant produced by a marine bacterium *3 Biotech.* **6** 177

Ponte Rocha M V, Gomes Barreto R V, Melo V M M and Rocha Barros Gonçalves L 2009 Evaluation of cashew apple juice for surfactin production by Bacillus subtilis LAMI008 *Appl. Biochem. Biotechnol.* **155** 366–78

Puri N and Gupta A 2023 Water remediation using titanium and zinc oxide nanomaterials through disinfection and photo catalysis process: a review *Environ. Res.* **227** 115786

Qin X, Tang J C, Li D S and Zhang Q M 2012 Effect of salinity on the bioremediation of petroleum hydrocarbons in a saline-alkaline soil *Lett. Appl. Microbiol.* **55** 210–7

Rahimi M 2022 Discovery of new oil-degrading bacteria with biosurfactant production ability from oily tailings pond waste, refinery-contaminated soil, light and heavy crude oils for remediation of crude oil in water *MSc Thesis* Concordia University

Reddy M S, Naresh B, Leela T, Prashanthi M, Ch Madhusudhan N, Dhanasri G and Devi P 2010 Biodegradation of phenanthrene with biosurfactant production by a new strain of Brevibacillus sp *Bioresour. Technol.* **101** 7980–3

Richards R G and Mullins B J 2013 Using microalgae for combined lipid production and heavy metal removal from leachate *Ecol. Modell.* **249** 59–67

Sakiyo J J and Németh Á 2023 The potential of Bacilli-derived biosurfactants as an additive for biocontrol against plant pathogenic fungi *Microorganisms* **11** 707

Sayed K, Baloo L and Sharma N K 2021 Bioremediation of total petroleum hydrocarbons (TPH) by bioaugmentation and biostimulation in water with floating oil spill containment booms as bioreactor basin *Int. J. Environ. Res. Public Health* **18** 2226

Sharma P, Bano A, Surendra Pratap Singh S P, Nawal Kishore Dubey N K, Ram Chandra R and Hafiz Iqbal H M 2022 Recent advancements in microbial-assisted remediation strategies for toxic contaminants *Clean. Chem. Eng.* **2** 100020

Shiomi N 2015 *Advances in Bioremediation of Wastewater and Polluted Soil* (Books on Demand)

Silva G C, Kitano I T, Ribeiro I A F and Lacava P T 2022 The potential use of actinomycetes as microbial inoculants and biopesticides in agriculture *Front. Soil Sci.* **2** 833181

Singh A, Singh B and Ward O 2012 Potential applications of bioprocess technology in petroleum industry *Biodegradation* **23** 865–80

Singh S, Kang S H, Mulchandani A and Chen W 2008 Bioremediation: environmental clean-up through pathway engineering *Curr. Opin. Biotechnol.* **19** 437–44

Spain J C 1995 Biodegradation of nitroaromatic compounds *Annu Rev Microbiol* **49** 523–55

Tadros T F 2013 *Emulsion Formation and Stability* (New York: Wiley)

Thompson I P, van der Gast C J, Ciric L and Singer A C 2005 Bioaugmentation for Bioremediation: The Bioremediation Options for Heavy Metal Pollution *Environ. Microbiol.* **7** 909–15

Tripathy D B and Mishra A 2016 Sustainable biosurfactants *Sustain. Inorg. Chem.* **1** 175–92

Vaksmaa A, Guerrero-Cruz S, Ghosh P, Zeghal E, Hernando-Morales V and Niemann H 2023 Role of fungi in bioremediation of emerging pollutants *Front. Mar. Sci.* **10** 1070905

Varjani S J 2017 Microbial degradation of petroleum hydrocarbons *Bioresour. Technol.* **223** 277–86

Wang H, Guo W, Zheng C, Wang D and Zhan H 2017 Effect of temperature on foaming ability and foam stability of typical surfactants used for foaming agent *J. Surfactants Deterg.* **20** 615–22

Weiss M, Geyer R, Russow R, Richnow H H and Kästner M 2004 Fate and metabolism of [15N] 2,4,6-trinitrotoluene in soil *Environ. Toxicol. Chem. /SETAC* **23** 1852–60

Yuan Y-Q, Zou X-W and Xiong P-F 2003 Effects of temperature on the emulsification in surfactant-water-oil systems *Int. J. Mod. Phys.* B **17** 2773–80

Zhang Y and Miller R M 1992 Enhanced octadecane dispersion and biodegradation by a Pseudomonas rhamnolipid surfactant (biosurfactant) *Appl. Environ. Microbiol.* **58** 3276–82

Zhong H, Yang X, Tan F, Brusseau M L, Yang L, Liu Z, Zeng G and Yuan X 2016 Aggregate-based sub-CMC solubilization of -alkanes by monorhamnolipid biosurfactant *New J. Chem.* **40** 2028–35

Chapter 9

Microbial surfactants in food, pharmaceuticals, and agriculture

Vivek Kumar Yadav, Anuradha Singh and Pradeep Kumar Singh

The use of microbial surfactants has become a revolution in different industries and this chapter presents an in-depth study of the various uses of microbial surfactants in food, pharma and agriculture. Microbial surfactants derived from various microbial sources have unique properties that distinguish them from their biological counterparts, paving the way for effective and environmental solutions. In the food industry, microbial surfactants play an important role in improving the emulsification, stabilization and foaming properties of foods. Their natural origin and biodegradability fit perfectly with the food industry's growing demand for green alternatives. This chapter explores how microbial surfactants can contribute to the development of new formulas that improve texture, shelf life and overall food quality. Furthermore, in the pharmaceutical industry, microbial surfactants have shown great potential in drug delivery and technology development. Their biocompatibility, low toxicity, and ability to increase solubility make them valuable assets for drug research and development. Besides, agriculture has experienced a revolution with the incorporation of microbial surfactants into agricultural chemical formulations. These surfactants effectively improve the spread and adhesion of pesticides and reduce environmental impact while ensuring crop protection.

Additionally, this chapter explores the challenges and future prospects associated with the widespread use of microbial surfactants in these industries. As microbial surfactants continue to be recognized as having a variety of solutions and benefits, this chapter provides valuable resources to help researchers, clinicians, and industry professionals to utilize the biosurfactants in industries like food, medicine, and agriculture.

9.1 Introduction: microbial surfactants in food, pharmaceuticals, and agriculture

Microbial surfactants are remarkable natural compounds produced by microorganisms like bacteria and fungi, and they are gaining significant attention in various

doi:10.1088/978-0-7503-5989-4ch9

industries, including food, pharmaceuticals, and agriculture [1]. These versatile molecules possess unique properties that make them valuable in enhancing the quality and sustainability of products and processes across these sectors.

In food, microbial surfactants have carved a niche for themselves as valuable additives. They improve emulsification, foaming, and texture in a wide range of food products [2]. For example, in producing dairy goods like ice cream, microbial surfactants can help create a smoother texture and prevent the formation of ice crystals, ultimately enhancing the overall sensory experience for consumers. Similarly, they are employed in the baking industry to improve dough consistency and the volume of baked goods like bread [3]. In essence, microbial surfactants contribute to the creation of better-tasting and more visually appealing food products.

In the pharmaceutical industry, microbial surfactants have found intriguing applications, particularly in drug formulation and delivery systems [4]. These compounds can improve the solubility and stability of poorly water-soluble drugs, which can be a significant challenge in pharmaceutical development [5]. By incorporating microbial surfactants into drug formulations, pharmaceutical companies can enhance drug bioavailability, allowing patients to benefit more effectively from medications. Additionally, these surfactants play a crucial role in designing drug delivery systems, such as micelles and nanoparticles, which enable targeted drug release, reducing side effects and improving therapeutic outcomes [6].

Agriculture, too, has seen the emergence of microbial surfactants as eco-friendly alternatives to traditional chemical pesticides [7]. The harmful environmental and health effects associated with chemical pesticides have led to a growing interest in sustainable farming practices. Microbial surfactants offer a promising solution. They can be used as biopesticides, disrupting the cell membranes of plant pathogens, such as fungi and bacteria, effectively controlling crop diseases [7]. Moreover, microbial surfactants can enhance soil health by promoting beneficial microbial communities and improving soil structure, increasing crop yields and sustainable agriculture practices. [8]

Beyond their specific applications, microbial surfactants align with the broader global drive toward sustainability and environmental responsibility [9]. Unlike many synthetic surfactants derived from petrochemicals, microbial surfactants are biodegradable and less environmentally harmful. This makes them an appealing choice for industries striving to reduce their ecological footprint and meet stringent regulatory standards. By adopting microbial surfactants, companies can demonstrate their commitment to sustainable practices and contribute to a cleaner and greener future. In summary, biosurfactants are versatile and environmentally friendly compounds with many applications in the food, pharmaceutical, and agricultural sectors. Their ability to enhance product quality, improve drug delivery, and provide sustainable alternatives in agriculture underscores their importance in modern industries. As we continue to explore and harness the potential of these natural compounds, we move one step closer to a more sustainable and responsible future in food production, healthcare, and farming.

9.2 Microbial surfactants in food

Surfactants have been used in the food industry for centuries to improve the texture, stability, and flavour of food products. Natural surfactants, such as lecithin from egg yolk and proteins from milk, have been used in the preparation of mayonnaise, salad creams, dressings, and desserts. Later, polar lipids, such as monoglycerides, were introduced as emulsifiers for food products. In recent years the use of biosurfactants in the food industry has gained significant attention as the demand for sustainable and environmentally friendly solutions continues to grow. They have a wide range of applications in the food industry. They can be used to improve the viscosity, texture, and stability of food products, as well as to inhibit the growth of harmful microorganisms. Biosurfactants are non-toxic and biodegradable, making them a safe and sustainable alternative to synthetic food additives [2]. They also have antimicrobial, antiadhesive, and antioxidant properties, which can further improve the safety and quality of food products [10]. As a result of their unique properties, biosurfactants are gaining increasing attention as potential ingredients for a variety of food products, including ice cream, baked goods, dairy products, and meats.

9.2.1 Types of microbial surfactants used in the food industry

In the food industry, various types of microbial surfactants play pivotal roles in improving food product quality and processing. These natural compounds, primarily produced by microorganisms like bacteria and fungi, can be categorized into several classes based on their chemical structures and properties. Rosenberg and Ron [11] categorized them into two groups: one comprising high molecular weight molecules (polymeric and particulate surfactants) and the other consisting of low molecular weight surfactants, which include glycolipids, lipopeptides, and phospholipids [11].

9.2.1.1 Glycolipids

Glycolipids are the most common biosurfactants, characterized by their structural composition of a fatty acid linked to a carbohydrate component. This group of biosurfactants varies based on the specific types of lipids and carbohydrates. Depending on the particular carbohydrate component, glycolipids can be further categorized into various subtypes, including rhamnose lipids, trehalose lipids, sophorose lipids, cellobiose lipids, mannosyl erythritol lipids, lipomannosylmannitols, lipomannans and lipoarabinomannanes, diglycosyl diglycerides, monoacylglycerol, and galactosyl-diglyceride [12].

9.2.1.2 Rhamnolipids

Rhamnolipids are the most studied type of biosurfactant made up of one or two molecules of rhamnose bound to one or two molecules of hydroxydecanoic acid. Rhamnolipids was fist isolated in 1949 from *Pseudomonas aeruginosa* [13].

9.2.1.3 Trehalose lipids

Trehalose is another group of glycolipids that consists of a sugar called trehalose linked to long-chain fatty acids. These compounds are commonly produced in many species of *Mycobacterium*, *Nocardia*, and *Corynebacterium*. These microorganisms are known for synthesizing trehalose lipids with diverse structures and sizes. These lipids contain a glycosidic linkage with the reducing end of the rhamnose disaccharide, and the hydroxyl (OH) group of the second hydroxydecanoic acid is implicated in the ester formation process [14]. Trehalose lipids from *Rhodococcus erythropolis* and *Arthrobacter* spp. have been observed to reduce surface and interfacial tension in culture broth by 25–40 mNm and 1–5 mNm, respectively. Production of rhamnose-containing glycolipids was first described in *P. aeruginosa* by Jarvis and Johnson [15].

9.2.1.4 Sophorolipids

Sophorolipids are a type of glycolipids mainly produced by yeast. Sophorolipids are composed of a dimeric carbohydrate known as sophorose, linked to hydroxyl fatty acid through a glycosidic bond. Typically, sophorolipids consist of a mixture of at least six to nine distinct hydrophobic sophorolipids

These glycolipids are valued for their remarkable emulsifying properties, making them ideal for stabilizing emulsions in salad dressings, mayonnaise, and sauces. Sophorolipids, produced by non-pathogenic yeasts, have attracted a lot of interest in recent years due to their unique environmental-friendly properties [16]. The sophorolipid market is expanding as a result of rising demand for sophorolipids across a number of end-use industries, including the food and health sectors.

9.2.1.5 Lipopeptides and lipoproteins

Lipopeptides and lipoproteins are compounds comprising a lipid linked to a polypeptide chain [11]. Many biosurfactants of the lipopeptides category such as surfactin have demonstrated antimicrobial effects against various microorganisms, including bacteria, algae, fungi, and viruses [17]. Lipopeptide biosurfactants formed by *Bacillus subtilis* strains, such as surfactin, fengycin, and mycosubtilin, have been shown to have antimicrobial activity against foodborne pathogens and spoilage microorganisms, even at different purities [18, 19]. Notably, iturin from *B. subtilis* retained its activity even after processes like autoclaving, exposure to a pH range of 5–11, and had a shelf life of 6 months at −18 °C. Furthermore, lipoproteins, such as lipoprotein B, contribute significantly to enhancing the foaming properties of dough in baking. This results in bread with improved texture and greater volume, meeting consumer expectations for quality and taste [20].

The wide variety of microbial surfactants available in the food industry highlights their adaptability to meet specific product needs, all within the context of the industry's increasing focus on sustainability and eco-friendly practices. As ongoing research reveals novel microbial surfactants and their uses, the possibilities for innovation in food processing and product development continue to hold great promise.

9.3 Applications of microbial surfactants in food processing and preservation

Microbial surfactants have discovered a wide array of uses within the domain of food processing and preservation, making significant contributions to the improvement of food quality, safety, and sustainability. These naturally occurring compounds, produced by microorganisms, possess distinctive characteristics that render them invaluable across various facets of the food industry. One prominent application involves their capacity as emulsifiers and stabilizers [2]. Notably, microbial surfactants such as rhamnolipids and sophorolipids excel in stabilizing emulsions, facilitating consistent blending of ingredients in products like salad dressings and sauces [12]. Furthermore, they enhance the foaming properties in baked goods such as bread, resulting in improved texture and volume. In addition to these advantages, specific microbial surfactants also exhibit antimicrobial properties, which assist in preserving food by inhibiting the growth of spoilage microorganisms and foodborne pathogens [13]. This action effectively extends the shelf life of perishable items. Moreover, the biodegradable and environmentally friendly characteristics of microbial surfactants align with the food industry's increasing dedication to sustainable practices

9.3.1 Microbial surfactants as emulsifiers, foaming agents, and antiadhesive agents in food industry

Food additives are substances designed to enhance the overall characteristics of food products. Throughout history, certain food additives like, sugar, salt and SO_2 have been employed to preserve various food items such as meats, fish, and beverages. In recent times, the evolution of food additives has taken them from household kitchens to large-scale commercial usage. These additives can be sourced from natural ingredients or created through chemical synthesis and are incorporated into food products to provide specific technological benefits. It is crucial that food additives are used solely to protect nutritional value of food without inducing any detrimental consequences. In order to categorize these additives based on their functions, the Food and Agriculture Organization (FAO) and World Health Organization (WHO) have categorized them into three groups: (1) enzyme preparations, (2) flavouring agents, and (3) other additives.

Microbial surfactants have emerged as versatile and eco-friendly additives in the food industry, serving crucial functions as emulsifiers, foaming agents, and antiadhesive agents. These natural compounds, produced by microorganisms, play a significant role in improving food product quality and processing. They also being used for encapsulation of fat-soluble compounds known as direct food additives, i.e. vitamins. Additionally, compounds known as antimicrobial, antiadhesive are categorized as 'indirect food additives.'

Microbial biosurfactants exhibit good antiadhesive and antimicrobial properties against numerous pathogens, as summarized in table 9.1. These biosurfactants interact with porins, which are proteins spanning cell membranes, potentially leading to the release of cellular contents and cell death [25]. However, before incorporating biosurfactants into food processing, they must undergo rigorous toxicological assessments to evaluate their compatibility with food compounds, to asses daily intake limits, and determine the potential protective effects.

Table 9.1. Biosurfactants used in the food industry.

Biosurfactant	Organism	Use in food industry	Benefits	References
Surfactin	*Bacillus subtilis*	Emulsifier, foaming agent, antimicrobial agent	Improves the texture and stability of food products, reduces the use of synthetic surfactants, enhances the flavour and aroma of food	[21]
Rhamnolipid	*Pseudomonas aeruginosa*	Emulsifier, foaming agent, detergent, antimicrobial agent	Improves the solubility of fats and oils in water, reduces the surface tension of liquids, removes dirt and grime from food processing equipment, inhibits the growth of foodborne pathogens	[13, 22]
Trehalolipid	*Rhodococcus erythropolis*	Emulsifier, foaming agent, stabilizer, antimicrobial agent	Stabilizes food emulsions and prevents them from breaking down, improves the shelf life of food products, has a mild taste and odour	[23]
Sophorolipid	*Candida bombicola*	Emulsifier, foaming agent, detergent, antimicrobial agent	Enhances the texture and flavour of food, reduces the use of artificial additives, has a broad spectrum of antimicrobial activity	[16]
Lipopeptide and lipoprotein	*Bacillus subtilis*	Emulsifier, foaming agent, antimicrobial agent	Improves the digestibility of food, has a low toxicity and environmental impact, can be used to develop biodegradable food packaging materials	[17, 18, 24]

As emulsifiers, surfactants like rhamnolipids and sophorolipids enable the stable blending of ingredients that would typically separate, enhancing the texture and appearance of products like salad dressings and sauces. They also contribute to the creation of smooth and stable foams in various food items, from frothy cappuccinos to fluffy bread [2].

Additionally, it has been reported that *Nesterenkonia* species biosurfactants enhance the texture of muffins [26]. *Nesterenkonia* sp. falls within the genus *Micrococcus* and is characterized as obligate aerobes that thrive within the temperature range of 25 °C–37 °C. Phylogenetic and chemotaxonomic analysis of these isolates revealed their relationship with other genera, including *Kocuria*, *Kytococcus*,

Dermacoccus, and *Nesterenkonia* [26]. Addition of *Nesterenkonia* sp. into muffin production offers various advantages, such as reducing hardness, chewiness, and gumminess compared to control treatments [27]. Figure 9.1 depicts and table 9.1 describes the functions of microbial biosurfactants/bioemulsifiers (e.g., rhamnolipids, lipopeptides, surfactin, emulsan and glycolipids), as emulsifying agents, bakery additives, bread improvers and taste enhancers.

Furthermore, mayonnaise formulations that included *Candida utilis*-derived bioemulsifiers as an essential component improve emulsion stability during storage. Incorporation of glycerol monostearate, a chemical emulsifier, at a concentration of 0.1% along with bioemulsifiers significantly improved dough properties and volume [28].

Surfactants derived from *B. subtilis* were also reported to improve the structural qualities of dough and the textural quality of cookies [29]. Similarly, Mnif *et al* [30]

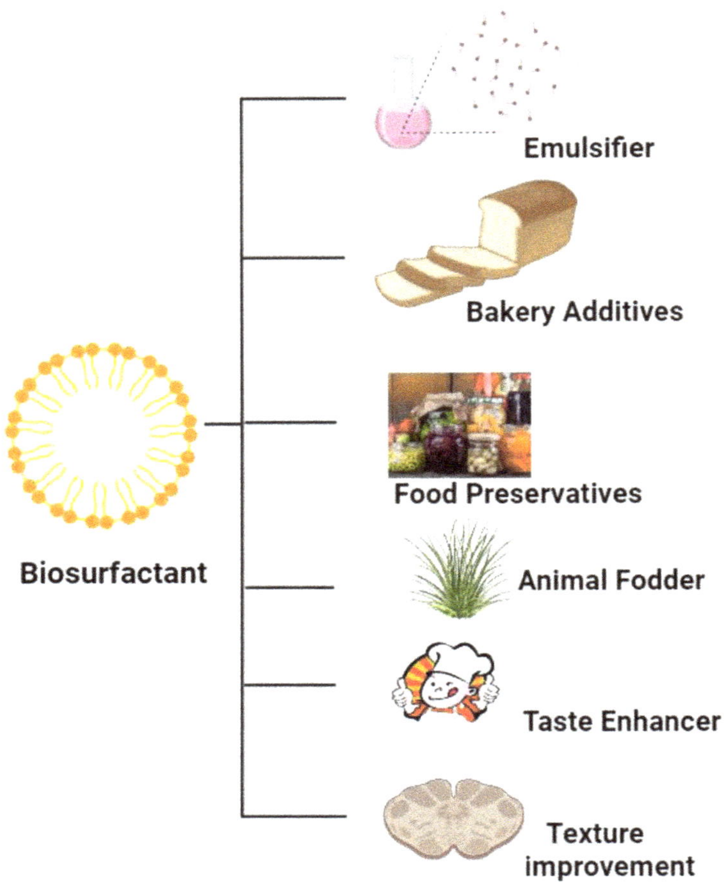

Figure 9.1. Application of biosurfactants in the food industry.

reported an increase in bread dough quality by using a *B. subtilis*-derived biosurfactant at a concentration of 0.075% (w/w) versus soya lecithin. These benefits also extended to the structural qualities of bread, resulting in improvements in chewiness, cohesiveness, and a reduction in stiffness.

Additionally, The inclusion of biosurfactants in cakes replaces 100% of vegetable oils and increases nutritional value by reducing fatty acids [31]. Biosurfactants have been used to improve the qualities of animal feed supplementing rapeseed meal with 'generally recognized as safe' (GRAS) microorganisms. It has long been known that surfactants promote the growth of chickens, and non-ionic surfactants can improve an animal's body weight, milk production and feed digestion. Fortification of rapeseed meal with GRAS-generated biosurfactants may provide probiotic benefits and serve as an alternative to antibiotics in animal feeds [32].

The emulsifying capacity of biosurfactant/bioemulsifier molecules plays an important role in the food industry and can be predicted by a detailed knowledge of their hydrophile–lipophile balance (HLB). This HLB value determines whether they are suitable for water-in-oil (W/O) or oil-in-water (O/W) emulsions. To investigate their applicability, biosurfactants/bioemulsifiers can be classified using the HLB scale (ranging from 0 to 20). HLB values of 3–6 are ideal for W/O microemulsions, while values of 8–18 are appropriate for O/W microemulsions [33]. For example, rhamnolipids, surfactin, and sophorolipids are chosen for O/W emulsion enhancement based on their HLB values. Representative HLB values for various biosurfactants, are shown in table 9.2.

The primary function of surfactants is to reduce interfacial tension, thereby forming small droplets in non-reactive liquids such as oil and water. Surfactants reduce negative interactions at the W/O interface and facilitate the dispersion of droplets from one phase to another. Small reductions in emulsions increase the stability of suspensions or liquids [39].

The ability of biosurfactants and bioemulsifiers to create microemulsions that can act as carriers for vitamins, oils, and other important substances is another possible use for them [40]. Furthermore, one study has reported that rhamnolipids derived from

Table 9.2. Hydrophilic–lipophilic balance (HLB) values for various biosurfactants.

Type of biosurfactants	HLB Value	References
Glycolipids	10–16	[34]
Lipopeptides	8–15	[35]
Saponins	10–15	[36]
Rhamnolipids	8–12	[36]
Sophorolipids	10–14	[37]
Trehalose lipids	9–13	[37]
Surfactin	10–14	[38]

P. aeruginosa, used in nano-biosurfactant preparation, enhance emulsification capacity for commercial baking [41]. Sophorolipids are well-known for their strong emulsifying abilities when used with vegetable oils in bakery preparations. In a study by Gaur *et al*, they were produced by *Candida* spp. and their potential as emulsifiers in the food industry was explored. The biosurfactant has a significant emulsifying capacity for many oils, including olive oil (51%), mustard oil (50%), soybean oil (39%), and almond oil (50%) [42]. It has been established that biosurfactants are effective emulsifiers for a variety of oils, implying their potential in a variety of food-related applications.

Furthermore, one of the most crucial applications of microbial surfactants within the food industry pertains to their role as antiadhesive agents [43]. The initial step in the development of biofilms, which are intricate communities of microorganisms enmeshed in extracellular polymeric substances (EPS), is microbial adhesion. These biofilms can form on numerous surfaces, including food processing equipment, food items, and materials used in food packaging. Biofilms on food processing equipment can lead to food contamination, spoilage, and foodborne illnesses, posing a significant risk to food safety. On the other hand, biofilms on food products can diminish their quality and shelf life. In this context, microbial surfactants play a pivotal role in averting biofilm formation and eliminating pre-existing ones. Microbial surfactants serve as effective antiadhesive agents through several mechanisms. They modify the hydrophobicity of microbial cells' surfaces, disrupt the EPS matrix that holds biofilms together, and compete with microbial adhesion proteins for binding sites on surfaces. Their ability to prevent adhesion and remove biofilms has been demonstrated across a wide spectrum of foodborne pathogens, including *Escherichia coli, Salmonella, Listeria monocytogenes*, and *Staphylococcus aureus* [43]. Thus, microbial surfactants offer a powerful means to enhance food safety and product quality within the food industry.

9.3.2 Safety and regulatory considerations of using microbial surfactants in food

The utilization of microbial surfactants in food products brings forth important safety and regulatory considerations. While these natural compounds offer numerous benefits, ensuring their safe use is paramount. The incorporation of biosurfactants and bioemulsifiers is important to establish them as food additives due to their non-toxic nature [44, 45]. Presently, no specific regulatory guidelines are in place to classify any biosurfactants or bioemulsifiers as official food additives. The journey from laboratory research to commercial market readiness, which includes conforming to the regulatory framework, is time-consuming and indispensable for commercialization. Prior to endorsing a substance as a food additive, thorough assessments for acute toxicity, often measured by the LD50 dose, and allergy potential are essential prerequisites [45]. The Organization for Economic Co-operation and Development (OECD) sets the framework for risk evaluation of food additives, considering key parameters encompassing acute toxicity, allergic reactions, reproductive toxicity, and mutagenic behaviour, which can be extrapolated for the safety evaluation of biosurfactants and bioemulsifiers. Campos *et al* [28] provided a concrete example of using glycolipids as food additives. In their research, glycolipids were incorporated as a food additive at a concentration of 0.7% (w/v) in mayonnaise production. This translates to the ingestion of approximately 0.10 g of biosurfactants when an adult weighing 50 kg consumes about 15 g of the treated food. This dosage, equivalent to 2 mg kg^{-1} of the consumer's total body weight, is significantly lower than the amount required to determine acute toxicity in laboratory animals (3600 mg kg^{-1}), indicating minimal acute risk. Regulatory bodies like the US Food and Drug Administration (FDA), the Food and Agriculture Organization (FAO), and the WHO recommend establishing an acceptable daily intake (ADI) to ensure an ample margin of safety and mitigate health risks for all consumer groups [46]. The ADI specifies the quantity of a food additive that can be ingested daily over a lifetime without any discernible health risks. It is calculated on a per-unit-of-body-weight basis and is assigned solely to food additives that are efficiently eliminated from the body within 24 h.

Notably, glycolipid biosurfactants such as rhamnolipids, which are primarily produced by opportunistic pathogenic *P. aeruginosa*, have prompted the search for alternative rhamnolipid-producing strains [47]. Tripathi *et al*, [48] discovered a rhamnolipid-like biosurfactant-producing non-pathogenic strain within the genus *Marinobacter.* Due to their cost-effective production from abundant renewable feedstocks, such strains extend the potential of rhamolipid biosynthesis to non-genetically engineered bacteria, offering prospects for food-related applications.

Non-pathogenic biosurfactant-producing organisms show promise because they are less pathogenic than *P. aeruginosa* strains. The production of commercial rhamnolipids with no identified risks is currently in progress in the United States through Jeneil Biotech in Saukville, WI, USA [49]. Food ingredients containing additives that have been approved by the United States Department of Agriculture (USDA) for specific use as biosufactants, with proper labelling, or that are GRAS, have the potential to be used in a variety of applications. For example, biosurfactants derived from *Candida utilis* are listed as GRAS organisms in the US FDA's Code of Federal Regulations Title 21 (21CFR-172.590).

This list includes approved food additives based on a history of safety, allowing their use in food products. In addition, microorganism-derived ingredients may undergo a GRAS notice evaluation process. Various lactic acid bacteria, recognized as GRAS strains, could serve as food additives to establish regulatory frameworks and streamline biosurfactant usage. The pursuit of eco-friendly, biodegradable, and safe materials is imperative for future applications [50, 51].

Surfactants originated from microbial sources are made up of hydrophobic elements (lipids) and hydrophilic compounds such as carbohydrates, proteins/peptides, and acid compounds These substances' straightforward and biodegradable structure makes them non-toxic. [51, 52]. However, the origin of some biosurfactants produced from pathogenic bacteria, such as *P. aeruginosa*, necessitates a thorough examination of their toxicity.

The evaluation of biosurfactant cytotoxicity must follow the recognized European standards (UNE EN ISO 10993-5:2009) and guidelines for the *in vitro* cytotoxicity biological assessment of medical devices. These guidelines, adopted for the cytotoxicity evaluation of biosurfactants produced by *Lactobacillus pentosus* [51, 53].

Studies on biosurfactants produced by marine strains of *Marinobacter and Pseudomonas* have also indicated a lack of cytotoxicity in *in vitro* models [54, 55]. Consequently, it is evident that low concentrations of biosurfactants do not pose a cytotoxic risk to human health, and the same is expected when adopting biosurfactants in food formulations. The utilization of biosurfactants at low concentrations, while still achieving high efficiency, can result in a low ADI value. A reduced ADI indicates a minimal need to attain the desired functionality of the molecule in food production, making the adoption of biosurfactants highly cost-effective. Overall, while microbial surfactants offer promising advantages in the food industry, their safety and regulatory compliance must always be rigorously assessed and maintained to safeguard consumer well-being.

9.3.3 Microbial surfactants in pharmaceuticals

Microbial surfactants have emerged as valuable compounds in the pharmaceutical industry, with many applications and benefits. These natural molecules, produced by microorganisms like bacteria and fungi, possess unique properties that make them increasingly attractive for pharmaceutical formulations and drug delivery systems. One of their primary roles is to improve the solubility and stability of poorly water-soluble drugs, a common challenge in drug development. By incorporating microbial surfactants into drug formulations, pharmaceutical companies can enhance drug bioavailability, ensuring patients can effectively benefit from medications. Furthermore, microbial surfactants have demonstrated promise in drug delivery systems, such as micelles and nanoparticles, which enable targeted drug release, reducing side effects and improving therapeutic outcomes. As research in this field continues to evolve, microbial surfactants hold the potential to revolutionize drug development and delivery, ultimately improving patient care and treatment efficacy.

9.4 Types of microbial surfactants used in the pharmaceutical industry

In the pharmaceutical industry, a diverse array of microbial surfactants has been employed to address various formulation and drug delivery challenges. These microbial surfactants come in several types, each with unique properties and applications. There are numerous potential biomedical and pharmaceutical applications for biosurfactants in existence, as shown in figure 9.2, which are encompassed.

9.4.1 Antimicrobial activity

Biosurfactants demonstrate antimicrobial effects against various pathogens, including bacteria, fungi, and viruses. These activities are mediated through diverse mechanisms, such as membrane disruption, biofilm interference, and modulation of cell surface characteristics. For instance, rhamnolipids, a glycolipid produced by *P. aeruginosa*, can disrupt bacterial cell membranes, resulting in cell death. Sophorolipids, another glycolipid biosurfactant produced by *Candida bombicola*, can disrupt biofilm structures, rendering them more susceptible to antibiotic

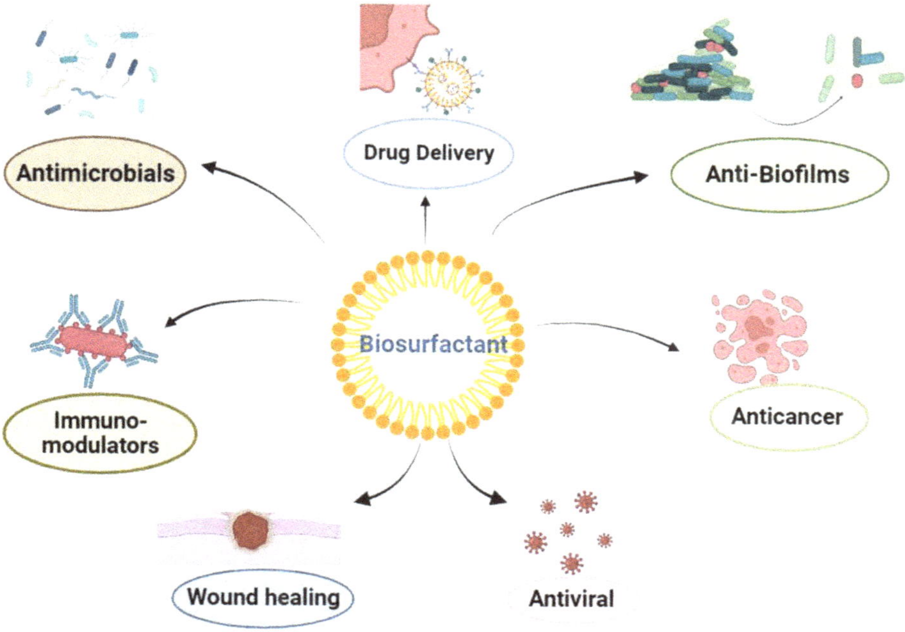

Figure 9.2. Application of biosurfactants in the biomedical and pharmaceutical industry.

treatments. Traditional antibiotics often struggle to treat biofilms, making biosurfactants a promising approach to biofilm management.

9.4.2 Modulation of cell surface properties

Biosurfactants can be employed to adjust the cell surface properties of microorganisms, which can have diverse applications, including drug delivery and vaccine development. Lipopeptides, a type of biosurfactant produced by *B. subtilis*, have been shown to enhance cell membrane permeability, facilitating drug delivery into cells, aiding in targeted drug release to improve therapeutic outcomes. Each type of microbial surfactant offers specific advantages, allowing pharmaceutical companies to tailor their use to meet the challenges posed by drug development and formulation. The diverse range of microbial surfactants available underscores their versatility and potential to drive innovation in the pharmaceutical industry, ultimately benefiting patient care and treatment efficacy.

9.5 Applications of microbial surfactants in drug delivery and formulation

Microbial surfactants have carved a significant niche in the pharmaceutical industry, particularly in drug delivery and formulation [4]. Their versatile applications span various functions crucial to drug development and patient care. One prominent application lies in enhancing the solubility and stability of poorly water-soluble drugs, a major challenge in drug development [6]. Microbial surfactants like rhamnolipids and sophorolipids serve as valuable tools to overcome this hurdle, enabling the formulation of more effective medications. Additionally, these surfactants play a pivotal role in drug delivery systems, such as micelles and nanoparticles, which facilitate targeted drug release to specific sites in the body, improving therapeutic outcomes and minimizing side effects [56]. Furthermore, biosurfactants are promising drug delivery systems (DDSs) for improving the oral bioavailability of poorly water-soluble drugs, a major challenge in pharmaceutical sciences. Their potential lies in their self-assembly and emulsification properties [57]. One DDS strategy, microemulsion drug delivery systems (MDDSs), uses biosurfactants to create systems that improve the oral bioavailability of hydrophobic drugs. MDDSs are globular substances that usually comprise lipids, surfactants, cosurfactants, and/or cosolvents [58]. Currently, MDDSs are being developed for use through various routes of delivery methods, including oral, nasal, ocular, topical, and intravenous [5]. Besides, microbial surfactants exhibit compatibility with various drug delivery routes, including oral, topical, and parenteral administration, making them adaptable to a wide range of pharmaceutical applications. As research continues to unravel new possibilities, microbial surfactants remain at the forefront of innovations in drug delivery and formulation, ultimately benefiting patients by improving drug efficacy and patient compliance [59].

9.5.1 Microbial surfactants as solubilizing agents, emulsifiers, and stabilizers in the pharmaceutical industry

In the pharmaceutical industry, microbial surfactants have proven to be indispensable solubilizing agents, emulsifiers, and stabilizers, playing a pivotal role in various aspects of drug formulation and manufacturing [59]. One of their primary functions is as solubilizing agents, addressing the challenge of low aqueous solubility often encountered with certain drugs. Microbial surfactants like rhamnolipids and sophorolipids can enhance drug solubility, allowing for the formulation of more effective drug products [60]. Additionally, these surfactants serve as excellent emulsifiers, facilitating the incorporation of oil-based drugs into water-based formulations. This is particularly valuable for intravenous drug delivery systems and injectable medications, where uniform dispersion is critical for patient safety [22]. Moreover, microbial surfactants contribute to the stability of pharmaceutical formulations by preventing phase separation and degradation of active ingredients, ensuring that the drug product maintains its efficacy over its shelf life [14]. These multifaceted roles of microbial surfactants in pharmaceuticals underscore their importance in improving drug solubility, formulation, and quality, ultimately benefiting patients by enhancing the efficacy and safety of medications.

9.5.2 Safety and regulatory considerations of using microbial surfactants in pharmaceuticals

Safety evaluation of surfactants in pharmaceutical applications is crucial, especially for nanoformulations, as they can interact with the drug substance and affect its stability, efficacy, and safety [61]. Another key aspect for safety evaluation involves the safety of the source microorganisms and the production process. It is imperative to confirm that the microorganisms used are safe and free from harmful contaminants, and that the production process is well-controlled to prevent the introduction of impurities or by-products that could compromise drug safety. Furthermore, comprehensive safety evaluations are necessary to assess any potential allergenicity, toxicity, or adverse effects associated with microbial surfactants in pharmaceutical formulations. Regulatory bodies in the US and Europe have published comprehensive guidelines on the safety evaluation of new excipients, including surfactants. These guidelines emphasize the importance of risk-benefit assessments and comprehensive safety studies, tailored to the specific excipient and its intended use. The US FDA guidelines require appropriate justification for the use of surfactants at a predefined concentration in drug development. They also recommend performing pivotal toxicology studies, including acute, repeat-dose, reproductive, and carcinogenicity studies. Safety pharmacology studies and genetic toxicology studies are also required [62]. The European Medicines Agency (EMA) guidelines are similar to the FDA guidelines, but they also require that the pharmaceutical development section of the dossier include an explanation of the choice of the excipient and its concentration, and how its properties can affect the drug product performance or manufacturability [63]. Additionally, compatibility studies of the excipient with the drug and, where relevant, other excipients should be established. For novel excipients, both the US FDA and EMA guidelines require full details of the manufacture, characterization, and controls, as well as data concerning the toxicology of the novel excipient. For biosurfactants, special impurity tests should also be considered, such as those described in the ICH Q3D (R1) and Q6B guidelines [61].

Stringent adherence to these regulations is essential to ensure that pharmaceutical products containing microbial surfactants meet the highest safety standards. Transparent labelling practices are also crucial to inform healthcare professionals and patients about the presence of these additives in medications, facilitating informed decisions and potential allergen avoidance. In essence, while microbial surfactants offer substantial advantages in pharmaceutical formulations, their safety and compliance with regulatory requirements must be rigorously upheld to guarantee the integrity and safety of pharmaceutical products and to prioritize patient health.

9.6 Microbial surfactants in agriculture

Microbial surfactants are making significant inroads in the field of agriculture, offering a sustainable and environmentally friendly alternative to traditional chemical pesticides [64]. These natural compounds, produced by microorganisms such as bacteria and fungi, have garnered attention for their eco-conscious applications. One of their primary roles is as biopesticides, where they exhibit antimicrobial properties against plant pathogens like fungi and bacteria [65]. By disrupting the cell membranes of these harmful organisms, microbial surfactants can

Table 9.3. List of biosurfactants in the agriculture industries.

Application	Microbial surfactant	Benefits
Improve soil quality	Rhamnolipid, surfactin	Improve soil drainage, increase nutrient availability, enhance microbial activity
Enhance plant growth	Lipopeptide, sophorolipid	Stimulate seed germination, promote root growth, increase nutrient uptake
Control pests and diseases	Rhamnolipid, surfactin, lipopeptide, sophorolipid	Disrupt pest and disease cell membranes, reduce pesticide use
Remove heavy metals	Rhamnolipid, surfactin, lipopeptide	Enhance bioavailability of heavy metals, facilitate phytoremediation
Improve bioremediation	Rhamnolipid, surfactin, lipopeptide, sophorolipid	Enhance degradation of pollutants, improve soil health

effectively control crop diseases while reducing the reliance on chemical pesticides that can harm the environment and human health [65]. Moreover, these surfactants contribute to improved soil health and plant growth. They also promote beneficial microbial communities in the soil, enhance nutrient uptake, and improve soil structure [66]. This not only results in increased crop yields but also aligns with the global shift toward sustainable farming practices [67]. As agriculture faces the challenge of reducing its ecological footprint, microbial surfactants offer a promising avenue to protect crops and improve agricultural sustainability simultaneously. A list of biosurfactants utilized in the agriculture industries is shown in table 9.3.

9.6.1 Types of microbial surfactants used in agriculture

In agriculture, a diverse array of microbial surfactants finds practical applications, each type offering distinct advantages for various agricultural needs (figure 9.3). One notable category is glycolipids, such as rhamnolipids and sophorolipids, known for their exceptional emulsifying properties [13, 16]. These glycolipids enable improved spreading and adherence of agricultural sprays, ensuring even coverage of pesticides, herbicides, and other agricultural inputs [67]. Additionally, they can aid in mitigating soil compaction, enhancing soil structure, and improving water retention, all of which contribute to healthier and more productive crops [68]. Another essential group is lipopeptides, including surfactin, which exhibit potent antimicrobial properties. These surfactants can effectively control plant diseases caused by pathogenic microorganisms, providing a natural and environmentally friendly alternative to chemical pesticides [24]. Lastly, lipoproteins, such as lipoprotein B, are utilized to enhance the wettability of plant leaves, ensuring efficient absorption of foliar fertilizers and other nutrients [69]. The versatility of these microbial surfactants in agriculture highlights their potential to address specific challenges, optimize agricultural practices, and promote sustainable farming methods while minimizing the ecological impact. As agriculture evolves towards more

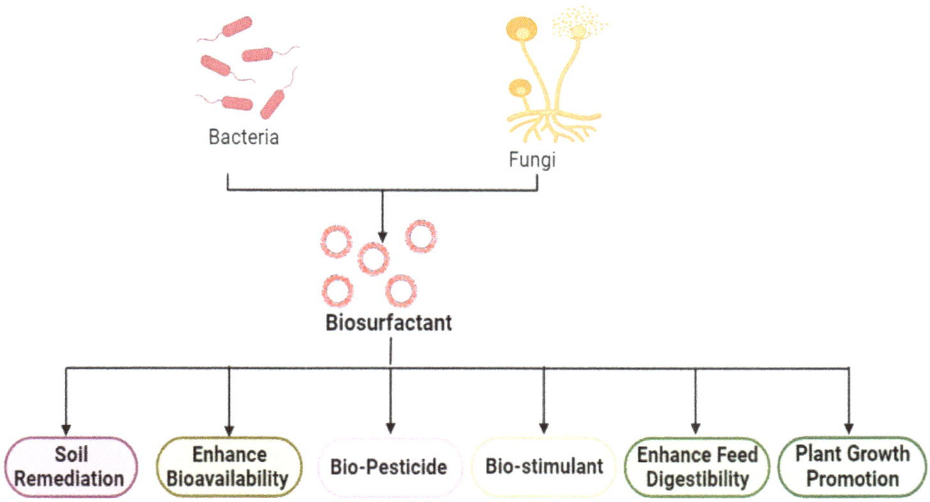

Figure 9.3. Application of biosurfactants in the agriculture.

environmentally responsible practices, microbial surfactants stand as valuable tools in the pursuit of sustainable and efficient crop production.

9.6.2 Applications of microbial surfactants in crop protection and soil remediation

Microbial surfactants have gained prominence in the realm of agriculture due to their multifaceted applications in crop protection and soil remediation [65]. In crop protection, these natural compounds, such as rhamnolipids and sophorolipids, serve as eco-friendly biopesticides and fungicides [65]. They effectively combat plant pathogens by disrupting their cell membranes, thereby reducing the reliance on chemical pesticides that can have detrimental effects on the environment and human health. Moreover, microbial surfactants enhance the adhesion and spreading of agricultural formulations, ensuring uniform coverage of crops and improving the efficacy of pest control [66]. In soil remediation, microbial surfactants play a crucial role in mitigating soil contamination. Their unique ability to mobilize and solubilize hydrophobic contaminants, such as petroleum hydrocarbons and heavy metals, makes them valuable in bioremediation efforts [67]. By increasing the bioavailability of these contaminants to soil microorganisms, microbial surfactants accelerate their degradation, facilitating the restoration of contaminated soils to a healthier state [8]. As agriculture increasingly embraces sustainable and environmentally responsible practices, microbial surfactants stand as versatile tools with the potential to enhance crop protection and soil remediation while reducing the ecological footprint of farming activities.

9.6.2.1 Microbial surfactants as biopesticides, biostimulants, and biofertilizers in agriculture

Microbial surfactants have emerged as versatile agricultural solutions, finding applications as biopesticides, biostimulants, and biofertilizers70]. As biopesticides, surfactants like rhamnolipids and sophorolipids exhibit potent antimicrobial properties, effectively combating plant pathogens without the environmental hazards associated with chemical pesticides. They provide a natural and eco-friendly approach to crop protection, promoting sustainable farming practices [67]. Additionally, microbial surfactants function as biostimulants by enhancing plant growth and vigour [71]. They can improve nutrient uptake, root development, and overall plant health. This leads to increased crop yields and improved stress tolerance, all while reducing the need for synthetic growth enhancers. Furthermore, these surfactants serve as biofertilizers by facilitating the nutrient solubilization and uptake processes [72]. They improve the availability of essential nutrients to plants, promoting nutrient-rich and healthy crops. The multifaceted roles of microbial surfactants in agriculture underscore their potential to optimize crop production, enhance sustainability, and minimize the environmental impact of farming activities.

9.6.2.2 Safety and regulatory considerations of using microbial surfactants in agriculture

The utilization of microbial surfactants in agriculture brings forth paramount safety and regulatory considerations essential for ensuring responsible and sustainable farming practices. Safety begins with the source microorganisms used in surfactant production, as it is crucial to confirm their safety and absence of harmful contaminants. Rigorous risk assessments are imperative to evaluate potential allergenicity, toxicity, and adverse effects on human health associated with the use of microbial surfactants in agricultural applications. Regulatory bodies, such as the Environmental Protection Agency (EPA) in the United States and corresponding agencies worldwide, play a pivotal role in establishing guidelines and evaluating the safety and efficacy of microbial surfactants used in agriculture [73]. Adherence to these regulations is vital to ensure that agricultural products containing microbial surfactants meet the stringent safety and environmental standards set forth by regulatory authorities. Transparent labelling practices are also crucial to inform farmers and consumers about the presence of these additives in agricultural products, enabling informed decisions and safe handling practices. As agriculture seeks to minimize its environmental footprint and promote sustainable practices, microbial surfactants offer promising solutions, but their safety and compliance with regulatory requirements must be rigorously upheld to ensure responsible and environmentally conscious agricultural practices.

9.7 Commercialization and future outlook

The commercialization of microbial surfactants represents a promising frontier in the food, pharmaceutical, and agricultural industries [70]. As sustainable and eco-friendly alternatives, these natural compounds have gained traction due to their versatile applications. In the food industry, microbial surfactants have the potential to revolutionize food processing, improving product quality, extending shelf life, and reducing the ecological impact of production. Likewise, in pharmaceuticals, their role in enhancing drug solubility and drug delivery systems holds immense promise for the development of more effective medications. In agriculture, microbial surfactants as biopesticides, biostimulants, and soil remediation agents can foster sustainable and environmentally responsible farming practices. The future outlook is optimistic, with ongoing research and innovation expected to uncover new microbial surfactant types and applications, further bolstering their commercial viability. As industries increasingly prioritize sustainability and environmentally responsible practices, microbial surfactants are poised to play an ever-expanding role in shaping the future of food, pharmaceuticals, and agriculture. Their commercialization represents not only a significant economic opportunity but also a means to address global challenges related to food security, healthcare, and environmental conservation.

9.7.1 Market trends and opportunities for microbial surfactants

Market trends and opportunities for microbial surfactants are on a notable upswing, reflecting a growing emphasis on sustainability and eco-friendly practices across various industries. In the food sector, the demand for natural and clean-label ingredients continues to rise, creating a niche for microbial surfactants as emulsifiers, stabilizers, and texture enhancers. Their biodegradable nature aligns with the industry's commitment to reducing ecological footprints. In pharmaceuticals, the need for innovative drug delivery systems and improved drug solubility opens doors for microbial surfactants. Their role in enhancing bioavailability and reducing side effects holds promise for the development of more effective medications. In agriculture, the shift towards sustainable farming practices bolsters the demand for microbial surfactants as biopesticides, biostimulants, and soil remediation agents. As global awareness of environmental issues grows, the market for microbial surfactants is poised to expand further. However, capitalizing on these opportunities requires overcoming challenges related to production scalability, cost-effectiveness, and regulatory compliance. As industries increasingly prioritize sustainability and consumers seek safer and more environmentally responsible products, microbial surfactants stand to gain a strong foothold in the market, representing a significant opportunity for both established players and emerging innovators in the field.

9.8 Conclusion

In conclusion, the exploration of microbial surfactants in the realms of food, pharmaceuticals, and agriculture reveals a world of potential and promise. These natural compounds, produced by microorganisms, have demonstrated their versatility and eco-friendliness, offering innovative solutions to critical challenges in each industry. In the food sector, they enhance product quality, improve stability, and enable cleaner label declarations. In pharmaceuticals, they revolutionize drug solubility and delivery systems, paving the way for more effective treatments. In agriculture, they contribute to sustainable farming practices, reducing the environmental impact of crop production. However, while the horizon appears bright, it is essential to acknowledge the importance of safety, regulatory compliance, and responsible practices in harnessing the full potential of microbial surfactants. The road ahead includes addressing scalability and cost-effectiveness challenges and embracing ongoing research and innovation to unlock new opportunities. As industries worldwide increasingly prioritize sustainability and environmental responsibility, microbial surfactants are poised to play a pivotal role in shaping the future of food, pharmaceuticals, and agriculture. By embracing these natural and versatile compounds, we embark on a journey toward more sustainable, efficient, and eco-conscious practices that benefit not only our industries but also our planet and its inhabitants.

References

[1] Borah D, Chaubey A, Sonowal A, Gogoi B and Kumar R 2021 Microbial biosurfactants and their potential applications: an overview *Microbial Biosurfactants* (Springer) 91–116

[2] Ribeiro B G, Guerra J M C and Sarubbo L A 2020 Biosurfactants: production and application prospects in the food industry *Biotechnol. Prog.* **36** e3030

[3] Liaqat A, Chughtai M F J, Khaliq A, Farooq U, Shahbaz M, Ali A *et al* 2023 Applications of biosurfactants in dairy industry *Applications of Next Generation Biosurfactants in the Food Sector* (Amsterdam: Elsevier) pp 509–26

[4] Ceresa C, Fracchia L, Fedeli E, Porta C and Banat I M 2021 Recent advances in biomedical, therapeutic and pharmaceutical applications of microbial surfactants *Pharmaceutics* **13** 466

[5] Ohadi M, Shahravan A, Dehghannoudeh N, Eslaminejad T, Banat I M and Dehghannoudeh G 2020 Potential use of microbial surfactant in microemulsion drug delivery system: a systematic review *Drug Des. Devel. Ther.* **14** 541–50

[6] Rodrigues L R 2015 Microbial surfactants: fundamentals and applicability in the formulation of nano-sized drug delivery vectors *J. Colloid Interface Sci.* **449** 304–16

[7] Raj A, Kumar A and Dames J F 2021 Tapping the role of microbial biosurfactants in pesticide remediation: an eco-friendly approach for environmental sustainability *Front. Microbiol.* **12** 791723

[8] Singh R, Glick B R and Rathore D 2018 Biosurfactants as a biological tool to increase micronutrient availability in soil: a review *Pedosphere* **28** 170–89

[9] Olasanmi I O and Thring R W 2018 The role of biosurfactants in the continued drive for environmental sustainability *Sustainability* **10** 4817

[10] Falk N A 2019 Surfactants as antimicrobials: a brief overview of microbial interfacial chemistry and surfactant antimicrobial activity *J. Surfactants Deterg.* **22** 1119–27

[11] Rosenberg E and Ron E Z 1999 High- and low-molecular-mass microbial surfactants *Appl. Microbiol. Biotechnol.* **52** 154–62

[12] Inès M and Dhouha G 2015 Glycolipid biosurfactants: potential related biomedical and biotechnological applications *Carbohydr. Res.* **416** 59–69

[13] Elshikh M, Moya-Ramírez I, Moens H, Roelants S, Soetaert W, Marchant R *et al* 2017 Rhamnolipids and lactonic sophorolipids: natural antimicrobial surfactants for oral hygiene *J. Appl. Microbiol.* **123** 1111–23

[14] Karanth N G K, Deo P G and Veenanadig N K 1999 Microbial production of biosurfactants and their importance *Curr. Sci.* **77** 116–26

[15] Jarvis F G and Johnson M J 1949 A glyco-lipide produced by Pseudomonas aeruginosa *J. Am. Chem. Soc.* **71** 4124–6

[16] de Oliveira M R, Magri A, Baldo C, Camilios-Neto D, Minucelli T and Celligoi M 2015 Sophorolipids a promising biosurfactant and its applications *Int. J. Adv. Biotechnol. Res.* **6** 161–74

[17] Meena K R and Kanwar S S 2015 Lipopeptides as the antifungal and antibacterial agents: applications in food safety and therapeutics *Bio. Med. Res. Int.* **2015** 473050

[18] Kourmentza K, Gromada X, Michael N, Degraeve C, Vanier G, Ravallec R *et al* 2021 Antimicrobial activity of lipopeptide biosurfactants against foodborne pathogen and food spoilage microorganisms and their cytotoxicity *Front. Microbiol.* **11** 561060

[19] Cameotra S S and Makkar R S 2004 Recent applications of biosurfactants as biological and immunological molecules *Curr. Opin. Microbiol.* **7** 262–6

[20] Antonioli Júnior R, Poloni J de F, Pinto É S M and Dorn M 2022 Interdisciplinary overview of lipopeptide and protein-containing biosurfactants *Genes (Basel)* **14** 76

[21] Naughton P J, Marchant R, Naughton V and Banat I M 2019 Microbial biosurfactants: current trends and applications in agricultural and biomedical industries *J. Appl. Microbiol.* **127** 12–28

[22] Yi G, Son J, Yoo J, Park C and Koo H 2019 Rhamnolipid nanoparticles for *in vivo* drug delivery and photodynamic therapy *Nanomed. Nanotechnol. Biol Med.* **19** 12–21

[23] Rawat G, Dhasmana A and Kumar V 2020 Biosurfactants: the next generation biomolecules for diverse applications *Environ. Sustain.* **3** 353–69

[24] Nanjundan J, Ramasamy R, Uthandi S and Ponnusamy M 2019 Antimicrobial activity and spectroscopic characterization of surfactin class of lipopeptides from *Bacillus amyloliquefaciens* SR1 *Microb. Pathog.* **128** 374–80

[25] Farias C B B, Almeida F C G, Silva I A, Souza T C, Meira H M, Rita de Cássia F *et al* 2021 Production of green surfactants: market prospects *Electron. J. Biotechnol.* **51** 28–39

[26] Stackebrandt E, Koch C, Gvozdiak O and Schumann P 1995 Taxonomic dissection of the genus micrococcus: kocuria gen. nov., nesterenkonia gen. nov., kytococcus gen. nov., dermacoccus gen. nov., and micrococcus Cohn 1872 gen. emend *Int. J. Syst. Evol. Microbiol.* **45** 682–92

[27] Kiran G S, Priyadharsini S, Sajayan A, Priyadharsini G B, Poulose N and Selvin J 2017 Production of lipopeptide biosurfactant by a marine *Nesterenkonia sp.* and its application in food industry *Front. Microbiol.* **8** 1138

[28] Campos J M, Stamford T L M, Rufino R D, Luna J M, Stamford T C M and Sarubbo L A 2015 Formulation of mayonnaise with the addition of a bioemulsifier isolated from *Candida utilis Toxicol. Rep.* **2** 1164–70

[29] Mnif I, Besbes S, Ellouze-Ghorbel R, Ellouze-Chaabouni S and Ghribi D 2013 Improvement of bread dough quality by *Bacillus subtilis* SPB1 biosurfactant addition: optimized extraction using response surface methodology *J. Sci. Food Agric.* **93** 3055–64

[30] Mnif I, Besbes S, Ellouze R, Ellouze-Chaabouni S and Ghribi D 2012 Improvement of bread quality and bread shelf-life by *Bacillus subtilis* biosurfactant addition *Food Sci. Biotechnol.* **21** 1105–12

[31] Silva I A, Veras B O, Ribeiro B G, Aguiar J S, Guerra J M C, Luna J M *et al* 2020 Production of cupcake-like dessert containing microbial biosurfactant as an emulsifier *PeerJ.* **8** e9064

[32] Konkol D, Szmigiel I, Domżał-Kędzia M, Kułażyński M, Krasowska A, Opaliński S *et al* 2019 Biotransformation of rapeseed meal leading to production of polymers, biosurfactants, and fodder *Bioorg. Chem.* **93** 102865

[33] De S, Malik S, Ghosh A, Saha R and Saha B 2015 A review on natural surfactants *RSC Adv.* **5** 65757–67

[34] Fenibo E O, Douglas S I and Stanley H O 2019 A review on microbial surfactants: production, classifications, properties and characterization *J. Adv. Microbiol.* **18** 1–22

[35] Sayyed R Z, El-Enshasy H A and Hameeda B 2021 *Microbial Surfactants: Volume I: Production and Applications* (Boca Raton, FL: CRC Press)

[36] Khoshdast H, Abbasi H, Sam A and Noghabi K A 2012 Frothability and surface behavior of a rhamnolipid biosurfactant produced by *Pseudomonas aeruginosa* MA01 *Biochem. Eng. J.* **60** 127–34

[37] Vaughn S F, Behle R W, Skory C D, Kurtzman C P and Price N P J 2014 Utilization of sophorolipids as biosurfactants for postemergence herbicides *Crop. Prot.* **59** 29–34

[38] Gudiña E J, Rangarajan V, Sen R and Rodrigues L R 2013 Potential therapeutic applications of biosurfactants *Trends Pharmacol. Sci.* **34** 667–75

[39] Goodarzi F and Zendehboudi S 2019 A comprehensive review on emulsions and emulsion stability in chemical and energy industries *Can. J. Chem. Eng.* **97** 281–309

[40] Sagalowicz L and Leser M E 2010 Delivery systems for liquid food products *Curr. Opin. Colloid Interface Sci.* **15** 61–72

[41] Farheen V, Saha S B, Pyne S and Chowdhury B R 2016 Production of nanobiosurfactant from *Pseudomonas aeruginosa* and it's application in bakery industry *Int. J. Adv. Res. Biol. Eng. Sci. Technol.* **2** 67

[42] Gaur V K, Regar R K, Dhiman N, Gautam K, Srivastava J K, Patnaik S *et al* 2019 Biosynthesis and characterization of sophorolipid biosurfactant by *Candida* spp.: application as food emulsifier and antibacterial agent *Bioresour. Technol.* **285** 121314

[43] Banat I M, De Rienzo M A D and Quinn G A 2014 Microbial biofilms: biosurfactants as antibiofilm agents *Appl. Microbiol. Biotechnol.* **98** 9915–29

[44] Partal P, Guerrero A, Berjano M and Gallegos C 1999 Transient flow of o/w sucrose palmitate emulsions *J. Food Eng.* **41** 33–41

[45] Kralova I and Sjöblom J 2009 Surfactants used in food industry: a review *J. Dispers. Sci. Technol.* **30** 1363–83

[46] Sharma D 2021 *Biosurfactants: Greener Surface Active Agents for Sustainable Future.* (Berlin: Springer)

[47] Twigg M S, Tripathi L, Zompra A, Salek K, Irorere V U, Gutierrez T *et al* 2018 Identification and characterisation of short chain rhamnolipid production in a previously uninvestigated, non-pathogenic marine pseudomonad *Appl. Microbiol. Biotechnol.* **102** 8537–49

[48] Tripathi L, Twigg M S, Zompra A, Salek K, Irorere V U, Gutierrez T *et al* 2019 Biosynthesis of rhamnolipid by a marinobacter species expands the paradigm of biosurfactant synthesis to a new genus of the marine microflora *Microb. Cell. Fact.* **18** 12

[49] Marchant R and Banat I M 2012 Biosurfactants: a sustainable replacement for chemical surfactants? *Biotechnol. Lett.* **34** 1597–605

[50] Sharma P, Madhyastha H, Madhyastha R, Nakajima Y, Maruyama M, Verma K S *et al* 2019 An appraisal of cuticular wax of *calotropis procera* (Ait.) R. Br.: extraction, chemical composition, biosafety and application *J. Hazard. Mater.* **368** 397–403

[51] Sharma V, Singh D, Manzoor M, Banpurkar A G, Satpute S K and Sharma D 2022 Characterization and cytotoxicity assessment of biosurfactant derived from Lactobacillus pentosus NCIM 2912 *Braz. J. Microbiol.* **53** 327–40

[52] Rodríguez-López L, López-Prieto A, Lopez-Álvarez M, Pérez-Davila S, Serra J, González P *et al* 2020 Characterization and cytotoxic effect of biosurfactants obtained from different sources *ACS Omega.* **5** 31381–90
[53] International Organization for Standardization 2009 *Biological evaluation of medical devices—Part 5: tests for* in vitro *cytotoxicity* (GenevA, Switzerland: International Organization for Standardization)
[54] Adu S A, Twigg M S, Naughton P J, Marchant R and Banat I M 2023 Characterisation of cytotoxicity and immunomodulatory effects of glycolipid biosurfactants on human keratinocytes *Appl. Microbiol. Biotechnol.* **107** 137–52
[55] Voulgaridou G P, Mantso T, Anestopoulos I, Klavaris A, Katzastra C, Kiousi D E *et al* 2021 Toxicity profiling of biosurfactants produced by novel marine bacterial strains *Int. J. Mol. Sci.* **22** 2383
[56] Ohadi M, Shahravan A, Dehghannoudeh N, Eslaminejad T, Banat I M and Dehghannoudeh G 2020 Potential use of microbial surfactant in microemulsion drug delivery system: a systematic *Drug Des. Devel. Ther.* **5** 541–50
[57] Kurozuka A, Onishi S, Nagano T, Yamaguchi K, Suzuki T and Minami H 2017 Emulsion polymerization with a biosurfactant *Langmuir* **33** 5814–8
[58] Das M, Patowary K, Vidya R and Malipeddi H 2016 Microemulsion synthesis of silver nanoparticles using biosurfactant extracted from *Pseudomonas aeruginosa* MKVIT3 strain and comparison of their antimicrobial and cytotoxic activities *IET Nanobiotechnol.* **10** 411–8
[59] Bjerk T R, Severino P, Jain S, Marques C, Silva A M, Pashirova T *et al* 2021 Biosurfactants: properties and applications in drug delivery, biotechnology and ecotoxicology *Bioengineering* **8** 115
[60] Nguyen T T and Sabatini D A 2011 Characterization and emulsification properties of rhamnolipid and sophorolipid biosurfactants and their applications *Int. J. Mol. Sci.* **12** 1232–44
[61] Ismail R, Baaity Z and Csóka I 2021 Regulatory status quo and prospects for biosurfactants in pharmaceutical applications *Drug Discov. Today* **26** 1929–35
[62] US FDA 2005 *Guidance for Industry: Nonclinical Studies for the Safety Evaluation of Pharmaceutical Excipients* (US Department of Health and Human Services, Food and Drug Administration)
[63] European Medicines Agency 2007 Guideline on excipients in the dossier for application for marketing authorisation of a medicinal product EMEA/CHMP/QWP/396951/2006 https://www.ema.europa.eu/en/documents/scientific-guideline/guideline-excipients-dossier-application-marketing-authorisation-medicinal-product-revision-2_en.pdf
[64] Sachdev D P and Cameotra S S 2013 Biosurfactants in agriculture *Appl. Microbiol. Biotechnol.* **97** 1005–16
[65] Gaikwad S 2023 Tapping significance of microbial surfactants as a biopesticide and synthetic pesticide remediator: an ecofriendly approach for maintaining the environmental sustainability *Insecticides - Advances in Insect Control and Sustainable Pest Management* (InTech Open)
[66] Sharma P, Sangwan S, Singh S and Kaur H 2022 Microbial biosurfactants: an eco-friendly perspective for soil health management and environmental remediation *New and Future Developments in Microbial Biotechnology and Bioengineering* (Amsterdam: Elsevier) pp 277–98
[67] Gayathiri E, Prakash P, Karmegam N, Varjani S, Awasthi M K and Ravindran B 2022 Biosurfactants: potential and eco-friendly material for sustainable agriculture and environmental safety—a review *Agronomy* **12** 662

[68] Raddadi N, Giacomucci L, Marasco R, Daffonchio D, Cherif A and Fava F 2018 Bacterial polyextremotolerant bioemulsifiers from arid soils improve water retention capacity and humidity uptake in sandy soil *Microb. Cell Fact.* **17** 12

[69] da Silva A F, Banat I M, Giachini A J and Robl D 2021 Fungal biosurfactants, from nature to biotechnological product: bioprospection, production and potential applications *Bioprocess. Biosyst. Eng.* **44** 2003–34

[70] Kumar A, Singh S K, Kant C, Verma H, Kumar D, Singh P P *et al* 2021 Microbial biosurfactant: a new frontier for sustainable agriculture and pharmaceutical industries *Antioxidants* **10** 1472

[71] Chinnasamy M, Rathanasamy R, Selvam S, Mohankumar H K, Anandraj M and Pal S K 2022 Biosurfactant as biostimulant: factors responsible for plant growth promotions *Applications of Biosurfactant in Agriculture* (Amsterdam: Elsevier) pp 45–68

[72] Karamchandani B M, Pawar A A, Pawar S S, Syed S, Mone N S, Dalvi S G *et al* 2022 Biosurfactants' multifarious functional potential for sustainable agricultural practices *Front. Bioeng. Biotechnol.* **10** 1047279

[73] Touart L W 2020 The federal insecticide, fungicide, and rodenticide act *Fundamentals of Aquatic Toxicology* (Boca Raton, FL: CRC Press) pp 657–68

IOP Publishing

Microbial Surfactants
A sustainable class of versatile molecules
Divya Tripathy and Anjali Gupta

Chapter 10

Potential applications of biosurfactants in the biomedical field

Pragati Saini, Pragati Sahai and Garima Bartariya

As research in this field continues to unfold, several key themes emerge. Biosurfactants, which are surface-active compounds produced by microorganisms, have shown great potential in various biomedical fields. These microbial-derived surface-active compounds offer unique properties that can be harnessed for various biomedical applications that are diverse and promising and present a compelling avenue for innovation and advancement. The quest for alternatives to traditional antibiotics finds a potential solution in biosurfactants. With their capacity to disrupt microbial cell membranes, these compounds offer a new frontier in the fight against antimicrobial resistance, paving the way for the development of innovative antimicrobial agents. This chapter elaborates on the application of biosurfactants in biomedical fields. Cancer therapy, dental care, and the development of biocompatible coatings for medical devices all stand as areas where biosurfactants can contribute significantly. In the realm of imaging techniques, biosurfactants serve as valuable contrast agents, improving the visualization of specific tissues. This has implications for advancing diagnostic imaging modalities, providing clinicians with more detailed and accurate information. Their ability to enhance the biocompatibility of scaffolds and improve immune responses in vaccines opens avenues for creating artificial tissues and developing more effective and enduring immunization strategies. As researchers delve deeper into understanding the intricacies of biosurfactant behaviour and applications, the biomedical landscape stands to benefit from these microbial marvels, ushering in a new era of healthcare innovation. Research in these areas is ongoing, and as our understanding of biosurfactants expands, new applications in the biomedical field may emerge, offering innovative solutions to various healthcare challenges.

doi:10.1088/978-0-7503-5989-4ch10

10.1 Introduction

The chemical substances known as biosurfactants are derived from various microorganisms and have the capacity to reduce the interfacial tension between two phases that are similar or different. The cosmetics, pharmaceutical, biotechnology, food, and oil industries can all benefit from the usage of biosurfactants. It is a useful method for biological and physicochemical applications (Abbot *et al* 2022).

Their antifungal, antiviral, and antibacterial properties render biosurfactants suitable for application against an array of ailments and medicinal substances (Panda *et al* 2023). Because of their ability to self-aggregate, biosurfactants engage with specific targets and have the potential to break the cell membrane of microbial pathogens. They are also highly stable biomolecules. These surface-active substances can also function as a great adjuvant to stimulate the immune system and have immunological modulatory properties to boost the immune system.

Several reports of biosurfactant uses over medical purposes have come to light in recent years (Rodrigues and Teixeira 2010). Because of their antibacterial, antifungal, and antiviral properties, biosurfactants have been considered to be important molecules for use as therapeutic agents and in the treatment of a variety of diseases. Additionally, their ability to function as antiadhesive coating agents for medical insertional materials against a variety of diseases demonstrates their appropriateness and helps to reduce hospital infections without the need for synthetic medications or chemicals.

Biosurfactants are surfactants produced from microbiological sources. Biosurfactants have received more attention from academia and industry in recent times because of their flexibility, ecological compatibility, and potential for fermentation-based production. Their stability, low toxicity, renewable source of energy, usefulness under extreme pH and temperature settings, and biodegradability are further benefits. Those features make biosurfactants useful for solubilization, emulsification, and separation. They also lessen surface and interfacial tension and encourage the adsorption of bioactive substances across biological membranes.

Biosurfactants have less toxicity and more degradability, properties that cannot be found in traditional surfactants. In the medical field, biosurfactants are attractive, because their products can be used effectively in small amounts.

Several new biosurfactants have been reported recently, and rapid developments in molecular and cellular biology should deepen our understanding of the variety of biosurfactant structures and uses. Biosurfactants are crucial for the natural motility of swarming microorganisms and are involved in the signalling, differentiation, and biofilm formation processes within cells. In addition to their ability to complex with heavy metals and play natural physiological roles in enhancing the bioavailability of hydrophobic molecules, biosurfactants can also have antimicrobial activity (Van Hamme *et al* 2006).

In recent years, there has been an increased interest in microbial surfactants, referred to as biosurfactants, due to their natural origin and environmental compatibility. These qualities satisfy society's and regulatory bodies' demands for the use of more environmentally friendly and long-lasting chemicals. Synthetic surfactants can be substituted with microbial-derived surfactants in a wide range of industrial applications, including as foaming agents, detergents, emulsifiers, solubilizers, and wetting agents. There is a growing need for new 'green' food additives due to consumer trends shifting away from synthetic and toward natural ingredients and growing concerns about environmental and health issues (Nitschke and Silva 2018).

10.2 Biosurfactants in biomedical applications

Biosurfactants, which are surface-active molecules produced by microorganisms including bacteria, fungi, and yeasts, have gained attention in various fields due to their unique properties and environmentally friendly nature. In the biomedical field, biosurfactants have shown promising applications in several areas due to their biocompatibility, biodegradability, and low toxicity (figure 10.1).

Biosurfactants exhibit antimicrobial, antifungal, and antiviral properties, as well as immune, neurological, and anticancer properties. They can also enhance the electrical conductance of bimolecular lipid membranes and prevent the development of fibrin clots. By inhibiting hydrogen bonding and enhancing hydrophilic/hydrophobic interactions, they can reduce surface tensions between immiscible or miscible liquids. They can also be employed as antiadhesive/antibiofilm agents on medical devices and in transplant applications (Sarma and Prasad 2021, Inamuddin *et al* 2022) They are used to improve a few physical–chemical aspects of pharmaceutical formulations, as well as the products' effectiveness and functionality. Additionally, they are crucial to the creation of liposomes, micro- and nano-based drug delivery systems, self-emulsifying drug delivery systems, and regulating particle size and drug stability or solubility in liquid, semi-solid, and solid formulations (Ismail *et al* 2021).

Figure 10.1. Biosurfactant applications in biomedical fields.

10.2.1 Drug delivery systems

Biosurfactants can be used to improve the delivery of drugs to their target sites. They can solubilize hydrophobic drugs, making them more bioavailable. Biosurfactants can also be used to encapsulate drugs in liposomes, which can protect the drugs from degradation and deliver them to specific cells. Biosurfactants, amphiphilic molecules produced by microorganisms, have emerged as promising candidates for drug delivery systems due to their unique properties and potential advantages over conventional synthetic surfactants.

10.2.1.1 Enhancing drug solubility and bioavailability

Biosurfactants can effectively solubilize poorly water-soluble drugs, a significant challenge in drug development. By encapsulating hydrophobic drugs into micellar structures or forming self-assembled nanocarriers, biosurfactants can enhance the solubility and dispersibility of drugs in aqueous environments. This improved solubility leads to increased bioavailability, ensuring that a higher proportion of the drug reaches its target site and exerts its therapeutic effect. These systems provide a number of benefits, such as maximum drug loading capacity without drug loss, aqueous solubility, enhanced bioavailability, and controlled and facilitated transport of the active ingredient across membranes to the targeted site, which maximizes efficacy (Jain 2008).

10.2.1.2 Targeted and controlled drug delivery

Biosurfactants can facilitate targeted drug delivery by incorporating ligands or targeting moieties onto their surface. These ligands can specifically bind to receptors or cell surface markers on the target cells, enabling the selective delivery of drugs to the desired site of action. This targeted approach minimizes off-target effects and enhances the therapeutic efficacy while reducing potential side effects.

By enhancing the substrate's bioavailability for microorganisms or by interacting with the cell surface to increase its hydrophobicity, biosurfactants can enhance hydrocarbon bioremediation and facilitate the easier association of the hydrophobic substrates with the bacterial cells (Mahjoubi *et al* 2018, Mnif *et al* 2018). To improve the oral bioavailability of hydrophobic medicines, various solutions have been implemented, including the use of microemulsion drug delivery systems. They are typically globular in shape and contain lipids, surfactants, cosurfactants, and/or cosolvents (Das *et al* 2016). Currently, microemulsion drug delivery systems are being developed for usage via a number of delivery methods, including intravenous, topical, ocular, nasal, and ocular (Ohadi *et al* 2020b).

10.2.1.3 Promoting cellular uptake and penetration

Biosurfactants can enhance cellular uptake and penetration of drugs by disrupting cell membranes, facilitating drug permeation through biological barriers, and promoting endocytosis. This improved cellular uptake ensures that drugs reach their intracellular targets, such as the nucleus or mitochondria, where they can exert their therapeutic effect. Surfactants have a significant effect on the phospholipid membrane of cells in addition to their considerable influence on the characteristics of the cell surface. The penetration of surfactant molecules is linked to modifications in the permeability of cell membranes in the presence of surfactants (Zhou *et al* 2016). This process also applies to molecules of biodegradable substances and may be involved in the uncontrollable release of gas molecules, ions, or tiny metabolites from the cell. The detrimental impact of surface-active chemicals on microorganisms is also determined by the way surfactants interact with the cell membrane (Koley and Bard 2010). The cellular metabolism can be disrupted by tiny molecules entering the cell due to an increase in membrane permeability. Conversely, a liquid membrane may allow enzymes or other materials to leave the cell (Zhou *et al* 2016).

10.2.1.4 Mucoadhesion and transmucosal delivery

Proteins, lipids, glycoproteins, minerals, and proteins are all present in the mucus, which is a hydrophilic layer. Mucins are the predominant glycoproteins, which add to the mucus layer's viscosity, negative charge, and mesh-like structure (Falavigna *et al* 2020, Murgia *et al* 2018). To circumvent these limitations, supplementary agents are required due to the challenges associated with nasal delivery. While various methods have been studied to boost drug penetration through the nasal mucosa, the most commonly used approach is adding absorption enhancers, including surfactants, to the formulation. These latter are amphiphilic compounds that can increase medication absorption in a variety of ways, such as by temporarily opening tight junctions, disrupting the cell membrane, or stopping the enzymatic breakdown of medicines (Ozsoy and Güngör 2011, Ghadiri *et al* 2019). Biosurfactants can promote mucoadhesion, the ability to adhere to mucosal surfaces, such as the gastrointestinal tract or respiratory tract. This mucoadhesive property can enhance the residence time of drugs at the target site, prolonging their therapeutic effect and reducing the frequency of dosing. Additionally, biosurfactants can facilitate transmucosal delivery, enabling drugs to permeate through mucosal barriers and reach their systemic circulation.

10.2.1.5 Overcoming biological barriers

Biosurfactants can overcome various biological barriers that hinder drug delivery, such as the blood–brain barrier (BBB) and the blood–tumour barrier (BTB). By modifying the surface properties of drug carriers or disrupting these barriers, biosurfactants can facilitate the delivery of drugs to the central nervous system or tumour sites, where they are typically difficult to reach using conventional methods.

10.2.1.6 Examples of biosurfactants in drug delivery systems

- **Rhamnolipids:** Produced by *Pseudomonas aeruginosa*, rhamnolipids have been used to encapsulate drugs into liposomes and nanoparticles for targeted delivery.
- **Sophorolipids:** Produced by *Candida albicans*, sophorolipids have shown promise in enhancing the solubility and bioavailability of poorly water-soluble drugs.
- **Sucrose esters:** Produced by *Bacillus subtilis*, sucrose esters have been used to prepare self-assembled nanocarriers for drug delivery.

- **Mannosylerythroside:** Produced by *Arthrobacter*, mannosylerythroside has been explored for its ability to promote cellular uptake and transmucosal delivery of drugs.

Biosurfactants offer a promising approach to overcoming the challenges of drug delivery, particularly for poorly water-soluble drugs and those targeting difficult-to-reach sites. Their ability to enhance solubility, facilitate targeted delivery, promote cellular uptake, and overcome biological barriers makes them valuable tools for developing more effective and efficient drug delivery systems. Ongoing research continues to explore the diverse applications of biosurfactants in the pharmaceutical industry, paving the way for innovative drug delivery strategies.

10.2.2 Wound healing:

The normal biological process of wound healing involves four separate phases, namely haemostasis, inflammation, proliferation, and remodelling (Guo and Dipietro 2010). Numerous variables might contribute to poor wound healing, such as the fact that chronic wounds typically take longer than 6 weeks to heal. These aspects can also provide significant challenges for treatment approaches, since any given therapy must be carefully timed to the proper stage (Sajid *et al* 2020). Regarding their ability to form emulsions, their antitumour and antibacterial qualities, and their impact on the immune system, lipopeptides are among the biosurfactants that have been investigated the most (Ohadi *et al* 2020a). Using an excision wound model, Ohadi *et al* (2017) found that a lipopeptide produced by *Acinetobacter junii* B6 helps shield mouse cells from damage caused by free radicals while also demonstrating cell recovery.

Biosurfactants are amphiphilic molecules produced by microorganisms, including bacteria, fungi, and yeasts. They have a wide range of potential applications in the biomedical field, including wound healing. Biosurfactants can promote wound healing by:

- stimulating the proliferation and migration of fibroblasts, keratinocytes, and endothelial cells, which are essential for wound healing.
- reducing inflammation by inhibiting the production of inflammatory mediators.
- promoting angiogenesis, the formation of new blood vessels, which is important for supplying oxygen and nutrients to the wound site.
- preventing bacterial adhesion and biofilm formation, which can delay wound healing and increase the risk of infection.
- enhancing the delivery of drugs and growth factors to the wound site, which can accelerate the healing process.

10.2.3 Antimicrobial agents

Glycolipids, rhamnolipids, lipopeptides, polysaccharide-protein complexes, phospholipids, fatty acids, and neutral lipids are examples of these complex compounds. In contrast to synthetic surfactants, biodegradable and diversified, biosurfactants have the capacity to perform extremely specialized and highly selective tasks. Many biosurfactants have antiviral, antifungal, and antibacterial properties, which makes them suitable candidates to fight infection. The most potent known biosurfactant, surfactin, is made by *B. subtilis* and is included in the list of known biosurfactants. Mannosylerythritol lipids from *Candida antarctica* rhamnolipids from *P. aeruginosa*, iturin, which is also made by *B. subtilis* and those isolated from probiotic bacteria *Streptococcus thermophilus* A and *Lactococcus lactis* are some other biosurfactants having antimicrobial action. Because probiotic *lactobacilli* produce a variety of antimicrobial compounds, including biosurfactants, they are known to be powerful interfering bacteria and to be a significant component of the natural microbiota (Sambanthamoorthy *et al* 2014).

Biosurfactants can disrupt the cell membranes of microorganisms, leading to their death. They can also interfere with the formation of biofilms, which are communities of microorganisms that are embedded in a sticky matrix. Biofilms are more resistant to antibiotics than individual cells, so biosurfactants can be helpful in treating infections that are caused by biofilms. Biosurfactants are being investigated for a variety of applications as antimicrobial agents, such as:

- Biosurfactants can be incorporated into coatings for surfaces that are prone to contamination, such as medical implants and catheters.
- Biosurfactants can be used as disinfectants to clean and disinfect surfaces.
- Biosurfactants can be used as preservatives in cosmetics, pharmaceuticals, and food products.

In many studies, the cell-bound biosurfactant from a strain of *Lactobacillus rhamnoses* isolated from human breast milk was shown to change the integrity and viability of bacterial cells as well as reduce the total exopolysaccharide matrix content, which in turn inhibited the adhesion of various bacterial pathogens (including *P. aeruginosa*, *Staphylococcus aureus*, and *Escherichia coli*) to surfaces and promoted the eradication of their pre-formed biofilms (Patel *et al* 2021). A *Lactobacillus plantarum* strain isolated from a yoghurt sample produced glycolipid that inhibited *P. aeruginosa* and *Chromobacterium violaceum*'s pathogenicity, deactivating their quorum-sensing regulatory mechanism. The investigated biosurfactant was shown to have a dose-dependent inhibitory effect on violacein, acyl homoserine lactone, pyocyanin and exopolysaccharide synthesis, swarming motility and biofilm formation, as well as a decrease in LasA protease and LasB elastase activities (Patel *et al* 2022).

Biosurfactants are a promising new class of antimicrobial agents with the potential to address the growing problem of antibiotic resistance. As research continues, we can expect to see even more innovative and effective applications of biosurfactants in the fight against infection.

10.2.4 Cancer therapy

In recent times, biosurfactants have surfaced as a potentially effective substitute molecule for managing multiple cancer kinds, such as pancreatic, breast, cervical, oral, colon, lung, and liver malignancies (Banat *et al* 2010, Naughton *et al* 2019). Their capacity to control certain processes in mammalian cells, which stops the cancer from progressing abnormally and inhibits cell survival, migration, and proliferation, has shown promise for the treatment of cancer (Kamalakannan *et al* 2020) According to research by Adu *et al*, glycolipids and lipopeptides are two examples of biosurfactants that can stop tumour cells from proliferating and surviving (Marchant and Banat 2012). Furthermore, these biosurfactants changed the potential of the mitochondrial membrane, which ultimately led to necrosis and cell death.

Biosurfactants exhibit various mechanisms of action that contribute to their anticancer potential. These mechanisms can work by disruption of the cell membrane of cancer cells, leading to leakage of cellular contents and causing cell death. These compounds also have the ability to induce apoptosis, a programmed cell death mechanism that selectively eliminates cancer cells. Biosurfactants can also be used to inhibit angiogenesis, the formation of new blood vessels that tumours need to grow and spread. In the previous section it was also demonstrated that biosurfactants can also be used as drug delivery vehicles to improve the uptake and efficacy of anticancer drugs.

Glycolipid biosurfactants have a great deal of commercial potential because they are much less toxic and more biocompatible than synthetic surfactant compounds derived from petrochemicals. This includes biomedical applications like skincare pharmaceutical formulations. Furthermore, the potential of glycolipid biosurfactants, some of which are generated by bacteria isolated from environmental sources like the open ocean, to influence the growth of tumours and hence function as anticancer agents has been studied (Adu *et al* 2022).

Human pancreatic (HPAC), liver (H7402), lung (A549), brain (LN229, HNCG-2), oesophageal (KYSE109, KYSE450), breast, cervical (HeLa), leukaemic (HL60, K562), and colonic (HCT116, CaCo-2) cell lines have all been demonstrated to be cytotoxically affected by sophorolipids *in vitro*. Additionally, studies using human promyelocytic leukaemia cell lines and breast cancer (MCF-7), colon cancer (CaCo-2), liver cancer (HepG2), and other cancers have demonstrated the anticancer properties of rhamnolipid biosurfactants (Thanomsub *et al* 2006, Christova *et al* 2013). Glycolipids have been demonstrated by Haque *et al* (2021) to have a detrimental effect on the viability of a murine melanoma cell line. Additionally, it has been reported that the anticancer effects of glycolipid biosurfactants either solely targeted one class of glycolipid or used uncharacterized combinations of multiple glycolipids on melanoma cells due to the distinct chemical structures of rhamnolipids and sophorolipids as opposed to their impact on healthy skin cells.

Research on the use of biosurfactants in cancer therapy is still in its early stages, but preclinical studies have shown promising results. Further research is needed to fully evaluate the efficacy and safety of biosurfactants in cancer treatments. Further research directions can include, developing novel biosurfactants with enhanced anticancer activity, investigating the mechanism of action of biosurfactants in cancer cells, optimizing the formulations and delivery of biosurfactants for cancer therapy and conducting clinical trials to evaluate the safety and efficacy of biosurfactants in cancer patients. Biosurfactants have emerged as promising alternatives for developing new cancer therapeutics due to their unique properties and potential to overcome the limitations of conventional cancer treatments. Continued research is essential to fully realize the potential of biosurfactants in cancer therapy and improve the treatment outcomes for cancer patients.

10.2.5 Imaging

Biosurfactants are amphiphilic molecules produced by microorganisms, plants and animals. They have a wide range of applications in various industries including imaging. Biosurfactants can be used as contrast agents, solubilizing agents and emulsifiers in imaging application. They have some properties that support their use in imaging. These molecules are biodegradable and biocompatible, which makes them safer than synthetic surfactants. They are often more effective than synthetic surfactants in certain applications. They can be used as:

- **Contrast agents:** Biosurfactants can be used to enhance the contrast of images in various imaging modalities, including ultrasound, MRI, and CT. For example, they can be used to encapsulate contrast agents, such as gadolinium, which can improve the visibility of tumours and other abnormalities in MRI.
- **Solubilizing agents:** Biosurfactants can be used to solubilize poorly soluble drugs and imaging agents, making them more bioavailable and easier to administer. For example, they can be used to dissolve Paclitaxel, a chemotherapeutic drug, which can improve the effectiveness of chemotherapy.
- **Emulsifiers:** Biosurfactants can be used to emulsify oil and water, which is important for some imaging applications. For example, they can be used to emulsify oil based contrast agents, which can improve the delivery of contrast agents to the target tissue.

Researchers are continuing to develop new biosurfactants for imaging applications. As the technology for producing and purifying biosurfactants improves, they are likely to become more widely used in imaging applications.

10.3 Advantages of biosurfactants

Biosurfactants, which are surface-active compounds produced by microorganisms, offer several advantages over chemical surfactants. In the year 2000, the global production of surfactants, which includes soaps, reached a substantial 17 million metric tonnes. Projections suggest a prospective annual growth rate of 3%–4% on a global scale in the coming years (Rahman and Gapke 2008). These surfactants are usually chemically synthesized surfactants that are mainly petroleum-based products mostly non-biodegradable and toxic to their native environment. Also, these compounds may bio-accumulate during their production and release of by-products that are equally hazardous. With a growing awareness of the imperative to safeguard the ecosystem, environmental scientists are intensifying environmental regulations. This has led to a heightened interest in surfactants derived from microbes as potential alternatives to those that are chemically synthesized (Benincasa 2007).

There are various advantages and applications of biosurfactants in agriculture, medicine, petroleum, and industry.

10.3.1 Biodegradability

Biosurfactants are generally more environmentally friendly than chemical surfactants because they are produced by living organisms and are biodegradable. They can be broken down by natural processes, reducing their impact on ecosystems compared to many synthetic surfactants. Biological surfactants are easily degraded by microorganisms (Mohan *et al* 2006).

10.3.2 Low toxicity

Biosurfactants are generally less toxic than their chemical counterparts. This lower toxicity can be beneficial in various applications, particularly in industries where the use of toxic substances is a concern. It was reported that biosurfactants depicted higher EC 50 than chemical surfactants that are highly toxic (Desai and Banat 1997).

10.3.3 Availability of raw materials

Biosurfactants are products of vast microbial flora of bacteria, yeast, and fungi. The biodiversity provides the different sources of extraction of biosurfactants with diversified properties and engaging applications. Thus, the range of raw materials varies from being cheapest to higher end depending on the need and application of the biosurfactants to be produced in large quantities. For instance, if a raw material is the carbon source, then it may come from hydrocarbons and carbohydrates of lower price compared to lipids of higher price, used separately or in combination with each other (Kosaric 2001).

10.3.4 Renewable source

Biosurfactants are often derived from renewable resources, such as microbial fermentation using agricultural by-products or waste materials. This contrasts with many chemical surfactants that are derived from non-renewable petrochemical sources.

10.3.5 Physical factors

Various environmental factors such as pH, temperature, and ionic concentration tolerance do not create an impact on most of the biosurfactants. As reported, the biosurfactant produced from *Bacillus licheniformis* known as lichenysin is unaffected by high temperatures up to 50 °C, can tolerate varying pH, in the broad band from acidic to basic covering 4.0–9.0, and is tolerant to high salt concentration of NaCl (50 g l^{-1}) and Ca (25 g l^{-1}) (Krishnaswamy *et al* 2008).

10.3.6 High specificity

Biosurfactants often exhibit high specificity in terms of the types of surfaces they can interact with. This specificity can be advantageous in certain applications, such as enhanced oil recovery, where targeted surface activity is desirable (Lin 1996).

10.3.7 Surface and interface activity:

Mulligan (2005) stated that a good surfactant can lower surface tension of water from 75 to 35 mN m^{-1} and the interfacial tension water/hexadecane from 40 to 1 mN M^{-1}. Surfactin possesses the ability to reduce the surface tension of water to 25 mN M^{-1} and the interfacial tension of water/hexadecane to <1 mN M^{-1} (Krishnaswamy *et al* 2008).

10.3.8 Effectiveness at extreme conditions

Some biosurfactants are effective under extreme conditions, such as high or low temperatures, high salinity, or extreme pH levels. This makes them suitable for use in environments where traditional chemical surfactants may not be effective (Darvishi *et al* 2011).

10.3.9 Synergy with chemical surfactants

Biosurfactants can be used in combination with chemical surfactants to enhance overall surfactant performance. This synergistic effect allows for improved surface activity and stability in various applications (Drakontis and Amin 2020).

10.3.10 Health and safety

Biosurfactants are often safer for human health compared to certain chemical surfactants that may cause irritation or other adverse effects. This makes them attractive for use in products that come into direct contact with humans, such as personal care items (Sharma and Sharma 2021).

Other advantages are biocompatibility and digestibility which allows their application in cosmetic, pharmaceuticals and as functional food additives (Marchant and Banat 2012).

10.4 Applications in bioremediation

Biosurfactants are employed in bioremediation processes to enhance the solubility and bioavailability of hydrophobic pollutants. Their ability to facilitate the degradation of pollutants makes them valuable in environmental clean-up efforts. While biosurfactants have these advantages, it is important to note that their production can sometimes be more expensive than that of chemical surfactants, and their properties may vary depending on the source organism and production conditions. Ongoing research aims to address these challenges and further optimize the use of biosurfactants in various industries (figure 10.2).

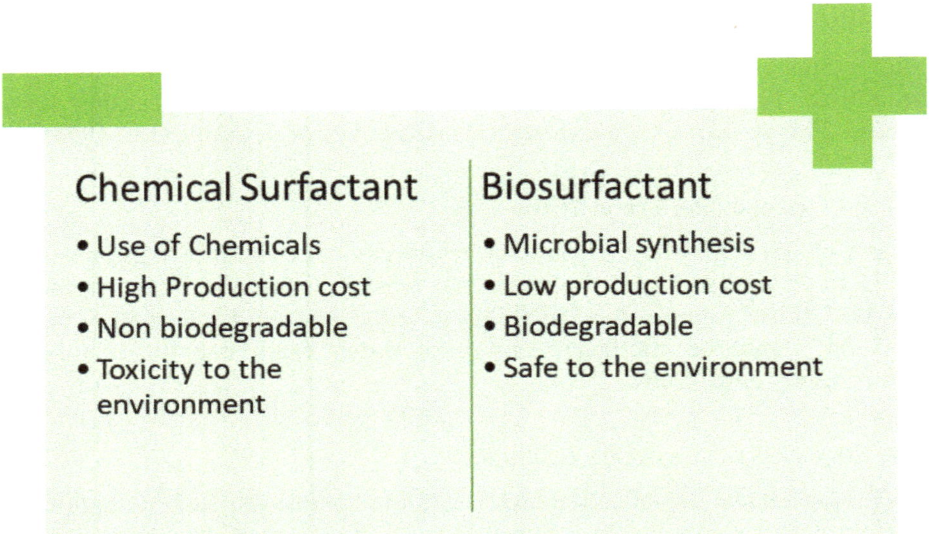

Figure 10.2. Advantages of biosurfactant over synthetic surfactant. Courtesy of A Karnwal (2023).

10.4.1 Biosurfactants in agriculture

Utilizing surfactants as mobilizing agents is a viable approach to enhance the solubility of bio-hazardous chemical compounds, such as polycyclic aromatic hydrocarbons (PAHs). This method effectively increases the apparent solubility of hydrophobic organic contaminants (HOCs). Additionally, surfactants play a crucial role in facilitating the adsorption of microbes to soil particles containing pollutants. This, in turn, reduces the diffusion path length between the point of absorption and the site where microorganisms take up the contaminants.

In the realm of agriculture, surfactants find application in the hydrophilization of heavy soils, ensuring optimal wettability and uniform distribution of fertilizers in the soil. They also serve to prevent the caking of specific fertilizers during storage and enhance the spreading and penetration of toxicants in pesticides, as highlighted by Makkar and Rockne (2003). Notably, rhamnolipid biosurfactants, primarily produced by the genus Pseudomonas, exhibit potent antimicrobial activity without posing anticipated adverse effects on humans or the environment through aggregate exposure.

Furthermore, Singh and Upadhyay (2021) stated that fengycins, recognized for their antifungal properties, present a promising avenue for biocontrol in plant disease management. This multifaceted role of surfactants and biosurfactants showcases their significance in various fields, ranging from environmental remediation to agricultural practices, with the potential to address challenges related to solubility, microbial interaction, and disease control.

10.4.2 Biosurfactants in commercial laundry detergents

The majority of surfactants, crucial components in contemporary commercial laundry detergents, are produced through chemical synthesis and pose a threat to freshwater organisms due to their toxicity. Increased public awareness of the environmental dangers associated with these chemical surfactants has prompted a quest for eco-friendly, natural alternatives in laundry detergents. Biosurfactants like cyclic lipopeptide (CLP) offer a sustainable solution, maintaining stability across a broad pH range (7.0–12.0) and retaining their surface-active properties even under high-temperature conditions (Kaur *et al* 2023). These biosurfactants exhibit efficient emulsion formation with vegetable oils and showcase excellent compatibility and stability when incorporated into commercial laundry detergents (El-Khordagui *et al* 2021).

10.4.3 Biosurfactants as biopesticide

Traditional methods of arthropod control rely on the use of broad-spectrum chemicals and pesticides, but these often result in unintended consequences. The development of pesticide-resistant insect populations, coupled with the escalating costs of new chemical pesticides, has prompted a quest for alternative, environmentally friendly vector control solutions. Lipopeptide biosurfactants, synthesized by various bacteria, have demonstrated insecticidal activity against the fruit fly *Drosophila melanogaster*. As a result, these biosurfactants hold promise as a biopesticide, offering a potentially eco-friendly approach to arthropod control (Nalini *et al* 2023).

10.4.4 Biosurfactants as antimicrobial agents

Biosurfactants find diverse applications in the field of medicine, particularly due to their antimicrobial properties. The varied structures of biosurfactants contribute to their ability to exhibit versatile performance. Their structural characteristics enable biosurfactants to impact cell membrane permeability, resembling a detergent-like effect and resulting in toxicity (Santos *et al* 2016). Numerous biosurfactants have been reported to possess robust antibacterial, antifungal, and antiviral activities. These surfactants also act as antiadhesive agents against pathogens, making them valuable for the treatment of various diseases and as probiotic and therapeutic agents. As reported, the lipopeptide produced from marine sp. of *Bacillus circulans* exhibited antimicrobial action against most of the Gram-negative and positive pathogens along with neutralizing semi-pathogenic strains including multi-drug-resistant strains (Gaur and Manickam 2021).

10.4.5 Biosurfactants possessing anticancer activity

Certain extracellular glycolipids are produced by a microbe's trigger cell differentiation rather than cell proliferation in the human promyelocytic leukaemia cell line. Additionally, when Rat pheochromocytoma PC12 cells (PC 12 cells) are exposed to mannose erythritol lipid (MEL), it enhances the activity of acetylcholinesterase and disrupts the cell cycle at the G1 phase. This disruption results in the excessive growth of neurites and partial cellular differentiation. These findings suggest that MEL has the ability to induce neuronal differentiation in PC12 cells, laying the foundation for considering microbial extracellular glycolipids as innovative agents for treating cancer cells (Jayashree and Murugaragavan 2021).

10.4.6 Biosurfactants acting as antiadhesive agents

Biosurfactants exhibit the ability to hinder the adhesion of pathogenic organisms to solid surfaces or infection sites. When vinyl urethral catheters were pre-coated with a surfactin solution before being inoculated with media, a noticeable reduction in biofilm formation occurred for *Salmonella typhimurium*, *Salmonella enterica*, *E. coli*, and *Proteus mirabilis*. In another instance, treating silicone rubber with the surfactant from *S. thermophilus* resulted in an 85% reduction in adhesion of *C. albicans*. Moreover, surfactants derived from *L. fermentum* and *L. acidophilus*, when adsorbed on glass, led to a 77% decrease in the number of adhering uropathogenic cells of *Enterococcus faecalis* (Silva *et al* 2021).

10.4.7 Biosurfactants as immunological adjuvants

When combined with conventional antigens, bacterial lipopeptides serve as powerful immunological adjuvants that are non-toxic and nonpyrogenic. The augmentation of the humoral immune response was evidenced when low molecular mass antigens such as Iturin AL and herbicolin A were utilized (Patel *et al* 2015).

10.4.8 Biosurfactants possessing antiviral activity

The literature has documented the antibiotic effects and growth inhibition of the human immunodeficiency virus in leucocytes through the use of biosurfactants (Ceresa *et al* 2021). Additionally, with the escalating incidence of HIV in women, there has been a growing demand for a female-controlled, effective, and safe vaginal topical microbicide. Among these, sophorolipid surfactants derived from *Candida bombicola*, particularly the sophorolipid diacetate ethyl ester and its structural analogues, have emerged as the most potent spermicidal and virucidal agents (Saikia *et al* 2021). It has been reported that this substance exhibits virucidal activity comparable to nonoxynol-9 against human semen (Thakur *et al* 2020).

10.4.9 Biosurfactants in gene delivery

The development of an effective and safe approach for introducing exogenous nucleotides into mammalian cells is deemed crucial for both basic sciences and clinical applications. In addition to this, surfactants have demonstrated medicinal advantages by serving as agents that stimulate stem fibroblast metabolism and exert immunomodulatory actions. Literature highlights that the deficiency of pulmonary surfactant, a complex of phospholipids and proteins, is linked to respiratory failure in prematurely born infants. However, the isolation of genes associated with the protein molecules of this surfactant and their subsequent cloning in bacteria has enabled the fermentation production of surfactants for medical applications (Bjerk *et al* 2021).

10.4.10 Biosurfactants in the food processing industry

The various aspects of food processing primarily including different food formulation and antiadhesive agents employs the use of biosurfactants. When they are part of food formulation, they reduce the surface and interfacial tension thereby acting as stabilizing agent for emulsions. They also act as anti-agglomerates, controlling agglomeration of lipid molecules by stabilizing the aeration and modifying agents for rheological properties of wheat products. This leads to enhancement in texture, taste, and shelf life of products with starch and fat as ingredients (Ribeiro *et al* 2020).

10.4.11 Biosurfactants in cosmetic industry

Biosurfactants act as substitutes for toxic chemical surfactants even in the cosmetic industry. Their properties of emulsification, water-binding, foaming, wetting and spreading make them a good emulsifier, foaming and wetting agent, solubilizer and cleanser. Apart from this, their antimicrobial property and as enzyme mediator, give them the place in the line of antacids, insect repellants, hygiene products, baby products, dentist cleanser and contact lens solution among the few to be listed (Benhur *et al* 2020).

10.4.12 Biosurfactants in petroleum

Biosurfactants and bioemulsifiers represent a novel category of molecules and stand out as some of the most potent and versatile by-products offered by modern microbial technology. These compounds play a crucial role in various applications, including the degradation of biocorrosion and biofouling in hydrocarbons within oil reservoirs, the enhancement of enzymes and biocatalysts for petroleum upgrading. Additionally, biosurfactants are significant contributors to the petroleum industry, impacting processes such as extraction, transportation, upgrading, refining, and petrochemical manufacturing (Xi *et al* 2021).

10.4.13 Biosurfactants in microbial-enhanced oil recovery

Microbial enhanced oil recovery involves leveraging microorganisms and their metabolic processes to boost oil production from marginally producing reservoirs. In recent times, microbial surfactants have gained widespread use in oil recovery. The mechanism facilitating oil release is attributed to the acidification of the solid phase. Specific microorganisms, including *B. subtilis*, *P. aeruginosa*, and *Torulosis bombicola*, have been documented to utilize crude oil and hydrocarbons as sole carbon sources, making them suitable for applications such as oil spill clean-ups (Geetha *et al* 2018).

10.5 Challenges and limitations

To restore the environmental loss due to over exploitation of synthetic surfactants, various sustainable measures have been taken that include bioremediation, biomineralization and biodegradation involving the microbial system. However, the above applications have certain constraints such as affinity to strong soil particles that makes their down-streaming process difficult. Secondly, low biological availability and low water solubility make them less approachable and more expensive (Ahamad *et al* 2018).

It has been reported that even with the recent biotechnological and environmental application advancements their use is very restricted. The main enigma is economy over sustainability. The higher cost of their recovery due to strong adsorption and pollutant hydrophobicity make their use less available. When compared with chemical surfactants that are cheap and readily available the market is still not open for the use of biosurfactants (Decesaro *et al* 2017).

Looking at the current scenario there is an urgent need for the development of the novel biosurfactant to meet the following challenges and overcome the limitations (figure 10.3.)

Production costs: One significant challenge in the use of biosurfactants is the production cost. Compared to chemical surfactants, the production of

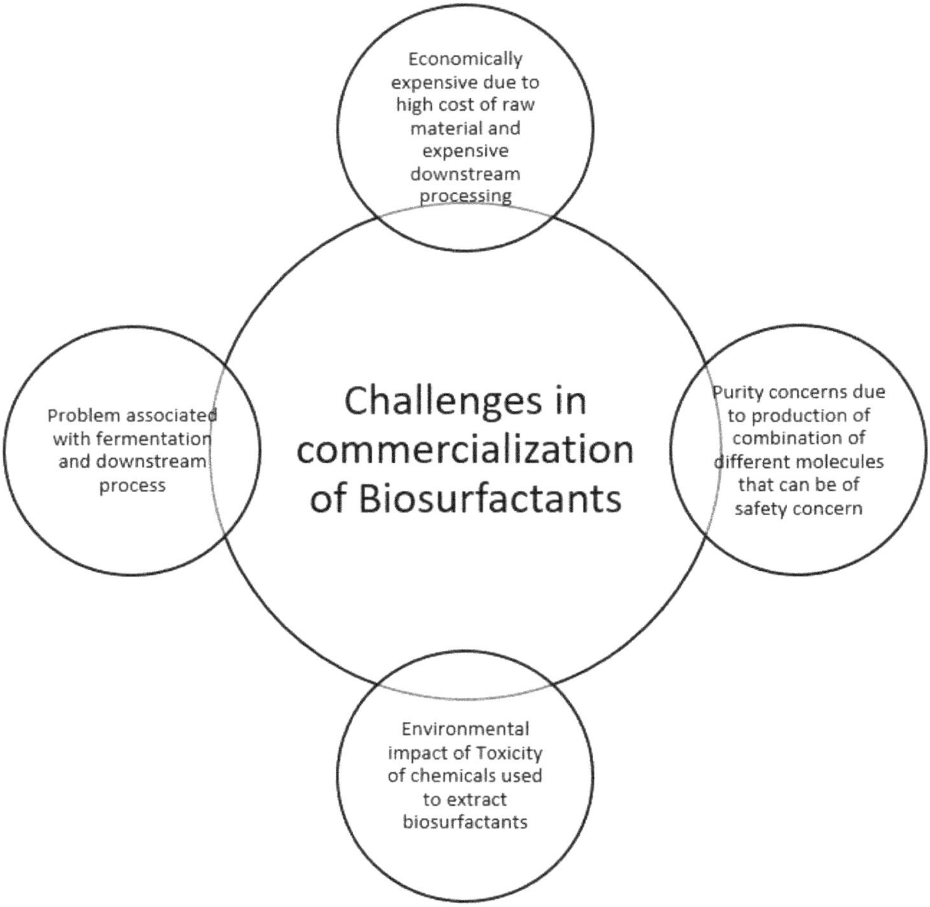

Figure 10.3. Challenges in commercialization of biosurfactants.

biosurfactants can be more expensive, which may limit their widespread adoption, especially in large-scale industrial applications. This cost challenge is often associated with the complex production processes and the need for specialized microbial strains (Banat *et al* 2014).

Scale-up issues: Scaling up the production of biosurfactants from laboratory to industrial levels can be challenging. Achieving consistent quality and yield on a large scale may pose difficulties, impacting the cost-effectiveness of biosurfactants. The quality yield on a large scale is a complex task that may require optimization (Kosaric and Sukan 2014).

Specificity: Some biosurfactants may exhibit specificity in terms of the types of surfaces they can interact with or the range of substances they can emulsify. This specificity may limit their utility in certain applications (Dolman *et al* 2019).

Strain specificity: The effectiveness of biosurfactants can be strain specific. Different microorganisms produce different types of biosurfactants, and finding the most suitable strain for a particular application can be a complex process (Dolman *et al* 2019).

Limited stability: Biosurfactants may have limited stability under certain conditions, such as extreme temperatures or pH levels. This can affect their performance and applicability in various industries (Gurkok 2021).

Purity and standardization: Achieving high purity and standardization of biosurfactants can be challenging. Impurities in biosurfactant preparations may affect their performance and safety in certain applications (Gurkok 2021).

Limited diversity: The range of biosurfactants available is not as extensive as that of chemical surfactants. This limited diversity may constrain their applicability in certain industrial processes. Compared to chemical surfactants, the structural diversity of biosurfactants is somewhat limited. This can impact their versatility and applicability in different industries and specific applications (Roelants *et al* 2019).

Market competition: Chemical surfactants dominate the market, and there may be resistance to adopting biosurfactants due to established practices and the ease of use of chemical alternatives. Chemical surfactants have a long history of use, and established products dominate the market. Introducing biosurfactants may face resistance or scepticism in industries accustomed to conventional surfactants (Roelants *et al* 2019).

Effectiveness in harsh conditions: Some biosurfactants may be less effective in harsh conditions, such as high salinity or extreme temperatures, limiting their application in certain industries like oil recovery from challenging environments (Schultz and Rosado 2020).

Fermentation issues: The fermentation process used to produce biosurfactants can be sensitive to variations in environmental conditions, affecting the yield and purity of the final product (Gurkok 2021).

Knowledge gaps: There are still gaps in our understanding of the production, properties, and applications of various biosurfactants. More research is needed to address these knowledge gaps and unlock the full potential of biosurfactants (Collins *et al* 2022).

Regulatory hurdles and approval: The regulatory landscape for biosurfactants may not be as well-established as that for chemical surfactants. This can pose challenges in terms of compliance and gaining approval for certain applications. The regulatory approval process for biosurfactants may be complex. Ensuring compliance with regulatory standards and obtaining approvals for the use of biosurfactants in specific industries can be time-consuming and may pose a barrier to adoption (Collins *et al* 2022).

Biodegradability variability: While biosurfactants are generally considered more environmentally friendly, the degree of biodegradability can vary among different types. Some biosurfactants may still raise environmental concerns (Roelants *et al* 2019).

Extraction and purification challenges: The extraction and purification processes for biosurfactants can be intricate and may involve multiple steps. Developing efficient and cost-effective methods for these processes is an ongoing challenge (Gurkok 2021).

Despite these challenges, ongoing research and technological advancements aim to overcome limitations and enhance the feasibility and applicability of biosurfactants in diverse industrial sectors.

10.6 Conclusion

In conclusion, the role of biosurfactants is significant and holds great promise across various industries and environmental applications. These naturally occurring surface-active molecules, produced by a range of microorganisms, exhibit unique properties that make them valuable in numerous contexts. Their role in environmental remediation, particularly in the biodegradation of hydrophobic pollutants and their ability to enhance the solubility and bioavailability of pollutants makes them effective agents in the clean-up of contaminated sites. In microbial-enhanced oil recovery and bioremediation efforts, biosurfactants facilitate the mobilization and degradation of hydrophobic compounds. This not only aids in the recovery of oil from reservoirs but also promotes the efficient removal of pollutants from soil and water (Roelants *et al* 2019).

Apart from their role in bioremediation and clean-up they act as antimicrobial, antiviral and immunological agents. Their application in medical and pharmaceuticals deploy their antimicrobial and emulsifying properties. They can be utilized in drug delivery systems, wound healing, and as agents against pathogenic microorganisms. From drug delivery systems to combating pathogens, biosurfactants showcase potential in improving healthcare technologies (Gaur and Manickam 2021).

Similarly, they hold important roles in commercial applications where biosurfactants offer sustainable alternatives to synthetic surfactants in various industrial processes, including food, cosmetics, and agriculture. Their biodegradable nature aligns with the growing demand for environmentally friendly products. In agriculture biosurfactants can enhance soil structure, nutrient availability, and water retention, contributing to improved agricultural practices. They have the potential to reduce the reliance on synthetic chemicals, promoting sustainable and eco-friendly farming supporting the circular economy. They find their unique place in cosmetics and personal care products as they can stabilize emulsions, enhance foaming, and contribute to the overall performance of such formulations providing functional benefits without the environmental drawbacks associated with some synthetic alternatives (Gurkok 2021).

The multifaceted roles of biosurfactants make them a valuable resource with diverse applications. As technology continues to advance and our understanding of these compounds deepens, biosurfactants are poised to play an increasingly vital role in addressing environmental, industrial, and medical challenges while contributing to a more sustainable and eco-friendly future. Biosurfactants offer a sustainable alternative to synthetic surfactants, contributing to efforts aimed at reducing the environmental impact of various industries. Their biodegradability and low toxicity make them attractive options for applications where eco-friendliness is a priority. Similarly, their ability to enhance the solubility and bioavailability of hydrophobic compounds is crucial in bioremediation processes. From oil spills to contaminated soil, biosurfactants play a vital role in facilitating the degradation and removal of pollutants, contributing to environmental clean-up.

Challenges and future directions: While the potential of biosurfactants is immense, challenges such as cost-effective production and large-scale extraction still need to be addressed. Ongoing research and technological advancements are crucial to overcoming these hurdles and unlocking the full potential of biosurfactants.

Future perspectives: Ongoing research into the isolation, production, and optimization of biosurfactants, along with advancements in biotechnological processes, will likely lead to increased commercialization and broader applications. Harnessing the potential of biosurfactants holds promise for addressing environmental challenges and promoting sustainable practices.

In summary, the exploration of biosurfactants opens up exciting possibilities for addressing environmental, industrial, and healthcare challenges. The interdisciplinary nature of biosurfactant research, combining microbiology, biotechnology, and engineering, positions these natural compounds as key players in the quest for sustainable and innovative solutions. As our understanding deepens and technology advances, once the cost and down-streaming process is stabilized, biosurfactants are poised to become integral components in creating a more sustainable and environmentally conscious future.

References

Abbot V, Paliwal D, Sharma A and Sharma P 2022 A review on the physicochemical and biological applications of biosurfactants in biotechnology and pharmaceuticals *Heliyon* **8** e10149

Adu S A, Twigg M S, Naughton P J, Marchant R and Banat I M 2022 Biosurfactants as Anticancer Agents: Glycolipids Affect Skin Cells in a Differential Manner Dependent on Chemical Structure *Pharmaceutics* **14** 360

Ahmad Z, Imran M, Qadeer S, Hussain S, Kausar R, Dawson L and Khalid A 2018 Biosurfactants for sustainable soil management *Advances in Agronomy* (New York: Academic) pp 81–130

Banat I M, Satpute S K, Cameotra S S, Patil R and Nyayanit N 2014 Cost effective technologies and renewable substrates for biosurfactants' production *Front. Microbiol.* **5** 697

Banat I M, Franzetti A, Gandolfi I, Bestetti G, Martinotti M G, Fracchia L, Smyth T J and Marchant R 2010 Microbial biosurfactants production, applications and future potential *Appl. Microbiol. Biotechnol.* **87** 427–44

Benhur A M, Pingali S and Amin S 2020 Application of biosurfactants and biopolymers in sustainable cosmetic formulation design *J. Cosmet. Sci.* **71** 455

Benincasa M 2007 Rhamnolipid produced from agroindustrial wastes enhances hydrocarbon biodegradation in contaminated soil *Curr. Microbiol.* **54** 445–9

Bjerk T R, Severino P, Jain S, Marques C, Silva A M, Pashirova T and Souto E B 2021 Biosurfactants: properties and applications in drug delivery, biotechnology and ecotoxicology *Bioengineering* **8** 115

Ceresa C, Fracchia L, Fedeli E, Porta C and Banat I M 2021 Recent advances in biomedical, therapeutic and pharmaceutical applications of microbial surfactants *Pharmaceutics* **13** 466

Christova N, Tuleva B, Kril A, Georgieva M, Konstantinov S, Terziyski I, Nikolova B and Stoineva I 2013 Chemical structure and *In Vitro* antitumor activity of rhamnolipids from *Pseudomonas aeruginosa* BN10 *Appl. Biochem. Biotechnol.* **170** 676–89

Collins T, Barber M and Rahman P K 2022 Biosurfactants industrial demand (market and economy) *Microbial Surfactants: Volume 2: Applications in Food and Agriculture* (CRC Press) ch 3

da Silva M D G C, Durval I J B, da Silva M E P and Sarubbo L A 2021 Potential applications of anti-adhesive biosurfactants *Microbial Biosurfactants* (Springer) pp 213–25

Darvishi P, Ayatollahi S, Mowla D and Niazi A 2011 Biosurfactant production under extreme environmental conditions by an efficient microbial consortium, ERCPPI-2 *Colloids Surf.*, B **84** 292–300

Das M, Patowary K, Vidya R and Malipeddi H 2016 Microemulsion synthesis of silver nanoparticles using biosurfactant extractedfrom *Pseudomonas aeruginosa* MKVIT3 strain and comparison of their antimicrobial and cytotoxic activities *IET Nanobiotechnol.* **10** 411–8

Decesaro A, Machado T S, Cappellaro Â C, Reinehr C O, Thomé A and Colla L M 2017 Biosurfactants during *in situ* bioremediation: factors that influence the production and challenges in evalution *Environ. Sci. Pollut. Res.* **24** 20831–43

Desai J D and Banat I M 1997 Microbial production of surfactants and their commercial potential *Microbiol. Mol. Biol. Rev.* **61** 47–64

Dolman B M, Wang F and Winterburn J B 2019 Integrated production and separation of biosurfactants *Process Biochem.* **83** 1–8

Drakontis C E and Amin S 2020 Biosurfactants: formulations, properties, and applications *Curr. Opin. Colloid Interface Sci.* **48** 77–90

El-Khordagui L, Badawey S E and Heikal L A 2021 Application of biosurfactants in the production of personal care products, and household detergents and industrial and institutional cleaners *Green Sustainable Process for Chemical and Environmental Engineering and Science* (Amsterdam: Elsevier) pp 49–96

Falavigna M, Stein P, Flaten G and di Cagno M 2020 Impact of mucin on drug diffusion: development of a straightforward in vitromethod for the determination of drug diffusivity in the presence of mucin *Pharmaceutics* **12** 168
Gaur V K and Manickam N 2021 Microbial biosurfactants: production and applications in circular bioeconomy *Biomass, Biofuels, Biochemicals* (Amsterdam: Elsevier) pp 353–78
Geetha S J, Banat I M and Joshi S J 2018 Biosurfactants: production and potential applications in microbial enhanced oil recovery (MEOR) *Biocatal. Agric. Biotechnol.* **14** 23–32
Ghadiri M, Young P and Traini D 2019 Strategies to enhance drug absorption via nasal and pulmonary routes *Pharmaceutics* **11** 113
Guo S and Dipietro L A 2010 Factors affecting wound healing *J. Dent. Res.* **89** 219–29
Gurkok S 2021 Important parameters necessary in the bioreactor for the mass production of biosurfactants *Green Sustainable Process for Chemical and Environmental Engineering and Science* (Amsterdam: Elsevier) pp 347–65
Haque F, Khan M S A and AlQurashi N 2021 ROS-Mediated necrosis by glycolipid biosurfactants on lung, breast, and skin melanoma cells *Front. Oncol.* **11** 253
Inamuddin , Adetunji C O and Ahamed M 2022 *Green Sustainable Process for Chemical and Environmental Engineering and Science Biomedical Application of Biosurfactant in the Medical Sector* (Cambridge, MA: Academic Amsterdam Elsevier) p 708
Ismail R, Baaity Z and Csóka I 2021 Regulatory status quo and prospects for biosurfactants in pharmaceutical applications *Drug Discov. Today* **26** 1929–35
Jain K K 2008 *Drug Delivery Systems—An Overview* **vol 437** (Berlin: Springer) p 150
Jayashree R and Murugaragavan R 2021 Biosurfactants and its applications *Sabujeema Int. Multi. e-Magaz* **4** 61–4
Kamalakannan S, Gopalakrishnan A V, Thangarasu R, Kumar N S and Vellingiri B 2020 Biosurfactants and anti-inflammatory activity: a potential new approach towards COVID-19 *Curr. Opin. Environ. Sci. Health* **17** 72–81
Karnwal A 2023 Prospects of microbial bio-surfactants to endorse prolonged conservation in the pharmaceutical and agriculture industries *ChemistrySelect* **8** e202300401
Kaur H, Kumar P, Cheema A, Kaur S, Singh S and Dubey R C 2023 Biosurfactants as promising surface-active agents: current understanding and applications *Multifunctional Microbial Biosurfactants* (Cham: Springer Nature) pp 271–306
Koley D and Bard A J 2010 Triton *X*-100 concentration effects on membrane permeability of a single HeLa cell by scanning electrochemical microscopy (SECM) *Proc. Natl. Acad. Sci. U.S. A.* **107** 16783–7
Kosaric N 2001 Biosurfactants and their applications for soil bioremediation *Food Technol. Biotechnol.* **39** 295–304
Kosaric N and Sukan F (ed) 2014 *Biosurfactants: Production and Utilization—Processes, Technologies, and Economics* **vol 159** (Boca Raton, FL: CRC Press)
Krishnaswamy M, Subbuchettiar G, Ravi T K and Panchaksharam S 2008 Biosurfactants properties, commercial production and application *Curr. Sci.* **94** 736–47
Lin S C 1996 Biosurfactants: recent advances *J. Chem. Technol. Biotechnol.: Int. Res. Process, Environ. Clean Technol.* **66** 109–20
Mahjoubi M, Cappello S, Souissi Y, Jaouani A and Cherif A 2018 Microbial bioremediation of petroleum hydrocarbon–contaminated marine environments *Petroleum Science and Engineering* ed M Zoveidavianpoor (London: IntechOpen)

Makkar R S and Rockne K J 2003 Comparison of synthetic surfactants and biosurfactants in enhancing biodegradation of polycyclic aromatic hydrocarbons *Environ. Toxicol. Chem.* **22** 2280–92

Marchant R and Banat I M 2012 Microbial biosurfactants: challenges and opportunities for future exploitation *Trends Biotechnol.* **30** 558–65

Marchant R and Banat I M 2012 Biosurfactants: a sustainable replacement for chemical surfactants? *Biotechnol. Lett.* **34** 1597–605

Mnif I, Ellouz-Chaabouni S and Ghribi D 2018 Glycolipid biosurfactants, main classes, functional properties and related potentialapplications in environmental biotechnology *J. Polym. Environ.* **26** 2192–206

Mohan P K, Nakhla G and Yanful E K 2006 Biokinetics of biodegradation of surfactants under aerobic, anoxic and anaerobic conditions *Water Res.* **40** 533–40

Mulligan C N 2005 Environmental applications for biosurfactants *Environ. Pollut.* **133** 183–98

Murgia X, Loretz B, Hartwig O, Hittinger M and Lehr C-M 2018 The role of mucus on drug transport and its potential to affecttherapeutic outcomes *Adv. Drug Deliv. Rev.* **124** 82–97

Nalini S, Sathiyamurthi S, Dhas T S and Revathi M 2023 Lipopeptide and rhamnolipid biosurfactant as biopesticides *Multifunctional Microbial Biosurfactants* (Springer) pp 171–87

Naughton P J, Marchant R, Naughton V and Banat I M 2019 Microbial biosurfactants: current trends and applications in agricultural and biomedical industries *J. Appl. Microbiol.* **127** 12–28

Nitschke M and Silva S S E 2018 Recent food applications of microbial surfactants *Crit. Rev. Food Sci. Nutr.* **58** 631–8

Ohadi M, Forootanfar H, Dehghannoudeh G, Eslaminejad T, Ameri A, Shakibaie M and Adeli-Sardou M 2020a Antimicrobial, anti-biofilm, and anti-proliferative activities of lipopeptide biosurfactant produced by *Acinetobacter junii* B6 *Microb. Pathog.* **138** 103806

Ohadi M, Forootanfar H, Rahimi H R, Jafari E, Shakibaie M, Eslaminejad T and Dehghannoudeh G 2017 Antioxidant potential and wound healing activity of biosurfactant produced by *Acinetobacter junii* B6 *Curr. Pharm. Biotechnol.* **18** 900–8

Ohadi M, Shahravan A, Dehghannoudeh N, Eslaminejad T, Banat I M and Dehghannoudeh G 2020b Potential use of microbialsurfactant in microemulsion drug delivery system: a systematic review *Drug Des. Devel. Ther.* **14** 541–50

Ozsoy Y and Güngör S 2011 Nasal route: an alternative approach for antiemetic drug delivery *Expert. Opin. Drug Deliv* **8** 1439–53

Panda A, Dewali K, Miglani S, Mishra R, Mandal M and Bisht S S 2023 Microbial biosurfactant: current trends and applications in biomedical metabolism *Recent Advancement and Future Perspective of Microbial Metabolites. Application in Biomedicine* (New York: Academic) pp 147–71

Patel M, Siddiqui A J, Ashraf S A, Surti M, Awadelkareem A M, Snoussi M, Hamadou W S, Bardakci F, Jamal A, Jahan S *et al* 2022 *Lactiplantibacillus plantarum*-derived biosurfactant attenuates quorum sensing-mediated virulence and biofilm formation in *Pseudomonas aeruginosa* and *Chromobacterium violaceum Microorganisms* **10** 1026

Patel M, Siddiqui A J, Hamadou W S, Surti M, Awadelkareem A M, Ashraf S A, Alreshidi M, Snoussi M, Rizvi S M D, Bardakci F *et al* 2021 Inhibition of bacterial adhesion and antibiofilm activities of a glycolipid biosurfactant from *Lactobacillus rhamnosus* with its physicochemical and functional properties *Antibiotics* **10** 1546

Patel S, Ahmed S and Eswari J S 2015 Therapeutic cyclic lipopeptides mining from microbes: latest strides and hurdles *World J. Microbiol. Biotechnol.* **31** 1177–93

Pattanathu K S M R and Gakpe E 2008 Production, characterisation and applications of biosurfactants-review *Biotechnology* **7** 360–70

Rahman P K S M 2008 Production, Characterization and application of Biosurfactants - review *Biotechnol.* **7** 360–70 XP002614719

Ribeiro B G, Guerra J M and Sarubbo L A 2020 Biosurfactants: production and application prospects in the food industry *Biotechnol. Progr.* **36** e3030

Rodrigues L R and Teixeira J A 2010 Biomedical and therapeutic applications of biosurfactants *Adv. Exp. Med. Biol.* **672** 75–87

Roelants S L K W, Van Renterghem L, Maes K, Everaert B, Redant E, Vanlerberghe B and Soetaert W 2019 Microbial biosurfactants: from lab to market *Microbial Biosurfactants and their Environmental and Industrial Applications* (CRC Press) pp 341–63

Roy A 2017 A review on the biosurfactants: properties, types and its applications *J. Fund. Renew. Energ. Appl.* **8** 1–5

Saikia R R, Deka S and Sarma H 2021 Biosurfactants from bacteria and fungi: perspectives on advanced biomedical applications *Biosurfactants for a Sustainable Future: Production and Applications in the Environment and Biomedicine* (Wiley) pp 293–315

Sajid M, Ahmad Khan M S, Singh Cameotra S and Safar Al-Thubiani A 2020 Biosurfactants: potential applications as immunomodulator drugs *Immunol. Lett.* **223** 71–7

Sambanthamoorthy K, Feng X, Patel R *et al* 2014 Antimicrobial and antibiofilm potential of biosurfactants isolated from lactobacilli against multi-drug-resistant pathogens *BMC Microbiol.* **14** 197

Santos V L, Drummond R N and Dias-Souza M 2016 Biosurfactants as antimicrobial and antibiofilm agents *Current Developments in Biotechnology and Bioengineering: Human and Animal Health Applications* (Elsevier) pp 371–402

Sarma H and Prasad M N 2021 *Biosurfactants for a Sustainable Future: Production and Applications in the Environment and Biomedicine* (Hoboken, NJ: Wiley) p 544

Schultz J and Rosado A S 2020 Extreme environments: a source of biosurfactants for biotechnological applications *Extremophiles* **24** 189–206

Sharma D and Sharma D 2021 Biosurfactants or chemical surfactants? *Biosurfactants: Greener Surface Active Agents for Sustainable Future: Microbial Surfactants* (Springer) pp 1–35

Singh S P and Upadhyay S K (ed) 2021 *Bioprospecting of Microorganism Based Industrial Molecules* (New York: Wiley)

Thakur S, Singh A, Sharma R, Aurora R and Jain S K 2020 Biosurfactants as a novel additive in pharmaceutical formulations: current trends and future implications *Curr. Drug Metab.* **21** 885–901

Thanomsub B, Pumeechockchai W, Limtrakul A, Arunrattiyakorn P, Petchleelaha W, Nitoda T and Kanzaki H 2006 Chemical structures and biological activities of rhamnolipids produced by *Pseudomonas aeruginosa* B189 isolated from milk factory waste *Bioresour. Technol.* **97** 2457–61

Van Hamme J D, Singh A and Ward O P 2006 Physiological aspects. Part 1 in a series of papers devoted to surfactants in microbiology and biotechnology *Biotechnol. Adv.* **24** 604–20

Xi W, Ping Y and Alikhani M A 2021 A review on biosurfactant applications in the petroleum industry *Int. J. Chem. Eng.* **2021** 1–10

Zhou C, Wang F, Chen H, Li M, Qiao F, Liu Z, Hou Y, Wu C, Fan Y, Liu L *et al* 2016 Selective antimicrobial activities and action mechanism of micelles self-assembled by cationic oligomeric surfactants *ACS Appl. Mater. Interface* **8** 4242–9

Chapter 11

Potential commercial applications of microbial surfactants

Ashlesha P Kawale, Nishant Shekhar, Pravat K Swain, S Y Bodakhe and Arti Srivastava

This comprehensive chapter delves into the versatile applications of microbial surfactants, specifically biosurfactants, across various industries, prioritizing sustainable and eco-friendly practices while highlighting their environmental impact. The discussion spans agriculture, cosmetics, personal care, detergents, textiles, and the leather industry. The chapter accentuates the crucial role of microbial surfactants in promoting sustainable agriculture, emphasizing their impact on plant growth promotion in the rhizosphere. It explores the dual hydrophobic/hydrophilic character and superior performance of microbial surfactants, comparing them favourably to synthetic counterparts in agriculture.

The chapter further dissects the mechanisms of action in plant growth promotion, encompassing biological nitrogen fixation, potassium and phosphorus solubilization, and phytohormone synthesis. It also delves into the biocontrol potential of microbial surfactants, showcasing their effectiveness in suppressing plant pathogens and addressing abiotic stresses like salinity. The production and application aspects of microbial surfactants in agriculture are discussed, highlighting their eco-friendly nature and suitability for various biotechnological applications, including bioremediation.

In the leather industry, the chapter explores applications such as cleaning, degreasing, oil residue removal, and bioremediation, emphasizing the eco-friendly practices in leather production. Sustainable leather production practices in India and advancements in leather processing and finishing are also covered, with a focus on enzymatic degreasing using microbial lipases. The chapter concludes with insights into reducing environmental pollution and waste in the leather industry.

The chapter underscores the substantial potential of microbial surfactants across industries and emphasizes collaborative efforts to achieve a greener and more sustainable future. Future perspectives and challenges in the application of microbial surfactants are outlined, emphasizing the need to address technical, regulatory, and consumer-related challenges for seamless integration into mainstream practices.

11.1 Agriculture and plant growth promotion

All nations are very concerned about increasing agricultural production to fulfil the world's expanding food needs. Reviving native soil systems is crucial for increased agricultural productivity, according to thematic research titled 'Sustainable Agriculture and Food Security in Asia and the Pacific' carried out by the United Nations Economic and Social Commission for Asia and the Pacific (ESCAP) in April 2009. Various biological modifications can be used to carry out such rejuvenation activities in an environmentally responsible manner. Numerous microorganisms that coexist in the rhizosphere—the soil that is affected by plant roots—share a mutualistic interaction with plants that has a positive impact on the plants. Rhizobacteria have been shown to support plant development through a number of processes [1–3]. As a result, rhizosphere biology is regarded as the most intense field of agricultural study. Surfactants have numerous functional qualities that are well recognized and used in a variety of business areas. Surfactants are also used in a variety of agricultural applications. The key areas where surfactants are used are listed in a review by Deleu and Paquot [4]. Surfactants were utilized in crop safety and agrochemical compositions at a rate of around 0.2 million tonnes in 2004. Several studies have shown that green surfactants (biosurfactants generated from microorganisms) outperform synthetic surfactants. Because there are few publications on the use of biosurfactants in agriculture, this study emphasizes the importance of biosurfactants and biosurfactant-generating bacteria from soil. A comparison of these green surfactants with chemically synthesized surfactants reveals that biosurfactants derived from microbial sources have a dual hydrophobic/hydrophilic character [5]. Given their antimicrobial activity and ability to increase beneficial plant–microbe interactions, these biosurfactants can be widely used in agricultural settings to promote indirect plant growth, enhance the biodegradation of pollutants, and improve soil quality.

11.2 Mechanisms of action of microbial surfactants in plant growth promotion

a) Mechanism of biological nitrogen fixation
   Among the elements that are most essential, nitrogen is required for plant development and a variety of metabolic processes. Although it makes up 78% of all atmospheric gases, plants cannot use nitrogen (N_2) in this form. Biological nitrogen fixation (BNF) [6, 7] is the process by which some bacteria and archaea convert nitrogen to ammonia (NH_3) by use of a nitrogenase protein complex [8]. Natural nitrogen fixers come in many forms, and bacteria with the capacity to fix nitrogen dioxide (N_2) can take the place of synthetic fertilizers [9, 10]. Two types of nitrogen-fixing microorganisms exist: (i) symbiotic and (ii) non-symbiotic [6, 11–13]. *Frankia* and

Rhizobium species are frequently mentioned as symbiotic N_2 fixers in soil, however, non-symbiotic nitrogen-fixing microorganisms include diazotrophic plant growth-promoting rhizobacteria (PGPR), such as *Cyanobacteria, Azospirillum, Pseudomonas, Azotobacter, Acetobacter*, and *Nostoc* [11]. In the natural world, non-symbiotic bacteria play a significant role; although they fix a smaller amount of nitrogen than symbiotic bacteria, they nonetheless supply enough for host plants to satisfy their needs. Root nodules are the product of complicated steps of infection and formation connecting symbionts and the roots of leguminous plants [14]. Nitrogen fixation requires energy in the form of ATP (adenosine triphosphate), which is produced by oxidative phosphorylation of bacterial carbon resources. The energy produced is then stored as glycogen [15]. The nif (nitrogenase complex) genes are responsible for nitrogen fixation throughout both symbionts and free-living non-symbionts [16]. The enzymes needed to fix atmospheric N_2 into the kind of nitrogen that plants can use are encoded by the nif genes. The nitrogenase complex is the principal enzyme that is encoded by nif genes. The nif genes generate regulatory proteins for N_2 fixation in addition to enzymes. The stimulation of nif gene expression is caused by the concentration of oxygen (O_2) and low nitrogen levels in the root environment of the host plant [15].

While oxygen is necessary for *Rhizobium* and bacteroid species respiration, it also inhibits the activity of the nitrogenase enzyme and is a negative regulator of nif genes. Within root nodules, bacteroid respiration is supported by enough O_2 because of bacterial haemoglobin's ability to bind free oxygen radicals. At the same time, O_2 is kept from impeding N_2 fixation [11].

b) Potassium solubilization

After nitrogen, potassium (K), the seventh most prevalent element found in the core of our planet, is another essential macronutrient. It is essential to many plant processes connected to growth, development, and metabolism. In addition, approximately 80 enzymes that regulate starch synthesis, nitrate reduction, photosynthetic activity, and other energy metabolic activities are activated by potassium [17–19]. An appropriate supply of K also regulates an increase in plant vigour resistant to numerous biotic stressors, diseases, and pests [20]. The potassium content in soil varies from 0.04% to 3%; however, plants can only use 1–2% of the overall concentration, with the remaining 98%–99% being fixed in their mineral state [21].

Mineral K, which is inaccessible to plants, makes up the rest, or 90%–98% of soil K. Exchangeable, non-exchangeable, solution, and mineral K are among the types of K that may be found in soil [21]. In certain soils, K in solution form may seep through and is quickly ingested by microorganisms and plants. The typical range of solution K in agricultural soils is 2 to 5 mg L −1. Feldspar and mica are two minerals that contain K. In some clay minerals, unexchangeable K combines to oxygen atoms in the interconnected layers to form a permanent state.

c) Phosphorus solubilization

The most important macroelement, phosphorus (P), is necessary for sufficient plant feeding. It is essential to several plant metabolic processes, energy transmission, and photosynthesis [22]. Most soils have an excess of P, but this surplus is not accessible to plants [23], with just 0.1% of the total phosphorus in the soil being used by plants [24–26]. Complex interactions between phosphorus and different cations, such as Fe^{3+} and Al+ in acidic soils and Ca^{2+} and Mg^{2+} into alkali soils, immobilize phosphorus in soil [6, 27].

P is soluble in soil due to the action of several microbial species, such as fungi and bacteria [24, 28]. Reducing a soil's C:P ratio and increasing the availability of P for crop plants can be achieved by integrating organic materials with a modest dosage of chemical fertilizer [29]. The ability to solubilize P is possessed by around 50% among all bacteria [23]. The most important PSBs include *Serratia*, *Burkholderia*, *Microbacterium*, *Bacillus*, *Rhizobium*, *Pseudomonas*, and *Azotobacter* [30]. Organic solutes are released by some soil bacteria, such as actinobacteria, and inorganic phosphate is dissolved [11]. A significant process in which COO− (carboxyl group) and OH− (hydroxyl ion) function as chelators of cations involving Fe, Al^{3+}, and Ca^{2+} and compete for P sites of adsorption in soil is allow to eliminate organic acid by microbes to dissolved inorganic P bound to soil colloids [22, 31, 32]. Low molecular weight organic acids known as chelators fight for P fixation sites with iron and aluminium oxides [33].

d) Phytohormones

Organic components known as phytohormones or regulatory substances for plant growth stimulate the development of plants [34]. Plant growth with fast development are stimulated because of PGPR through the synthesis of phytohormones. Gibberellins, cytokinin, abscisic acid, ethylene, and auxin are examples of phytohormones that can stimulate root cell blossoming by having neighbouring roots produce more than they should, which increases the intake of nutrients and water. Plants have an elaborate and complicated mechanism in place to withstand stress. To enhance plant development, microorganisms employ several biochemical and molecular strategies. By modifying hormone levels, increasing food availability, and promoting resistance to pathogens, PGPR promotes the development of plants [35, 36].

Additionally, they produce certain metabolites that regulate plant infections in their roots zone. For instance, by reducing the synthesis of ethylene, rhizobitoxine promotes plant growth and expansion under stressful circumstances [37]. Furthermore, several bacteria possess sigma factors that alter plant gene expression to refine their capability to survive under stressful surroundings [38].

11.2.1 Effect on plant growth and development

Soil can become contaminated with substantial amounts of metalloids and residual metals due to various factors such as pollution from the emerging industrial sector, petrochemical spills, combustible fuel, mine waste, pesticides, sewage solids, atmospheric deposition, and residues from coal combustion [39, 40]. In industrial waste, the most prevalent inorganic contaminants are arsenic, chromium, cadmium, lead, nickel, copper, mercury, and zinc [41]. In addition to causing insect resistance, pesticides can alter plants' morphology, physiology, biochemistry, and molecular make-up in ways that negatively impact growth and yield [42]. Recent studies have shown that among the microorganisms utilized in bioremediation, growth-promoting rhizobacteria (PGPR) are increasingly utilized because of their diverse capacities to both detoxify and breakdown toxins and to significantly promote plant development [43, 44]. Their interactions may facilitate PGPR's capacity to enhance plant development and mitigate trace metal toxicity [40]. This is due to its richness in complex organic compounds and trace metals. As a result, it is critical to use PGPR to promote plant development having normal circumstances, abiotic stress, and plant-borne pathogen assault, as these results may not be reached in the absence of PGPR.

11.2.2 Enhancement of nutrient uptake and utilization

Although the term 'nutrient uptake' has many different meanings, it always refers to an estimate of productivity per unit of consumption or loss of nutrients, as a minimum it is often assessed in perennial plants [45]. This estimate is based on the plants' capacity for effective intake of nutrients from the soil. It is also dependent on the environment, transportation, storage, mobilization, and utilization inside the facility. The reduced quantity of nutrients in the litterfall is often indicative of efficient nutrient usage [46, 47]. Since it incorporates a number of physiological processes that determine how many nutrients ingested into plants are typically used for the creation of biomass, nutrient absorption is a crucial ecological metric [48]. However, highly varied nutrient uptake estimates can be obtained depending on the techniques employed to evaluate productivity, nutrient absorption or loss, or both [49]. Utilizing vital plant nutrients during crop production is crucial in intensive agriculture to boost yield and preserve the long-term viability of the cropping system.

Considerations for the sources, fluxes, and destiny of nutrients include native nutrient budgets, inputs from biologically fixed or fertilizer sources, and the roles they play in plant absorption, different loss mechanisms, or soil build-up to increase fertility [50].

11.2.3 Promotion of root growth and development

The primary factor influencing root growth is exudates, which can be either lysates from the autolysis of epidermal and cortical cells or secretions that are actively released from the root or diffusates that are passively released as a result of osmotic differences between the soil solution and the cell [51, 52]. The majority of root exudates include lignins, proteins, carbohydrates, amino acids, organic acids, flavonols, and enzymes; they also include ions, free oxygen and water, enzymes, mucilage, and a variety of primary and secondary metabolites that contain carbon [50].

The amount and composition of root exudates are influenced by the genotype of the plant, the stage of growth, and parameters such as temperature, pH of the soil, atmospheric CO_2 concentration, nutrients, and the water-holding capacity (WHC) of the rhizosphere [50, 53].

Conversely, plants may either promote or suppress the growth of particular rhizospheric bacteria by discharging secondary metabolites into the rhizosphere [54]. This demonstrates how the interaction that is active in the rhizosphere is complex and happens in several directions for mutual gain.

11.2.4 Protection against environmental stressors

The main abiotic stresses that impair development and growth is salinity [55]. Activities carried out either by humans or by Nature that raise the degree of saturated salts—most notably sodium chloride—in the soil are what lead to soil salinization. Primary sources of salinity include airborne salts from oceans and sand dunes that accumulate inland, precipitation that washes these salts downstream, weathering of rocks and minerals that releases soluble salts, and the inflow of seawater followed by succeeding retreat [56]. It is well known that inoculating plants with growth promoters, particularly rhizosphere, can regulate abiotic stress through both direct and indirect mechanisms, hence inducing systemic tolerance [57].

11.3 Microbial surfactants as biocontrol agents

Stanghellini and Miller [58] showed that rhamnolipids may rupture zoospore membranes and induce lysis of zoospores of several oomycete plant diseases, so demonstrating the potential of biosurfactants for biological control. Numerous strains of *Bacillus* have been commercially sold for biocontrol purposes, and as such, they comprise a significant portion of the available economic biocontrol agents [59]. In the majority of instances, however, it is unknown whether the disease-suppressive effects are related to the formation of CLP. Ongena and Jacques [59] have provided a summary of the majority of research indicating that *Bacillus* CLPs have a function in biocontrol. However, a number of the biosurfactants that pseudomonads create have potent antibiotic properties when examined *in vitro*, and the strains that produce them are frequently effective biocontrol agents. Nevertheless, it is frequently unclear how biosurfactants contribute to the antagonistic potential of biocontrol strains *in situ*.

It is evident that the *in vitro* pathogen inhibition method cannot be applied to the field, a far more complex setting whereby the biocontrol mechanism may alter greatly. Furthermore, many environmental factors may have an impact on the significance of a particular metabolite for the effective use of a biocontrol agent. *In situ* biosurfactant production levels can also be influenced by a diversity of variables, including the host plant, the properties of the soil, the existence of organisms that can either promote or hinder the synthesis of certain chemicals, etc. Application techniques employed in infection studies, such as the creation of bacterial inoculum, might also affect how well research aims to understand biocontrol processes.

11.3.1 Suppression of plant pathogens

Plant pathogen suppression is acknowledged to exist in a certain area or region, although crops in specific areas do not appear to be impacted. The following physical and chemical aspects of soil will influence suppressiveness: pH, moistness content, texture, organic matter content, and mineral or combination of nutrients. Enhancing soil structure and texture, adding organic manures, restricting, and using certain inorganic fertilizers can all promote suppressiveness. [60].

11.3.2 Induction of plant defence mechanisms

Heterotrophic organisms rely on the presence of organic nutrients and can either feed on live things or make use of dead organic materials. Plants provide a nutrition supply for a wide variety of microbial infections. It is clear that a plant has to be able to identify an intruder and mount a suitable defence in order to protect itself against such a diverse array of infections. Prior physical or chemical barriers, such as an abundance of trichomes and a waxy cuticle or poisonous secondary metabolites, create the initial line of defence [61]. Usually, these built-in defences are enough to keep viruses from damaging the plant. But when this first line of defence fails, plants trigger other defence mechanisms, which are commonly referred to as 'induced defence.' Physical defences like thicker cell walls and chemical defences like the synthesis of antibacterial chemicals are examples of this. Therefore, passive defences that are continuously present or active defence mechanisms that are only generated upon assault might mediate resistance against microbial infections.

11.3.3 Enhancement of beneficial microbial communities in the rhizosphere

Plant growth promoters specially rhizosphere that are efficient and critical components of soil nutrient cycling in the soil–plant–microbe system [50]. The structure and functions of the microbial community in the soil system influence the state of soil and plant health, as well as the richness of the soil nutrient pool [62]. These microorganisms play a crucial role in nutrient solubilization, mobilizing, mineralization, dissolving, and absorption [63]. Their varied functions enhance plant development and disease suppression. The nutrients solubilization (P, K, Zn, and S), siderophore synthesis, denitrification, immunological regulation, signal transmission, and infection control are some of the well-known rhizospheric microbial community-mediated mechanisms that enhance plant growth.

11.4 Production and application of microbial surfactants in agriculture

In the event of the presence of water-soluble and/or water-insoluble substrates, microorganisms produce microbial surfactants extracellularly in aqueous media or intracellularly via *de novo* pathway and/or assembly from other substrates, resulting in variations in the structure or production domain within the organisms [64, 65]. The right choice of microbial strains, kind of substrate, and fermentation technique are necessary for the creation of these biomolecules [66].

Due to their multipurpose qualities and microbial generation, biosurfactants have become more in demand in recent years. The environmental friendliness of these biomolecules is also crucial since they are safer and more biodegradable than synthetic surfactants. Because of these exceptional qualities, biosurfactants can be used in biotechnological applications as alternatives to chemical surfactants, such as the petroleum, agricultural, food processing, cosmetics, detergent, leather, textile, paper, and medicinal product sectors [67, 68]. Additionally, oil spill clean-up, bioremediation of polluted soil and water, and recovery of oil residue from storage tanks are all done with the help of biosurfactants [69, 70].

11.4.1 Production methods of microbial surfactants

Below is a detailed discussion of the several physicochemical and nutritional factors that affect the productivity of microbial surfactants:

(a) Carbon: Hydrocarbon groups, fats and oils, and carbohydrates are popular carbon sources added to fermentation medium for the microbial surfactant synthesis process [70]. These carbon sources impact the composition of biosurfactants and are categorized as water-soluble (such as glucose, sucrose, and glycerol) and water-insoluble (such as vegetable and crude oils) [71, 72]. However, glucose is the carbon source that bacteria use the most because it is simpler to metabolize via the glycolytic pathway, producing the required metabolites [70].

(b) Nitrogen: Another crucial ingredient for microbial development and the synthesis of biosurfactants is nitrogen [73]. The synthesis of biosurfactants uses a variety of organic and inorganic nitrogen sources [74]. *Pseudomonas aeruginosa* is grown on mineral salt medium by the effects of various nitrogen sources, such as $NaNO_3$, $(NH_4)_2SO_4$, and CH_4N_2O. In comparison to other nitrogen sources, sodium nitrate was more efficient and produced 3.16 g l^{-1} of rhamnolipids.

(c) Physiochemical criterion: Physical and chemical factors, including pH, temperature, metal ions, oxygen needs, and agitation speed, are important in determining how bacteria grow and/or activate metabolically in order to produce biosurfactants. *Pseudomonas* sp. produced a surprising maximum amount of rhamnolipids [75]. In addition, because the biomolecules are created at different times, the incubation duration is important for the generation of microbial surfactants.

11.4.2 Formulation and delivery of microbial surfactants

Different scholars have provided different definitions for formulations. Formulations, according to Arora *et al* [76], are biologically active products that include one or more advantageous microbial strains in affordable, user-friendly carrier materials. According to a comprehensive definition, a bioformulation is a ready-to-use formulation that contains live cells or metabolites of them (from one or more strains), strengthened by inert and nontoxic substances to preserve the viability and effectiveness of the cells or metabolites and lengthen their shelf life.

Many microorganisms that demonstrate strong biocontrol activity against the pests they targeted in lab settings are difficult to apply in the field with the same effectiveness due to unknown environmental factors and intra- and inter-specific competition with other organisms in their niche, that affects the organisms' growth, physiology, metabolism, and gene activity in multiple ways [77]. It is vital to provide a supporting medium that can preserve live cells' vitality and viability throughout the whole life cycle, from creation to utilization, in order to get around these genetic and metabolic differences of living cells. The likelihood of bacteria performing at their best and being commercially successful in agro-food production is increased by well-designed bacterial preparations [78].

Seed injection and soil inoculation are two extensively used methods of delivery. Seed treatment is used to introduce bacterial formulations in powdered or liquid form to agricultural systems. Seeds are precoated in powdered formulations, however, adequate coating is difficult to obtain with small seed size. One significant downside of this strategy is the direct exposure of the bacterial formulation to severe environmental conditions prior to the establishment of the root system, which raises the danger of bioinoculant loss from the soil [78].

One more practical but more expensive way to administer bioformulation in the field is by soil inoculation. One way to accomplish this is to combine soil with granular, powdered, or encapsulated bioformulation. Soil inoculation takes a lot of inoculants to work well, even if it doesn't require any special equipment to administer in the field [79]. A substantial amount of bioinoculant applied to the soil eliminates a number of obstacles and improves the bioinoculant's ability to reach plant roots. Granular formulations are well-known in North America for improving field performance when applied in the seedbed during the time of sowing [78].

11.4.3 Application strategies of microbial surfactants in agriculture

Over the past few decades, there has been a notable surge in the worldwide market for microbial surfactants. This resulted from these biomolecules' exceptional functional qualities, which laid the groundwork for research in a variety of industries, including bioremediation, medicine, cosmetics, and food. The kind of substrate, fermentation process, and appropriate selection of microbial strains all affect the synthesis of microbial surfactants. From an economic perspective, microbial surfactant synthesis for commercial use was a significant impediment.

Profitable and economical surfactant manufacturing may be possible through the optimization of fermentation settings utilizing less expensive and renewable substrates in conjunction with reliable downstream processing techniques. Furthermore, significant quantities of surfactant synthesis that would be advantageous for commercial applications might be obtained by using hybrid and mutated hyper-producing microbes.

11.5 Cosmetics and personal care products

With a complicated structure, human skin is the biggest organ in the body. The skin's main job is to act as a barrier of defence against infections and dangerous chemicals that enter the body from the outside. It also keeps bodily fluids from being lost excessively. Although some bacteria help to trigger the body's natural defences, others generate antimicrobial substances (such bacteriocins) that stop infections and the spread of pathogens. Since the microflora in our skin are subject to external stresses including naturally occurring pollutants and the use of certain cosmetics and grooming products, it is imperative that we maintain a microbiome that is balanced for the entirety of human health. As a result, it is critical to make sure the microflora is present under the right environmental conditions. [80, 81]

11.5.1 Importance of microbial surfactants in cosmetics and personal care products

Several microorganisms, including fungus, yeasts, and bacteria, have been shown to be able to effectively produce biosurfactants. The amount and quality of biosurfactants generated during microbial culture development, however, are mostly determined by the kind of microbe utilized, the medium composition, the substrate properties, and other external and internal variables [82, 83]. Selecting the right strain of microbe is the first stage in the production of biosurfactants. Biosurfactant production can occur both within and outside of the microbe's cells when nutritional circumstances are not suitable [84]. Either the stagnant phase or the rapid phase of growth might see it happen. On the other hand, certain microbes cannot survive or produce surfactants [85]. Biosurfactants not only increase soil carbon absorption but also have physiological effects by increasing the bioavailability of hydrophobic molecules involved in cellular communication or maturation processes. Although the underlying physiological processes are still being investigated, biosurfactant synthesis in contaminated settings is considered to promote nutrition absorption from hydrophobic substances, boost cellular motility, and lower surface tension near the phase barrier [85].

11.5.2 Advantages of microbial surfactants over chemical surfactants in cosmetics and personal care products

The most obvious consequences of chemicals include disruptions to the skin's microbiota, inflammation, and other acute responses, which can be caused by the interaction of chemical surfactants with the epidermal layer of the skin [81, 86]. The precise processes behind chemical surfactants' possibly dangerous effects are not entirely understood. Conversely, it is hypothesized that these unfavourable outcomes result from the chemical surfactant characteristics, the concentration employed, and the duration of contact for the material to interact with the epidermis [87]. Chemical surfactants have the ability to upset the intracellular lipid equilibrium, leading to denaturation of proteins in skin cell membranes—a process referred to as delipidation [88]. Moreover, this may cause the stratum corneum to inflate acutely before deswelling. In addition to having an adverse effect on immunological responses, chemical surfactants can also harm skin-dwelling immune cells such as keratinocytes and Langerhans cells [89–91]. The concentrations and physical characteristics of the surfactants in solution—that is, whether they are monomers or micelles—have a significant impact on the possibility that chemical surfactants will permeate through skin layers and cause decomposition of proteins, hypersensitivity, or skin inflammation [89]. Therefore microbial surfactants are more helpful and harmless compared to chemical surfactants.

11.5.3 Applications of microbial surfactants in cosmetics and personal care products

A few of the primary roles of biosurfactants that are relevant to cosmeceuticals will be discussed here. Biosurfactants can be advantageous to the body in both direct and indirect ways, acting as transporters and promoters of cosmeceutical components.

11.5.3.1 Emulsifiers and stabilizers

The most common physical form in cosmetics for combining substances that are water-soluble, soluble in fat, and even insoluble into stable systems is an emulsion. Emulsions facilitate the solubilization of several active substances and improve their absorption into the skin and hair [92].

Emulsions are crucial delivery systems for the topical delivery of bioactive compounds. Emulsifiers and stabilizers are enhanced with ingredients such vitamins, anti-aging agents, humectants, moisturizers, antioxidants, and photoprotectors, providing advantages in the form of lotions, creams, and conditioners [93].

11.5.3.2 Foaming agents

One well-known function of surfactants is their capacity to foam. Adsorption of surfactant molecules onto the liquid layer enveloping the gaseous component is a dynamic occurrence. The structure of the molecule is precisely what determines the foaming, density, and flexibility of the foam [94]. For the interface to expand, a foam's elasticity needs to be low enough, yet high enough to preserve stiffness and, consequently, stability. A desirable quality for many cosmetic products, including shaving cream, soaps, and shampoos, is foaming. The intended application can change based on how stable the foam that a surfactant forms is. Certain situations call for a very stable foam, like bubble baths, while others call for an unstable foam for cleaning [95].

11.5.3.3 Cleansers and shampoos

In terms of economy, society, and culture, hair is very important. Among the most popular personal care products are hair products; shampoos in particular are used to cleanse the scalp and hair follicles. In addition, alopecia, dermatitis, pityriasis, dandruff, and lice are among the conditions that are treated with a lot of shampoos [96, 97]. Shampoos have been made over the years using a variety of plant extracts that contain biosurfactants. *Phyllanthus emblica*, *Ziziphus spina-christi*, *Citrus aurantifolia*, *Acacia concinna*, and *Sapindus mukorossi* extracts were used by Badi and Khan [98] to create a 'green' shampoo that demonstrated good cleansing and detergency, with a small bubble size, low surface tension, and stable foam after five minutes.

11.5.3.4 Conditioners and hair treatments

Conditioners are one type of hair product that moisturizes hair, giving it back its natural sheen and smoothness while also replenishing lost or damaged keratin and sebum [99]. Emollients (oily chemicals), thickeners, cationic surfactants, and supplementary emulsifiers make up conditioners [100]. Biosurfactants are active ingredients that may be used as humectants, thickeners, emulsifiers, carriers for moisturizing treatments, and even as conditioner preservatives. Ferreira *et al* [101] discuss the possible application of a biosurfactant made by *L. paracasei* in creams because of how well it works in O/W systems as an emulsifying agent when combined with antioxidant extract and essential oils.

11.5.3.5 Moisturizers and lotions

In the context of skincare, mannosylerythritol lipids (MELs), glycolipids derived from yeasts and fungi, are the most extensively researched biosurfactants. It has been observed that this kind of biosurfactant has antioxidant action, a moisturizing impact on dry skin, activates fibroblasts and papilla cells, and protects skin cells [102].

11.5.3.6 Anti-aging and skin rejuvenation products

Additionally, MELs brighten human melanocytes, suggesting a possible application in antimelanogenic cosmeceuticals to treat skin hyperpigmentation, including melasma and freckles [94]. According to a recent study, *Panax* saponins inhibited the growth of scars in rats and aided in the healing of skin wounds, suggesting that these natural products could be potential therapeutic candidates for reducing the formation of scars on skin and hastening the healing process [103].

11.5.3.7 Sunscreen products

Research has indicated that biosurfactants have the ability to offer physical shielding from the Sun's damaging rays and to improve the efficacy of other substances intended for the same reason.

Scientists examined the relationship between mica powder and a biosurfactant extract from maize steep liquor in order to raise the Sun protection factor (SPF) of a 'green' compound. Due to the biosurfactant extract's inherent ability to shield against UV rays and the formation of an emulsion with the mica powder, which enhanced the protective qualities of the mica by creating a stronger barrier and favouring physical protection, the addition of the biosurfactant to the aqueous emulsion containing mica powder increased the SPF [104].

11.5.4 Future perspectives and challenges

The significance of microbial surfactants in cosmetics and personal care products is underscored by their growing prominence. These biologically derived compounds offer distinct advantages over traditional chemical surfactants in the realm of beauty and personal care.

Microbial surfactants present notable benefits compared to their chemical counterparts. Their renewable sources align with the increasing demand for sustainability, contributing to environmentally friendly formulations. Additionally, the customization potential of microbial surfactants allows for tailoring formulations to specific functionalities, addressing diverse consumer needs.

The applications of microbial surfactants in cosmetics and personal care products are diverse. They serve as emulsifiers and stabilizers, enabling the creation of stable formulations. These surfactants also function as foaming agents, enhancing the sensory experience of products. In cleansers and shampoos, they contribute to effective cleaning, while in conditioners and hair treatments, they aid in product performance. Furthermore, microbial surfactants find application in moisturizers and lotions, offering skin hydration benefits. Their role extends to anti-aging and skin rejuvenation products, where they contribute to the overall efficacy. Additionally, microbial surfactants play a crucial role in sunscreen products, contributing to both formulation stability and skin protection.

Looking ahead, the future perspectives for microbial surfactants in the cosmetics industry are promising. Ongoing research endeavours focus on enhancing their properties, making them even more attractive for formulations. The integration of microbial surfactants with nanotechnology holds potential for innovative product development, providing more effective and targeted solutions. Collaborations between biotechnology and cosmetic industries may further streamline production processes, making microbial surfactants economically competitive.

However, challenges persist. Cost-effectiveness on a large scale remains a hurdle that necessitates continued research efforts. Ensuring the stability and shelf life of cosmetic products containing microbial surfactants, especially under varying environmental conditions, requires ongoing attention. Addressing regulatory complexities and educating consumers about the benefits of microbial surfactants are essential for wider acceptance.

In conclusion, microbial surfactants stand at the forefront of a sustainable and innovative future for cosmetics and personal care products. Overcoming challenges and capitalizing on evolving perspectives will be instrumental in realizing the full potential of these biologically derived compounds in the beauty industry.

11.5.5 Potential future applications of microbial surfactants in cosmetics and personal care products

The evolving landscape of microbial surfactants in cosmetics and personal care products presents promising avenues for future applications. As research and technology advance, these biologically derived compounds are poised to play diverse roles in shaping the beauty industry.

Microbial surfactants may find application in targeted drug delivery systems, enhancing the penetration of active ingredients and improving the efficacy of skincare products. Their potential to support a healthy skin microbiome opens doors to formulations designed to maintain a balanced microbial environment on the skin.

Beyond product formulations, microbial surfactants could contribute to sustainable packaging solutions, aligning with the industry's increasing focus on environmental impact. Their antimicrobial properties make them potential candidates for wound healing and scar reduction formulations, aiding in skin regeneration.

Advancements in customization and biotechnology may lead to the development of personalized skincare products, with microbial surfactants offering tailored functionalities to address individual skin needs. In hair care, these surfactants may enhance hair colour treatments and other formulations, contributing to improved colour retention and overall hair health.

Looking ahead, microbial surfactants could be integrated into wearable skincare technology, synergistically enhancing skincare routines and user experiences. The dynamic nature of microbial surfactants suggests a future where their applications continue to diversify, contributing to innovative solutions in the cosmetics and personal care industry.

11.5.6 Challenges and limitations in the use of microbial surfactants in cosmetics and personal care products, including regulatory requirements and consumer acceptance

It is a challenging effort to formulate cosmetics using only natural raw components. The choice of material is a problem since it must have accessibility and functionalities identical to those of its artificial equivalents [105]. Biosurfactants may be able to compete with commercial surfactants [73]. Still, the major obstacle is often the low level of concentration in manufacturing, particularly for microbial surfactants, which makes large-scale production challenging and expensive. Furthermore, it might be challenging for researchers to look for alternatives to reach a high concentration in the manufacturing due to a lack of funding and investments in some situations, particularly in third-world nations.

11.6 Detergents and cleaning agents

Detergents and cleaning agents play a crucial role in keeping our spaces clean and safe. These chemical formulations are designed to effectively get rid of dirt, stains, and germs from various surfaces, ensuring a sanitized and visually pleasing result. Detergents, commonly used for laundry, work by breaking down and getting rid of stains, making them easy to wash away. Cleaning agents, which are used in households, industries, and public spaces, are formulated to tackle specific types of dirt, grease, or microbes, providing a complete solution for cleanliness. The growing awareness of the environmental impact of traditional cleaning agents has led to the development of more eco-friendly and biodegradable formulations. As essential tools in our daily lives, detergents and cleaning agents not only contribute to our health and well-being but also help maintain the cleanliness of our surroundings.

In healthcare settings, detergents and cleaning agents are crucial for preventing the spread of pathogens and ensuring public health. They are recommended for disinfection and sterilization protocols, working alongside other antimicrobial agents. Surfactants, commonly found in detergents, are effective in lifting soils from surfaces and are considered essential for maintaining cleanliness in healthcare environments. Antimicrobial cleaning products, including detergents, play a vital role in preventing infections in hospitals, reducing costs, and following recommended disinfecting protocols.

In classrooms, the use of disinfectants and hand sanitizers has been proven to reduce illness-related absenteeism among children. Emphasizing the regulatory aspects of antimicrobials in cleaning products highlights the importance of following guidelines to ensure their effectiveness and safety.[73]

11.6.1 Performance and benefits of microbial surfactants in detergents and cleaning agents

Microbial surfactants in detergents and cleaning agents offer a range of performance benefits, contributing to the effectiveness and sustainability of these products. Surfactants are surface-active agents that play a crucial role in breaking down surface tension, facilitating the removal of dirt, grease, and stains. Here, the focus is on microbial-derived surfactants, which are produced by microorganisms and offer distinct advantages.

Biodegradability: Microbial surfactants are often biodegradable, making them more environmentally friendly compared to synthetic counterparts. Their natural origin and ability to break down into harmless byproducts contribute to reduced environmental impact.

Versatility: These surfactants exhibit versatility in their ability to function effectively across a broad range of pH levels and temperatures. This adaptability enhances their applicability in various cleaning scenarios, ensuring reliable performance in different environments.

Enhanced cleaning power: Microbial surfactants can enhance the cleaning power of detergents by improving the wetting, emulsifying, and dispersing capabilities. This leads to more efficient removal of stubborn stains, oils, and other contaminants from surfaces.

Low toxicity: In comparison to some synthetic surfactants, microbial surfactants often exhibit lower toxicity. This characteristic is particularly advantageous in terms of safety for both users and the environment, reducing potential health hazards associated with exposure.

Foaming properties: Some microbial surfactants exhibit excellent foaming properties, which can contribute to the perception of enhanced cleaning efficacy. Foaming aids in the mechanical removal of dirt and facilitates the overall cleaning process.

Renewable source: Microbial surfactants are derived from renewable resources, as they are produced by microorganisms. This contrasts with traditional surfactants derived from petroleum-based sources, aligning with the growing demand for sustainable and eco-friendly cleaning solutions.

Resistance to hard water: Microbial surfactants often show resistance to the effects of hard water, maintaining their effectiveness even in regions with water that has a high mineral content. This ensures consistent cleaning performance regardless of water quality.

Surfactants are the key ingredients in detergent formulations and play a crucial role in their performance. The use of surfactant mixtures, such as a combination of nonionic and anionic surfactants, has been found to enhance cleaning performance and reduce surface tension. Nonionic surfactants, in particular, contribute to making the surfactant system less sensitive to water hardness and increase solubility. The critical micelle concentration (CMC) of a mixed surfactant system can be lower than that of a single anionic surfactant, leading to improved cleaning power. The design and optimization of detergent products can be achieved using computational tools and methodologies, which can aid in the selection of surfactant mixtures and other additives. The final selection of surfactant mixtures and product composition should be verified experimentally to ensure desired properties and eliminate any undesirable interaction effects [106].

11.6.2 Comparison of microbial surfactants with chemical surfactants in detergents and cleaning agents

Comparison between microbial surfactants and chemical surfactants in detergents and cleaning agents reveals distinctions in origin, properties, and environmental impact (figure 11.1) [107].

The following points briefly highlight the differences.

11.6.2.1 Origin

Microbial surfactants: Derived from microorganisms through fermentation processes, offering a natural and sustainable source from renewable resources.

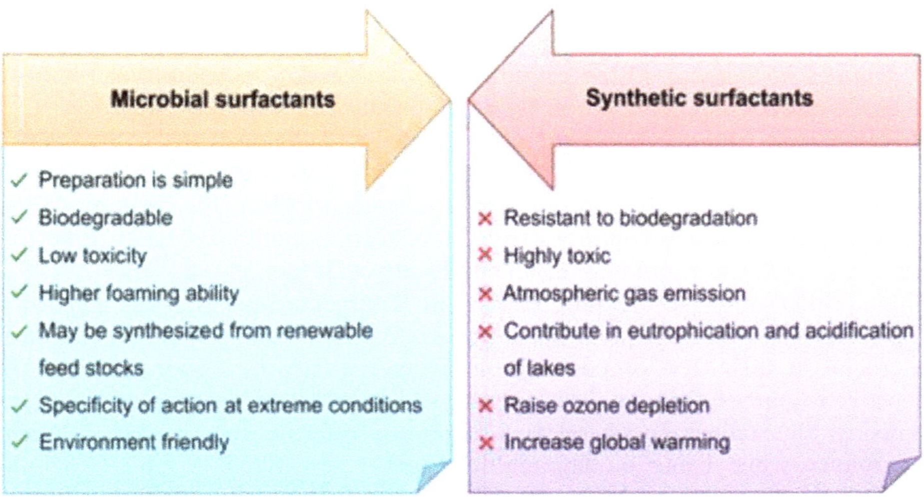

Figure 11.1. Basic differences between microbial and synthetic surfactants [107].

Chemical surfactants: Typically synthesized from petrochemicals or chemical processes, often originating from non-renewable resources.

11.6.2.2 Biodegradability

Microbial surfactants: Generally more biodegradable, undergoing natural processes and minimizing environmental impact.

Chemical surfactants: Some may have lower biodegradability, potentially persisting in the environment.

11.6.2.3 Environmental impact

Microbial surfactants: Tend to have a lower environmental impact due to their natural origin and biodegradability, offering a sustainable alternative.

Chemical surfactants: Production may involve non-renewable resources, contributing to environmental pollution with potential adverse effects on aquatic life.

11.6.2.4 Performance

Microbial surfactants: Exhibit comparable or superior performance in wetting, emulsification, and dispersion across a range of conditions.

Chemical surfactants: Historically widely used, providing effective cleaning, but performance may vary based on formulation.

11.6.2.5 Toxicity

Microbial surfactants: Generally lower toxicity, enhancing safety for users and the environment.

Chemical surfactants: Some may pose health and safety risks, especially with high exposure levels.

11.6.2.6 Cost

Microbial surfactants: Production costs may be higher, influencing the overall product cost.

Chemical surfactants: Often economically viable due to established industrial processes, but costs vary based on type and production method.

In conclusion, the selection between microbial and chemical surfactants depends on environmental considerations, desired performance, and cost. With a growing focus on sustainability, microbial surfactants gain attention for their eco-friendly properties. The industry continues to explore formulations that balance performance, cost-effectiveness, and environmental impact [108].

On another note, surfactants, crucial in detergents and cleaning agents, are categorized based on their polar head groups. Economic and environmental factors are crucial in their development, emphasizing eco-friendly options derived from renewable resources. Microbial surfactants, produced by microorganisms, emerge as promising alternatives to chemical surfactants, offering comparable detergency performance while being biodegradable. Further research is essential to unlock the full potential of microbial surfactants in commercial detergent formulations [109].

11.6.3 Applications of microbial surfactants in detergents and cleaning agents

Microbial surfactants exhibit diverse applications in the detergent and cleaning agent industry, owing to their distinctive properties and environmental benefits. These versatile compounds find utility in various sectors, showcasing their effectiveness and sustainability. Some key applications include:

Household detergents: Microbial surfactants are integral components in household detergents like laundry detergents, dishwashing liquids, and all-purpose cleaners. They augment wetting and emulsifying capabilities, improving the removal of dirt, grease, and stains.

Industrial cleaners: In industrial settings, microbial surfactants are incorporated into heavy-duty cleaners for equipment, machinery, and facility maintenance. Their efficacy in breaking down complex contaminants is valuable in manufacturing and automotive industries.

Personal care products: Microbial surfactants find use in personal care products such as shampoos, body washes, and facial cleansers due to their gentle yet effective cleaning action, making them suitable for direct contact with skin and hair.

Oil spill clean-up: In oil spill clean-up efforts, microbial surfactants reduce water surface tension, dispersing and breaking down oil. This facilitates its degradation by microorganisms, easing the clean-up process.

Agricultural and crop protection: Certain microbial surfactants enhance the effectiveness of pesticides and herbicides in agriculture by improving spreading and adherence to plant surfaces, leading to more efficient pest and weed control.

Bioremediation: Microbial surfactants play a crucial role in bioremediation by aiding in the breakdown of pollutants in soil and water. They enhance contaminant bioavailability to microorganisms, promoting degradation.

Food industry cleaning: In the food industry, microbial surfactants are integrated into cleaning agents for equipment, surfaces, and utensils. Their natural origin is advantageous where avoiding chemical residues is a priority.

Medical equipment sterilization: Microbial surfactants are applicable in cleaning solutions for medical equipment, ensuring effective contaminant removal and contributing to aseptic conditions in healthcare facilities.

Wastewater treatment: In wastewater treatment, microbial surfactants assist in separating oil and grease from water, improving treatment efficiency by facilitating the removal of hydrophobic contaminants.

Textile industry: The textile industry utilizes microbial surfactants in cleaning and finishing processes, aiding in impurity removal and enhancing fabric wetting properties during manufacturing.

The broad spectrum of applications underscores the versatility and efficacy of microbial surfactants in addressing diverse cleaning challenges across industries. As the demand for sustainable solutions grows, microbial surfactants offer a promising alternative to traditional chemical surfactants in cleaning applications. Biosurfactants can be applied as novel elements useful in managing and preventing outbreaks, or as alternative sustainable solutions in a number of contexts. Figure 11.2 below provides a summary of several uses.

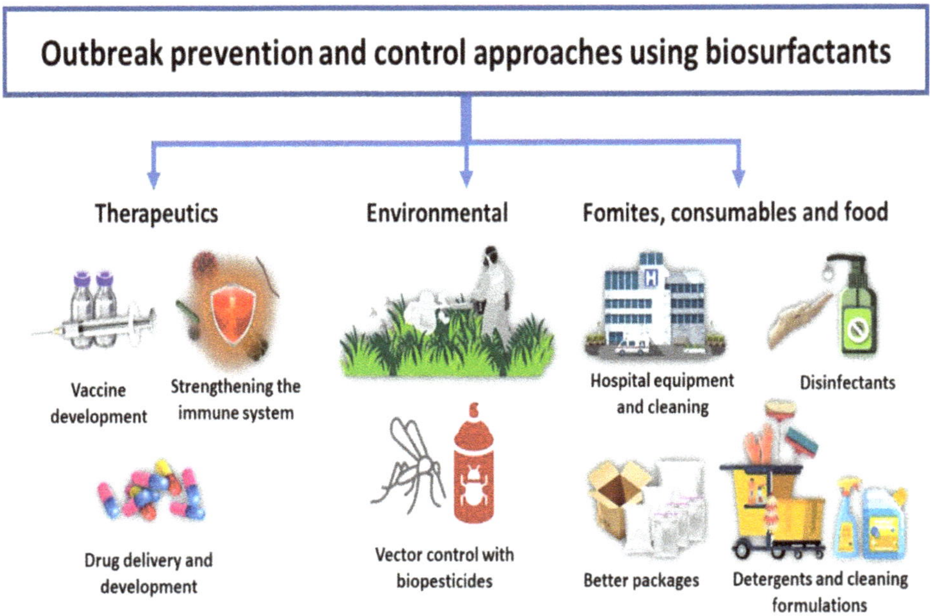

Figure 11.2. Revolutionizing outbreak defence: future-forward strategies with biosurfactants [111].

Additionally: Biosurfactants, a type of microbial surfactant, show promise in the detergent industry. They can serve as additives in laundry detergents, enhancing performance and stain removal. Biosurfactants, being multifunctional and environmentally compatible, are preferred over synthetic surfactants. Their use contributes to the biotreatment of oily wastewater, improving oil removal efficiency. Factors influencing microbial surfactant production include microbial strains, substrate type, and fermentation technology. Optimization of fermentation parameters and the use of recombinant and mutant hyperproducing microorganisms can boost surfactant production for industrial applications [109].

11.6.3.1 Laundry detergents and fabric softeners

Laundry detergents are formulated to eliminate dirt, stains, and odours from fabrics, employing surfactants, enzymes, and various cleaning agents. These detergents are available in powder, liquid, and pod forms, catering to different preferences and washing machine types, including specialized options for high-efficiency machines and colour-safe fabrics [110].

Fabric softeners, applied during the rinse cycle, enhance the fabric's texture, reduce static cling, and introduce a pleasing fragrance. Comprising quaternary ammonium compounds and emollients, fabric softeners are offered in liquid form or as dryer sheets. Users should consider potential skin sensitivity and the impact on towel absorbency when using these products.

Key points regarding cationic surfactants and fabric softeners include:

Cationic surfactants: Cationic surfactants like quaternary ammonium salts and imidazolinium salts are prevalent in fabric softening agents and laundry detergents. These compounds exhibit fabric softening and antistatic properties.

Fabric softener composition: Fabric softeners commonly contain aqueous dispersions of distearyl dimethyl ammonium chloride or tallow alkylated imidazolinium derivatives, with concentrations ranging from 1%–9%. The market has also introduced concentrated fabric softeners for consumer convenience.

Wash-cycle fabric softeners: Some fabric softeners are designed for use in the wash bath instead of the rinse bath. These formulations contain a higher concentration of active materials compared to rinse-cycle fabric softeners. This variation is prominent in the US market due to the absence of automatic fabric softener dispensers in many washing machines.

Solid fabric softeners: Solid fabric softeners, in forms such as spray-dried powder, ground product, granules, or flakes, are less commonly utilized. Commercial products typically incorporate compounds like DSDMAC or fatty acid amidoethyl substituted imidazolinium compounds.

Functions of fabric softeners: Fabric softeners address issues such as fibre pile disorder, electrostatic charging, and impart fluffiness and antistatic properties to fabrics. Additionally, they facilitate quicker drying and ironing, providing a pleasant sensation and long-lasting freshness to textiles [112].

11.6.3.2 Dishwashing detergents

Dishwashing detergents are designed for both manual handwashing and automatic dishwashers, with liquid versions being particularly popular for their quick dissolution and convenient dosing. These detergents employ surfactants and enzymes for efficient grease and food particle breakdown.

Introduction:

- Liquid detergents, including those for dishwashing, have gained favour due to their rapid dissolution and ease of dosing.
- Dishwashing detergents are indispensable for daily household surface care.
- Since the 1990s, alkylpolyglucosides (APGs), nonionic surfactants derived from renewable raw materials, have been integrated into dishwashing liquids. APGs boast desirable physical–chemical properties, skin mildness, and rapid biodegradability.

Formulation and optimization: The formulation of liquid dishwashing detergents involves a blend of anionic, nonionic, and zwitterionic surfactants, along with NaCl for viscosity control. Statistical analysis software, employing central composite experimental design, is utilized to explore the impacts of different components on key properties. The response surface methodology, employing a second-order design and analysis of experiments, helps pinpoint concentrations that meet the desired properties.

Investigated properties: Properties explored in the formulation and optimization of dishwashing detergents encompass washing performance, viscosity, cloud point, and emulsion stability [113].

11.6.3.3 Surface cleaners and disinfectants

Surface cleaners, equipped with surfactants, effectively remove dirt and stains from various surfaces. Disinfectants, containing quaternary ammonium compounds or alcohol, eradicate pathogens. Some products combine both functions for a comprehensive approach. When selecting and using these cleaners, it is crucial to consider surface compatibility, adhere to safety precautions, and ensure regulatory compliance.

Surface disinfectant cleaners (SDCs): SDCs, effective against gram-negative bacteria, demonstrate efficacy against multi-drug-resistant strains. Alcohol-based SDCs, incorporating amphoteric substances (AAS), oxygen-releasers (OR), and surface-active substances (SAS), exhibit effectiveness against clinically relevant bacterial species.

Quaternary ammonium compounds (QACs): QACs are commonly used in SDCs as surface-active ingredients, though resistance can develop. While QACs are more active against gram-positive bacteria, their efficacy against gram-negative bacteria is limited.

Preventing spread of multi-drug-resistant gram-negative bacteria (MRGN):

- Effective hygiene measures, including targeted surface disinfection, are crucial in preventing the transmission of MRGN.
- The primary influence on antibiotic drug resistance development is the use of antibiotics, emphasizing the importance of targeted surface cleaning and disinfection in healthcare settings.

Alcohols as surface disinfectants in healthcare:

- Isopropyl and ethyl alcohols serve as low-level disinfectants in healthcare settings.
- Recent studies highlight ethyl alcohol's efficacy against various viruses and bacteria, with 60%–70% solutions proving effective.
- Alcohol-containing products, often combined with other active agents like quaternary ammonium or phenolic compounds, are widely used for disinfecting healthcare facility surfaces.

Regulatory compliance: Out of 56 EPA-registered formulations with ethyl and/or isopropyl alcohol, 49 are designed for use in hospitals and healthcare settings.

In summary, the dual-action capabilities of surface cleaners and disinfectants play a vital role in maintaining hygiene. Understanding their efficacy, considering surface compatibility, and following safety guidelines are essential for optimal utilization, particularly in healthcare settings where preventing the spread of infections is paramount [113, 114].

11.6.3.4 Carpet cleaners

Carpet cleaners play a crucial role in eliminating dirt and stains from carpets, employing cleaning agents and surfactants. These products are available in various forms such as sprays and powders, with some specifically designed for use with carpet cleaning machines. Spot cleaners are tailored to target specific stains. Considerations for compatibility with carpet materials and ensuring proper ventilation are essential when utilizing these cleaning products.

Carpet cleaners with wood oils:

- Novel carpet cleaners were formulated using wood oils extracted from *Thujopsis dolabrata*, *Chamaecyparis obtusa*, and *Chamaecyparis taiwanensis*.
- At a concentration of 0.1%, these wood oil-containing carpet cleaners demonstrated significant effectiveness in rendering *D. pteronyssinus* inactive.
- Incorporating wood oils not only provided efficient mite control but also imparted a pleasant aroma, presenting a new potential for utilizing these wood oils in carpet cleaners.
- Research on wood oils has focused on their dual effects on humans and mites, considering both human comfort and health in shared living spaces.
- Wood oils are emerging as promising natural alternatives for mite control, potentially replacing chemical agents, addressing health concerns, and responding to the prevalent issue of mites in carpets [115].

In essence, the evolving landscape of carpet cleaners encompasses traditional efficacy considerations, innovative formulations, and a growing interest in natural solutions like wood oils, showcasing the industry's commitment to effectiveness, sensory experience, and health-conscious choices.

11.6.3.5 Industrial cleaning agents

Industrial cleaning agents are potent formulations engineered for robust cleaning in manufacturing and industrial environments. These formulations incorporate potent surfactants, solvents, and detergents to effectively combat stubborn grime, grease, and oils on equipment, machinery, and surfaces. Available in various forms, including specialized types like degreasers and acidic or alkaline cleaners, these agents prioritize safety precautions and regulatory compliance. Some formulations also emphasize environmental friendliness through the inclusion of biodegradable components.

Contact allergens in cleaning agents: Cleaning agents, whether for industrial or household use, undergo registration in the Danish Product Register Database (PROBAS) under special notification rules set by the Danish Environmental Protection Agency. As of February 1992, approximately 2350 registered washing and cleaning agents, containing around 1250 different chemical substances, were available in the Danish market. Major product types registered with contact allergens included general cleaners, skin cleaners, hair shampoos, and floor polishes.

Examples of contact allergens in various cleaning agents:

- General cleaners: Contained surface-active agents like coconut diethanolamide, 2-propanol, and polyethylene glycol mono(nonylphenyl) ether;

preservatives such as isothiazolinones, formaldehyde, bromonitropropanediol; corrosion inhibitor triethanolamine; and colourant tartrazine.

- Hair shampoos: Featured surface-active agents like coconut diethanolamide and propylene glycol; preservatives like chloromethyl and methylisothiazolinone, formaldehyde, formaldehyde releasers, and parabens; and colourant tartrazine.
- High-pressure cleaning agents: Primarily included polyethylene glycol mono (nonylphenyl) ether and other surface-active agents, along with corrosion inhibitor triethanolamine.
- Skin cleaners: Comprise surface-active agents like coconut diethanolamide and propylene glycol; preservatives including chloromethyl and methylisothiazolinone; antioxidant butylated hydroxytoluene; formaldehyde releasers bromonitrodioxane, methyl, and propyl paraben; and colourant tartrazine.
- Sprinkler liquids: Feature surface-active agent 2-propanol.
- Toilet cleaners: Include surface-active agent 2-propanol [116].

In summary, industrial cleaning agents stand out for their potent formulations tailored for heavy-duty applications, with due consideration for safety and regulatory compliance. Understanding contact allergens in these agents reinforces the importance of informed choices in both industrial and household cleaning product selection. Different types of industrial cleaning agents are illustrated briefly in table 11.1.:

Table 11.1. Different types of industrial cleaning agents, along with their typical uses: applications of these cleaning agents may vary based on the manufacturer and the intended use.

Cleaning agent	Description	Typical uses
Alkaline cleaners	Elementary solutions often contain sodium hydroxide or potassium hydroxide.	Removing grease, oils, and heavy-duty industrial cleaning
Acidic cleaners	Solutions with acidic properties, such as hydrochloric acid or phosphoric acid.	Removing mineral deposits, scale, and rust from surfaces
Solvent cleaners	Organic solvents like acetone, toluene, or petroleum-based compounds.	Removing oil, grease, and adhesive residues
Surfactant cleaners	It contains surfactants to lower surface tension and improve wetting and penetration.	General-purpose cleaning, emulsifying oils and greases
Enzyme cleaners	Contain enzymes to break down organic matter into simpler compounds.	Removing protein-based stains, such as blood or food stains
Disinfectants	Chemicals that kill or inhibit the growth of microorganisms.	Sanitizing and disinfecting surfaces to control pathogens
Abrasive cleaners	Contain abrasive particles to scrub physically and polish surfaces.	Removing tough stains, rust, and scale from hard surfaces
Degreasers	Specifically designed to remove grease and oil residues.	Cleaning machinery, engines, and industrial equipment
Oxidizing cleaners	Use oxygen-based compounds like hydrogen peroxide to clean and disinfect.	Removing stains, mould, and mildew from various surfaces

11.7 Application of microbial surfactants in the textile industry

Microbial surfactants play a pivotal role in the textile industry, offering multifaceted benefits throughout various stages of textile processing. These bio-based surfactants serve as highly effective wetting agents, facilitating essential processes like scouring and dyeing. Their exceptional emulsifying properties make them invaluable for the removal of stains and oils during textile manufacturing. In addition, microbial surfactants contribute significantly to dye dispersion, ensuring uniform and vibrant coloration.

Key Contributions of microbial surfactants in the textile industry:

Wetting agents: Microbial surfactants act as efficient wetting agents, enhancing processes such as scouring and dyeing in textile manufacturing.

Emulsification properties: Their emulsifying capabilities play a crucial role in effectively removing stains and oils from textiles, contributing to the overall cleanliness and quality of the fabric.

Dye dispersion: Microbial surfactants aid in the dispersion of dyes, ensuring even colouration and enhancing the overall aesthetic appeal of the textiles.

Fabric softening: In fabric softening applications, microbial surfactants contribute to creating a smoother texture in textiles, adding to the comfort and quality of the finished products.

Biodegradability and sustainability: The inherent biodegradability of microbial surfactants aligns with sustainable practices, minimizing environmental impact and making them compatible with various textile fibres.

Advantages of biosurfactants in textile processes:

Solubilization and emulsification: Biosurfactants are utilized for the solubilization and emulsification of chemicals employed in textile processing, enhancing the efficiency of various manufacturing procedures.

Cleaning effectiveness: In textile cleaning, biosurfactants contribute to improving the effectiveness of detergents and laundry supplies, ensuring thorough and efficient cleaning of textiles.

Environmental impact: Biosurfactants offer distinct advantages over chemical surfactants, boasting biodegradability, effectiveness across a range of temperatures or pH levels, and lower toxicity. Their use in the textile industry contributes significantly to reducing the environmental footprint of textile manufacturing processes [117].

In essence, the versatile applications of microbial surfactants in the textile industry underscore their significance in enhancing efficiency, quality, and sustainability across diverse textile processing stages.

11.7.1 Textile processing and finishing

Textile processing encompasses crucial stages such as pre-treatment, dyeing, printing, and finishing, where various techniques enhance fabric properties and contribute to aesthetic and functional characteristics. Finishing processes, including softening, anti-pilling, and water repellency, along with mechanical methods like calendering and chemical treatments such as resin application, play a pivotal role in determining the overall quality and sustainability of textiles. Notably, specialty finishes like UV protection and aromatherapy have been introduced to cater to evolving consumer preferences.

In response to the growing emphasis on sustainability and eco-friendly practices, textile processing and finishing are undergoing transformations. The incorporation of natural biomolecules and ingredients aligns with this trend, aiming to reduce costs, minimize environmental impact, and eliminate harmful chemicals. Sustainable techniques contribute to decreased water and energy consumption, negating the need for additional wastewater treatment chemicals.

Specific sustainable techniques in textile finishing:

Pad-dry-cure method: Utilizes natural ingredients to enhance textiles without harmful chemicals, contributing to reduced environmental impact.

Microencapsulation: Incorporates sustainable practices in encapsulating active ingredients, ensuring the desired properties of textiles without compromising eco-friendly standards.

Exhaust method: Sustainable finishing through exhaust methods minimizes resource consumption and eliminates the use of harmful chemicals in the treatment process.

Spraying technique: Adopts a spraying technique that uses natural products for finishing, ensuring desirable textile properties while adhering to sustainable principles.

Challenges and considerations:

Wash durability: While natural finishes showcase satisfactory results in property analysis after finishing, it is crucial to consider the wash durability of textiles. Typically, these finishes serve well for a limited number of washes, approximately 15–30.

Specific binding agents: The use of specific binding agents, such as polyacylates, has demonstrated improved results in attaching aroma microcapsules to fabrics, specifically jute blended fabric [118].

In summary, the integration of sustainable techniques in textile finishing not only enhances fabric properties but also aligns with the global push towards eco-friendly practices. These advancements prioritize reducing environmental impact, promoting cost efficiency, and addressing consumer preferences for textiles that harmonize aesthetics with sustainable principles.

11.7.2 Dyeing and printing of textiles

Textile industries employ diverse dyes and techniques in the dyeing process, introducing colour at different production stages. Concurrently, printing methods, such as screen printing and digital printing, bring intricate patterns and designs to the fabric's surface. Considerations for environmental impact and sustainability underscore the adoption of eco-friendly dyeing practices. Both dyeing and printing processes contribute significantly to the aesthetic diversity of textiles.

Integration of UV technology in textile dyeing and printing:

- UV technology presents a valuable application in textile dyeing and printing, offering the utilization of UV-curable dyeing liquors and finishing chemicals. This not only saves time and energy but also facilitates the creation of innovative textile products.
- Adoption of UV radiation curing technology: The textile industry's interest in UV radiation curing technology has surged due to its several advantages, including lower energy consumption, shorter start-up periods, faster and more reliable curing, reduced environmental pollution, room temperature curing, and space efficiency.
- UV-assisted screen printing: A notable method within UV technology involves UV-assisted screen printing. This technique utilizes a printing paste containing UV-curable compounds, achieving coloration and dye fixation on textiles through a polymerization process.

Advancements and benefits of UV technology:

- Rapid development: The application of UV technology in textile dyeing and printing is a relatively recent development, progressing swiftly. New methods, such as UV-curable inks and polymer-encapsulated pigment applications, have been introduced. These advancements offer excellent fastness, enabling the production of high-quality textiles with quick response times and enhanced flexibility [119].

In summary, the integration of UV technology in textile dyeing and printing represents a transformative process, introducing energy-efficient and environmentally conscious practices. This innovative approach not only enhances the efficiency of traditional methods but also opens avenues for high-quality, flexible, and responsive textile production.

11.7.3 Fabric softening and wrinkle reduction

Fabric softening, a process enhancing fabric comfort, involves both chemical and mechanical methods. Similarly, wrinkle reduction techniques aim to minimize creases through chemical finishes or mechanical processes like calendering. Considerations encompass environmental impact, fabric compatibility, and a growing inclination towards sustainable practices. Innovations in this domain include the development of eco-friendly formulations and smart textiles designed with self-wrinkle reduction properties. Both fabric softening and wrinkle reduction contribute significantly to the overall comfort and aesthetics of textiles.

Eco-friendly finishing regime for fabric softening and wrinkle reduction:

- A study explored the achievement of fabric softening and wrinkle reduction in cotton fabric through an eco-friendly finishing regime.
- The investigation focused on factors such as pre-treatment methods (carboxymethylation or ionic-crosslinking), post-treatment using amino-functional

silicone softener, its concentration, degree of carboxymethylation, and thermofixation conditions.
- Results demonstrated that post-treatment with an amino-based silicone microemulsion (SiE) at a concentration of up to 30 g l^{-1}, pH 4, and a wet pickup of 100%, followed by drying and curing, significantly improved fabric resiliency and softness without compromising strength.

Innovations and benefits:

Formation of semi-IPN: Fixation of the amino-functional silicone softener onto the modified cellulose structure resulted in the formation of a semi-inter and/or intra-penetrated network (semi-IPN), enhancing crosslinking, networking, and providing high softness.

Si–O–Si-cellulose complex: Fourier transform infrared analysis confirmed the formation of a Si–O–Si-cellulose complex, indicating the integration of the silicone softener into the fabric structure.

Surface characteristics: Scanning electron micrographs revealed that treated fabrics exhibited higher surface smoothness with reduced protruding fibres, ditches, and grooves compared to untreated fabrics [120].

In summary, the integration of eco-friendly finishing regimes in fabric softening and wrinkle reduction showcases a commitment to sustainability and innovation. These advancements not only enhance the tactile qualities of textiles but also contribute to reduced environmental impact, aligning with the contemporary shift towards eco-conscious practices in the textile industry.

11.7.4 Improvement of fabric colour fastness and stability

Natural dyes are gaining prominence across various sectors, including clothing, driven by the demand for commercial products featuring natural ingredients. However, challenges arise in maintaining the stability and brightness of natural dyes, especially the red colour, due to factors such as light, temperature, and pH. To address these concerns, the copigmentation method, leveraging additives, emerges as a promising solution.

Types of copigmentation methods:

1. Liquid–liquid copigmentation
2. Powder copigmentation

Factors affecting colour stability: Studies indicate that the copigmentation of dyes in powder form may impact the stability of the resulting colour, while natural dyes in liquid form tend to produce brighter colours.

Optimizing colour quality in powder form: To enhance colour brightness in powder form, a strategic combination of copigmentation additives and the utilization of the spray dryer process configuration can be employed.

Benefits and outlook: By leveraging the copigmentation method and optimizing the spray dryer process, there is significant potential to improve the colourfastness and stability of natural dyes when applied to fabrics [121].

Conclusion: In conclusion, the integration of the copigmentation approach, along with thoughtful adjustments in the spray dryer process, offers a pathway to overcome challenges associated with the stability and brightness of natural dyes. This advancement aligns with the growing preference for natural ingredients in textiles, ensuring that the colours derived from natural sources meet the desired standards for commercial products.

11.8 Application of microbial surfactants in the leather industry

Microbial surfactants, with biosurfactants being a notable type, hold promising applications across diverse industries, including the environmentally sensitive leather industry. Biosurfactants, derived from microorganisms, exhibit environmental compatibility, biodegradability, and lower toxicity in comparison to synthetic surfactants.

Applications in the leather industry:

Cleaning and degreasing: Microbial surfactants find utility in the leather industry for cleaning and degreasing leather products, ensuring effective maintenance without the use of harsh synthetic chemicals.

Oil residue removal: Addressing common challenges in leather processing, microbial surfactants prove beneficial in the removal of oil residue from storage tanks, contributing to the overall cleanliness and quality of the leather.

Oil spill clean-up: In scenarios of oil spills, a prevalent issue in the leather industry, microbial surfactants play a vital role in facilitating clean-up efforts by dispersing and breaking down oil, aligning with environmental sustainability. Bioremediation support: Microbial surfactants contribute to the bioremediation of contaminated soil and water, offering an essential solution for maintaining the environmental sustainability of the leather industry.

Environmental considerations: The use of microbial surfactants in the leather industry not only addresses specific cleaning and degreasing needs but also aligns with broader environmental goals, emphasizing sustainable practices in leather processing.

Overall impact: In conclusion, the diverse applications of microbial surfactants in the leather industry extend beyond conventional cleaning and degreasing. They play a pivotal role in environmental remediation efforts, promoting sustainable practices crucial for the longevity and ecological responsibility of the leather industry [109].

Outline of the applications of microbial surfactants in the leather industry is discussed thoroughly in table 11.2.

11.8.1 leather processing and finishing

Leather, a resilient and pliable material, is crafted through the tanning of animal rawhide and skin, predominantly sourced from cattle.

Sustainable leather production: Contrary to misconceptions, leather production is committed to utilizing available raw materials without promoting unnecessary harm or killing of animals. The industry strives to ensure that every part of the animal is put to use, minimizing waste.

Table 11.2. Outlines the applications of microbial surfactants in the leather industry: the adoption of microbial surfactants in the leather industry may depend on various factors, including cost, performance, and regulatory considerations.

Application	Description
Leather degreasing	Microbial surfactants can be used to replace chemical degreasers in the leather industry. They help in removing natural fats and oils from the hides and skins, preparing them for further processing.
Biodegradable tanning agents	Microbial surfactants can serve as eco-friendly alternatives to traditional tanning agents. They assist in the tanning process, helping to convert raw hides into durable and stable leather while minimizing environmental impact.
Emulsification of oils	Microbial surfactants have excellent emulsification properties, making them effective in breaking down and emulsifying oils present in the leather. This is particularly useful in cleaning and conditioning leather surfaces.
Cleaning leather products	Microbial surfactants can be used in the formulation of leather cleaning products. They help in lifting dirt, stains, and contaminants from leather surfaces without causing damage or leaving residues.
Surface modification	Microbial surfactants can aid in modifying the surface properties of leather, improving water repellency, softness, and other desirable characteristics. This contributes to the enhancement of leather quality.
Environmental sustainability	Microbial surfactants are often biodegradable and environmentally friendly, aligning with the leather industry's increasing focus on sustainability. They offer a greener alternative to certain chemical processes.

India's leather industry: economic contributor and export hub: In India, the leather industry holds significant economic importance, contributing significantly to the country's growth. The sector places a strategic emphasis on maximizing the export of leather goods.

Post-tanning operations: Following the tanning process, various operations in leather processing unfold, including samming, splitting, skiving, neutralizing, dyeing, greasing, and drying.

Enhancing leather quality through finishing processes: Finishing processes play a crucial role, involving surface treatments like coating with pigments or dyes, and embossing. These techniques serve to conceal defects, elevate product quality, and enhance the variety of leather goods.

Diverse drying methods: Leather processing employs different drying techniques, with options like oven drying and vacuum drying. These methods contribute to achieving specific qualities in the final leather product.

Evolution of the leather industry: Over time, the leather industry has undergone rapid transformations, witnessing improvements in processing time, product quality, and the array of available products. These advancements highlight the industry's adaptability and commitment to meeting evolving consumer expectations [122].

Conclusion: In conclusion, the world of leather production embraces sustainability, economic significance, and continual advancements in processing techniques. From India's export-focused leather industry to the intricacies of post-tanning operations and finishing processes, the sector demonstrates resilience and evolution in meeting the demands of a dynamic market.

11.8.2 Improvement of leather quality and characteristics

The utilization of microbial lipases in enzymatic degreasing has emerged as a transformative practice, showcasing improvements in leather quality while minimizing reliance on chemical compounds and surfactants.

Efficiency gains in sheepskin treatment: Enzymatic treatment during the deliming-bating and pickling stages demonstrated superior degreasing efficiency in sheepskin compared to conventional methods.

Lipase F from *Rhizopus oryzae*: A key player: Among the microbial lipases, Lipase F from *Rhizopus oryzae* stood out for its exceptional hydrolysis, yielding 58.3% and 37.2% for delimed and pickled skins, respectively.

Industrial promise of Lipase F: An enzymatic degreasing process utilizing 0.125% (w/v) of Lipase F for 3.5 h on pickled leather showcased a remarkable degreasing efficiency of 76.03%, positioning it as a frontrunner for industrial applications.

Enhanced mechanical properties: Enzymatic treatment significantly improved various mechanical properties of leather, including tensile strength, tearing strength, stitch tear strength, and percentage elongation.

Comparable performance with environmental benefits: In a noteworthy revelation, the overall performance of both chemical- and enzyme-based leathers demonstrated comparability, with the enzyme degreasing process notably enhancing the strength properties of wet blue sheep leather.

Seamless integration into conventional processes: A key advantage lies in the seamless integration of enzymatic treatment into the conventional leather manufacturing process, highlighting its practical viability.

Viable and eco-friendly approach: The findings strongly suggest that enzymatic degreasing not only holds promise for enhancing leather quality and characteristics but also addresses environmental concerns. This approach stands as a viable and eco-friendly alternative within the realm of leather processing [123].

11.8.3 Reduction of environmental pollution and waste generated by the leather industry

The leather industry generates a significant amount of solid and liquid waste, leading to pollution. Cleaner preservation techniques using chemicals and biological agents have been developed to reduce pollution problems in leather processing operations. Enzymatic dehairing has been implemented to reduce biochemical-oxygen-demand (BOD) and chemical-oxygen-demand (COD) levels in the effluent. Improved biological methods have been developed for the biodegradation of dyes and azo-dyes, reducing the dye pollution load in the effluent streams. Nanoparticle polymers and improved retanning materials have been synthesized for high exhaustion of dyeing and retanning properties, contributing to waste reduction. Mathematical models have been used to predict kinetics and growth in the leather processing operations, aiding in the development of cleaner production strategies. By adopting these technologies, it is possible to achieve a reduction in pollution loads such as BOD, COD, TDS, and TSS in the leather industry[124, 125].

11.9 Future perspectives and challenges

In exploring the application landscape of microbial surfactants across various industries, several future perspectives and challenges emerge.

In the realm of agriculture and plant growth promotion, ongoing research may uncover more intricate mechanisms, optimizing production methods and application strategies. Challenges include formulating standardized products and navigating agricultural regulatory frameworks.

Within the cosmetics and personal care sector, the future may witness a shift towards specialized applications like anti-aging products. However, challenges persist in ensuring regulatory compliance and gaining consumer acceptance.

The textile industry might see continued exploration of microbial surfactants, particularly in enhancing fabric colourfastness and stability. Challenges involve addressing the durability of finishes and compatibility with various textile materials.

In the leather industry, future prospects centre on refining processes for environmental sustainability, reducing pollution, and enhancing the quality of finished leather products. Challenges include achieving widespread adoption and fostering collaboration among researchers, industry stakeholders, and regulatory bodies.

Overall, while the future holds promise for microbial surfactant applications, addressing challenges—whether technical, regulatory, or consumer-related—will be crucial to realizing their full potential across diverse industries.

11.10 Conclusion

In summary, the extensive exploration of microbial surfactants in diverse commercial applications heralds a promising era for sustainable and eco-friendly practices across various industries. The versatility and potential impact of these surfactants in agriculture, cosmetics, personal care products, detergents, textiles, and the leather industry underscore their multifaceted roles.

Microbial surfactants have proven instrumental in agriculture, contributing to plant growth, nutrient uptake, and root development while serving as effective biocontrol agents against plant pathogens. Their application in cosmetics and personal care products spans emulsifiers, foaming agents, cleansers, and moisturizers, offering a natural alternative to chemical surfactants. Despite their advantages, challenges related to regulatory requirements and consumer acceptance warrant careful consideration for broader adoption.

In the realm of detergents and cleaning agents, microbial surfactants demonstrate performance benefits, showcasing their potential to transform the cleaning industry with eco-friendly and biodegradable alternatives. Their influence extends to the textile and leather industries, where they enhance fabric softening, wrinkle reduction, and colourfastness, aligning with the growing emphasis on sustainable practices.

Looking to the future, microbial surfactants hold significant promise, with anticipated advancements in production methods, formulations, and delivery strategies. However, addressing challenges such as regulatory compliance and consumer acceptance is crucial for a seamless integration into mainstream practices.

In conclusion, the cumulative findings presented in this chapter highlight the substantial potential of microbial surfactants as environmentally conscious alternatives across industries. As research and innovation progress, the successful incorporation of these bio-based surfactants into commercial processes will play a pivotal role in achieving sustainability goals. Collaborative efforts involving researchers, industries, and regulatory bodies are imperative to overcome challenges and usher in a greener and more sustainable future.

References

[1] Gamalero E and Glick B R 2011 Mechanisms used by plant growth-promoting bacteria *Bacteria in Agrobiology: Plant Nutrient Management* (Berlin: Springer) pp 17–46

[2] Zahir Z A, Arshad M and Frankenberger W T 2003 Plant growth promoting rhizobacteria: applications and perspectives in agriculture *Advances in Agronomy* **81** 97–168

[3] Glick B R, Todorovic B, Czarny J, Cheng Z, Duan J and McConkey B 2007 Promotion of plant growth by bacterial ACC deaminase *Crit. Rev. Plant Sci.* **26** 227–42

[4] Deleu M and Paquot M 2004 From renewable vegetables resources to microorganisms: new trends in surfactants *C.R. Chim.* **7** 641–6

[5] Singh A, Van Hamme J D and Ward O P 2007 Surfactants in microbiology and biotechnology: part 2. Application aspects *Biotechnol. Adv.* **25** 99–121

[6] Li M A, Naveed M, Mustafa A and Abbas A 2017 The good, the bad, and the ugly of rhizosphere microbiome *Probiotics and Plant Health* (Singapore: Sprinter Nature) pp 253–90

[7] Kim K Y, McDonald G A and Jordan D 1997 Solubilization of hydroxyapatite by enterobacter agglomerans and cloned *Escherichia coli* in culture medium *Biol. Fertil. Soils* **24** 347–52

[8] Dixon R and Kahn D 2004 Genetic regulation of biological nitrogen fixation *Nat. Rev. Microbiol.* **2** 621–31

[9] Ladha J K, de Bruijn F J and Malik K A 1997 Introduction: Assessing opportunities for nitrogen fixation in rice - a frontier project *Plant Soil* **194** 1–10

[10] Raymond J, Siefert J L, Staples C R and Blankenship R E 2004 The natural history of nitrogen fixation *Mol. Biol. Evol.* **21** 541–54

[11] Bhattacharyya P N and Jha D K 2011 Plant growth-promoting rhizobacteria (PGPR): emergence in agriculture *World J. Microbiol. Biotechnol.* **28** 1327–50

[12] Ahemad M and Khan M S 2012 Effects of pesticides on plant growth promoting traits of *Mesorhizobium* strain MRC4 *J. Saudi Soc. Agric. Sci.* **11** 63–71

[13] Zahran H 2001 Rhizobia from wild legumes: diversity, taxonomy, ecology, nitrogen fixation and biotechnology *J. Biotechnol.* **91** 143–53

[14] Giordano W and Hirsch A M 2004 The expression of MaEXP1, a melilotus alba expansin gene, is upregulated during the sweetclover-Sinorhizobium meliloti interaction *Mol. PlantMicrobe Interact.* **17** 613–22

[15] Marroqui S, Zorreguieta A, Santamaria C, Temprano F, Soberon M, Megias M and Downie J A 2001 Enhanced symbiotic performance by rhizobium tropici glycogen synthase mutants *J. Bacteriol.* **183** 854–64

[16] Glick B R 1995 The enhancement of plant growth by free-living bacteria *Can. J. Microbiol.* **41** 109–17

[17] Wang Y and Wu W-H 2017 Regulation of potassium transport and signaling in plants *Curr. Opin. Plant Biol.* **39** 123–8

[18] Adams E and Shin R 2014 Transport, signaling, and homeostasis of potassium and sodium in plants *J. Integr. Plant Biol.* **56** 231–49

[19] Zahoor R, Zhao W, Dong H, Snider J L, Abid M, Iqbal B and Zhou Z 2017 Potassium improves photosynthetic tolerance to and recovery from episodic drought stress in functional leaves of cotton (*Gossypium hirsutum* L.) *Plant Physiol. Biochem.* **119** 21–32

[20] Cecílio Filho A B, Dutra A F and Silva G S D 2017 Phosphate and potassium fertilization for radish grown in a Latosol with a high content of these nutrients *Revista Caatinga* **30** 412–9

[21] Munson R D, Sparks D L and Huang P M 1985 *Physical Chemistry of Soil Potassium* (Wiley)

[22] Billah M, Khan M, Bano A, Hassan T U, Munir A and Gurmani A R 2019 Phosphorus and phosphate solubilizing bacteria: keys for sustainable agriculture *Geomicrobiol. J.* **36** 904–16

[23] Sharma S B, Sayyed R Z, Trivedi M H and Gobi T A 2013 *Phosphate Solubilizing Microbes: Sustainable Approach for Managing Phosphorus Deficiency in Agricultural Soils* (SpringerPlus) Vol. 2 587

[24] Alori E T, Glick B R and Babalola O O 2017 Microbial phosphorus solubilization and its potential for use in sustainable agriculture *Front. Microbiol.* **8** 971

[25] Zhu J, Li M and Whelan M 2018 Phosphorus activators contribute to legacy phosphorus availability in agricultural soils: a review *Sci. Total Environ.* **612** 522–37

[26] Adnan M, Fahad S, Khan I A, Saeed M, Ihsan M Z, Saud S and Wu C 2019 Integration of poultry manure and phosphate solubilizing bacteria improved availability of Ca bound P in calcareous soils *3 Biotech.* **9** 368

[27] Alaylar B, Egamberdieva D, Gulluce M, Karadayi M and Arora N K 2020 Integration of molecular tools in microbial phosphate solubilization research in agriculture perspective *World J. Microbiol. Biotechnol.* **36** 93

[28] Kafle A, Cope K, Raths R, Krishna Yakha J, Subramanian S, Bücking H and Garcia K 2019 Harnessing soil microbes to improve plant phosphate efficiency in cropping systems *Agronomy* **9** 127

[29] Singh D and Prasanna R 2020 Potential of microbes in the biofortification of Zn and Fe in dietary food grains: a review *Agron. Sustain. Dev.* **40** 15

[30] Kalayu G 2019 Phosphate solubilizing microorganisms: promising approach as biofertilizers *Int. Agron.* **2019** 1–7

[31] Dash S, Borah S S and Kalamdhad A S 2020 Study of the limnology of wetlands through a one-dimensional model for assessing the eutrophication levels induced by various pollution sources *Ecol. Modell.* **416** 108907

[32] Pradhan A, Pahari A, Mohapatra S and Mishra B B 2017 Phosphate-solubilizing microorganisms in sustainable agriculture: genetic mechanism and application *Microorganisms for Sustainability* (Springer) pp 81–97
[33] Castagno L N, Sannazzaro A I, Gonzalez M E, Pieckenstain F L and Estrella M J 2021 Phosphobacteria as key actors to overcome phosphorus deficiency in plants *Ann. Appl. Biol.* **178** 256–67
[34] Damam M, Kaloori K, Gaddam B and Kausar R 2016 Plant growth promoting substances (phytohormones) produced by rhizobacterial strains isolated from the rhizosphere of medicinal plants *Inter. J. Pharm. Sci. Rev. Res.* **37** 130–6
[35] Nazli F, Mustafa A, Ahmad M, Hussain A, Jamil M, Wang X and El-Esawi M A 2020 A review on practical application and potentials of phytohormone-producing plant growth-promoting rhizobacteria for inducing heavy metal tolerance in crops *Sustainability* **12** 9056
[36] Spence C and Bais H 2015 Role of plant growth regulators as chemical signals in plant–microbe interactions: a double-edged sword *Curr. Opin. Plant Biol.* **27** 52–8
[37] Kumar K V, Srivastava S, Singh N and Behl H M 2009 Role of metal-resistant plant growth-promoting bacteria in ameliorating fly ash to the growth of *Brassica juncea J. Hazard. Mater.* **170** 51–7
[38] Gupta A, Gopal M, Thomas G V, Manikandan V, Gajewski J, Thomas G and Gupta R 2014 Whole genome sequencing and analysis of plant growth-promoting bacteria isolated from the rhizosphere of plantation crops coconut, cocoa, and arecanut *PLoS One* **9** e104259
[39] 2019 Microbes and enzymes in soil health and bioremediation *Microorganisms for Sustainability* ed A Kumar and S Sharma (Springer)
[40] Phulpoto A H, Maitlo M A and Kanhar N A 2020 Culture-dependent to culture-independent approaches for the bioremediation of paints: a review *Int. J. Environ. Sci. Technol.* **18** 241–62
[41] Haider F U, Ejaz M, Cheema S A, Khan M I, Zhao B, Liqun C and Mustafa A 2021 Phytotoxicity of petroleum hydrocarbons: sources, impacts, and remediation strategies *Environ. Res.* **197** 111031
[42] Varjani S, Kumar G and Rene E R 2019 Developments in biochar application for pesticide remediation: current knowledge and future research directions *J. Environ. Manage.* **232** 505–13
[43] Ali M H, Sattar M T, Khan M I, Naveed M, Rafique M, Alamri S and Siddiqui M H 2020 Enhanced growth of mungbean and remediation of petroleum hydrocarbons by *Enterobacter sp.* MN17 and biochar addition in diesel contaminated soil *Appl. Sci.* **10** 8548
[44] Haider F U, Liqun C, Coulter J A, Cheema S A, Wu J, Zhang R and Farooq M 2021 Cadmium toxicity in plants: impacts and remediation strategies *Ecotoxicol. Environ. Saf.* **211** 111887
[45] Aerts R and Chapin F S 1999 The mineral nutrition of wild plants revisited: a re-evaluation of processes and patterns *Adv. Ecol. Res.* **30** 1–67
[46] Vitousek P M 1984 Litterfall, nutrient cycling, and nutrient limitation in tropical forests *Ecology* **65** 285–98
[47] 2017 *Agriculturally Important Microbes for Sustainable Agriculture* ed V S Meena, P K Mishra, J K Bisht and A Pattanayak (Singapore: Sprinter Nature)
[48] Ramesh A, Sharma S K, Sharma M P, Yadav N and Joshi O P 2014 Inoculation of zinc solubilizing *Bacillus* aryabhattai strains for improved growth, mobilization, and

biofortification of zinc in soybean and wheat cultivated in vertisols of central India *Appl. Soil Ecol.* **73** 87–96
[49] Harrington R A, Fownes J H and Vitousek P M 2001 Production and resource use efficiencies in N- and P-limited tropical forests: a comparison of responses to long-term fertilization *Ecosystems* **4** 646–57
[50] Kumar A, Maurya B R, Raghuwanshi R, Meena V S and Tofazzal Islam M 2017 Co-inoculation with enterobacter and rhizobacteria on yield and nutrient uptake by wheat (*Triticum aestivum* L.) in the alluvial soil under indo-gangetic plain of India *J. Plant Growth Regul.* **36** 608–17
[51] Wu S C, Cao Z H, Li Z G, Cheung K C and Wong M H 2005 Effects of biofertilizer containing N-fixer, P and K solubilizers, and AM fungi on maize growth: a greenhouse trial *Geoderma* **125** 155–66
[52] Casieri L, Ait Lahmidi N, Doidy J, Veneault-Fourrey C, Migeon A, Bonneau L and Wipf D 2013 Biotrophic transportome in mutualistic plant–fungal interactions *Mycorrhiza* **23** 597–625
[53] Ryan P, Delhaize E and Jones D 2001 Function and mechanism of organic anion exudation from plant roots *Annu. Rev. Plant Physiol. Plant Mol. Biol.* **52** 527–60
[54] Meena S K, Rakshit A and Meena V S 2016 Effect of seed bio-priming and N doses under varied soil type on nitrogen use efficiency (NUE) of wheat (*Triticum aestivum* L.) under greenhouse conditions *Biocatal. Agric. Biotechnol.* **6** 68–75
[55] Läuchli A and Lüttge U 2004 *Salinity: Environment—Plants—Molecules* (Dordrecht: Springer)
[56] Rengasamy P 2002 Transient salinity and subsoil constraints to dryland farming in Australian sodic soils: an overview *Aust. J. Exp. Agric.* **42** 351–61
[57] Yang J, Kloepper J W and Ryu C-M 2009 Rhizosphere bacteria help plants tolerate abiotic stress *Trends Plant Sci.* **14** 1–4
[58] Stanghellini M E and Miller R M 1997 Biosurfactants: their identity and potential efficacy in the biological control of zoosporic plant pathogens *Plant Dis.* **81** 4–12
[59] Ongena M and Jacques P 2008 *Bacillus* lipopeptides: versatile weapons for plant disease biocontrol *Trends Microbiol.* **16** 115–25
[60] Broadbent P and Baker K F 1974 Behaviour of phytophthora cinnamomi in soils suppressive and conducive to root rot *Aust. J. Agric. Res.* **25** 121
[61] Gohre V and Robatzek S 2008 Breaking the barriers: microbial effector molecules subvert plant immunity *Annu. Rev. Phytopathol.* **46** 189–215
[62] Bulgarelli D, Schlaeppi K, Spaepen S, van Themaat E V L and Schulze-Lefert P 2013 Structure and functions of the bacterial microbiota of plants *Annu. Rev. Plant Biol.* **64** 807–38
[63] Nath D, Maurya B R and Meena V S 2017 Documentation of five potassium- and phosphorus-solubilizing bacteria for their K and P-solubilization ability from various minerals *Biocatal. Agric. Biotechnol.* **10** 174–81
[64] Gautam K K and Tyagi V K 2006 Microbial surfactants: a review *J. Oleo Sci.* **55** 155–66
[65] Satpute S K, Banpurkar A G, Dhakephalkar P K, Banat I M and Chopade B A 2010 Methods for investigating biosurfactants and bioemulsifiers: a review *Crit. Rev. Biotechnol.* **30** 127–44
[66] Marchant R and Banat I M 2012 Microbial biosurfactants: challenges and opportunities for future exploitation *Trends Biotechnol.* **30** 558–65
[67] Rodrigues L, Banat I M, Teixeira J and Oliveira R 2006 Biosurfactants: potential applications in medicine *J. Antimicrob. Chemother.* **57** 609–18

[68] Banat I M, Franzetti A, Gandolfi I, Bestetti G, Martinotti M G, Fracchia L and Marchant R 2010 Microbial biosurfactants production, applications and future potential *Appl. Microbiol. Biotechnol.* **87** 427–44
[69] Sobrinho H B, Luna J M, Rufino R D, Porto A L F and Sarubbo L A 2013 Biosurfactants: classification, properties and environmental applications *Recent Development in Biotechnology* (Houston, TX: Studium Press LLC) 1st ednvol 11 pp 1–29
[70] Nurfarahin A, Mohamed M and Phang L 2018 Culture medium development for microbial-derived surfactants production—an overview *Molecules* **23** 1049
[71] Prabhu Y and Phale P S 2003 Biodegradation of phenanthrene by *Pseudomonas sp.* strain PP2: novel metabolic pathway, role of biosurfactant and cell surface hydrophobicity in hydrocarbon assimilation *Appl. Microbiol. Biotechnol.* **61** 342–51
[72] Cunha C D, do Rosário M, Rosado A S and Leite S G F 2004 Serratia sp. SVGG16: a promising biosurfactant producer isolated from tropical soil during growth with ethanol-blended gasoline *Process Biochem.* **39** 2277–82
[73] Santos D, Rufino R, Luna J, Santos V and Sarubbo L 2016 Biosurfactants: multifunctional biomolecules of the 21st century *Int. J. Mol. Sci.* **17** 401
[74] Abdel-Mawgoud A M, Lépine F and Déziel E 2010 Rhamnolipids: diversity of structures, microbial origins and roles *Appl. Microbiol. Biotechnol.* **86** 1323–36
[75] Guerra-Santos L H, Kappeli O and Fiechter A 1984 *Pseudomonas aeruginosa* biosurfactant production in continuous culture with glucose as carbon source *Appl. Environ. Microbiol.* **48** 301–5
[76] Arora N K, Khare E and Maheshwari D K 2010 Plant growth promoting rhizobacteria: constraints in bioformulation, commercialization, and future strategies *Microbiology Monographs* (Berlin: Springer) pp 97–116
[77] Khare E and Arora N K 2014 Effects of soil environment on field efficacy of microbial inoculants *Plant Microbes Symbiosis: Applied Facets* (New Delhi: Springer) pp 353–81
[78] Bashan Y, de-Bashan L E, Prabhu S R and Hernandez J-P 2013 Advances in plant growth-promoting bacterial inoculant technology: formulations and practical perspectives (1998–2013) *Plant Soil* **378** 1–33
[79] Bashan Y 1998 Inoculants of plant growth-promoting bacteria for use in agriculture *Biotechnol. Adv.* **16** 729–70
[80] Ahle C M, Stodkilde K, Poehlein A *et al* 2022 Interference and co-existence of staphylococci and cutibacterium acnes within the healthy human skin microbiome *Commun. Biol.* **5** 923
[81] Adu S A, Twigg M S, Naughton P J, Marchant R and Banat I M 2023 Characterisation of cytotoxicity and immunomodulatory effects of glycolipid biosurfactants on human keratinocytes *Appl. Microbiol. Biotechnol.* **107** 137–52
[82] Patel P, Patel R, Mukherjee A and Munshi N S 2023 Microbial biosurfactants for green agricultural technology *Sustainable Agriculture Reviews 60: Microbial Processes in Agriculture* (Cham: Springer) pp 389–413
[83] Vu K A and Mulligan C N 2023 Remediation of organic contaminated soil by Fe-based nanoparticles and surfactants: a review *Environ. Technol. Rev.* **12** 60–82
[84] Ines M, Mouna B, Marwa E and Dhouha G 2023 Biosurfactants as emerging substitutes of their synthetic counterpart in detergent formula: efficiency and environmental friendly *J. Polym. Environ.* **31** 2779–91

[85] Dos Santos A V, Simonelli G and Dos Santos L C L 2023 Review of the application of surfactants in microemulsion systems for remediation of petroleum contaminated soil and sediments *Environ. Sci. Pollut. Res.* **30** 32168–83
[86] Kumari A, Kumari S, Prasad G S and Pinnaka A K 2021 Production of sophorolipid biosurfactant by insect-derived novel yeast metschnikowia churdharensis f.a., sp. nov., and its antifungal activity against plant and human pathogens *Front. Microbiol.* **12** 678668
[87] Ciurko D, Czyznikowska Z, Kancelista A, Laba W and Janek T 2022 Sustainable production of biosurfactant from agro-industrial oil wastes by Bacillus subtilis and its potential application as antioxidant and ACE inhibitor *Int. J. Mol. Sci.* **23** 10824
[88] Molaei S, Moussavi G, Talebbeydokhti N and Shekoohiyan S 2022 Biodegradation of petroleum hydrocarbons using an anoxic packed-bed biofilm reactor with *in situ* biosurfactant-producing bacteria *J. Hazard. Mater.* **421** 126699
[89] González-Penagos C E, Zamora-Briseño J A, Améndola-Pimenta M, Pérez-Vega J A, Montero-Muñoz J, Cañizares-Martínez M A and Rodríguez-Canul R 2022 The surfactant dioctyl sodium sulfosuccinate (DOSS) exposure causes adverse effects in embryos and adults of zebrafish (Danio rerio) *Toxicol. Appl. Pharmacol.* **443** 116019
[90] Li H, Miao M X, Jia C L *et al* 2022 Interactions between *Candida albicans* and the resident microbiota *Front. Microbiol.* **13** 930495
[91] Mapelli M, Mattavelli I, Salvioni E *et al* 2022 Impact of sacubitril/valsartan on surfactant binding proteins, central sleep apneas, lung function tests, and heart failure biomarkers: hemodynamic or pleiotropism? *Front. Cardiovasc. Med.* **9** 971108
[92] Pereira M J L, Rodrigues Neto E M, Girão Júnior F J, Macedo M L B D and Araujo T G 2018 Evaluation of the behavior of violet acid pigment 43 in formulations of shampoos and conditioners in bleached hair *J. Young Pharm.* **10** 282–7
[93] Kale S N and Deore S L 2017 Emulsion, microemulsion, and nanoemulsion: a review *Sys. Rev. Pharm.* **8** 39
[94] Bae I H, Lee E S, Yoo J W *et al* 2019 Mannosylerythritol lipids inhibit melanogenesis via suppressing ERK–CREB–MiTF–tyrosinase signaling in normal human melanocytes and a three-dimensional human skin equivalent *Exp. Dermatol.* **28** 738–41
[95] Mahmoodabadi M, Khoshdast H and Shojaei V 2019 Efficient dye removal from aqueous solutions using rhamnolipid biosurfactants by foam flotation *Iran. J. Chem. Chem. Eng.* **38** 127–40
[96] Abdullah N A and Kaki R 2017 Lindane shampoo for head lice treatment among female secondary school students in Jeddah, Saudi Arabia: an interventional study *J. Infect. Dis. Prev. Med.* **5** 2
[97] Gupta A K, Mays R R, Versteeg S G *et al* 2019 Efficacy of off-label topical treatments for the management of androgenetic alopecia: a review *Clin. Drug Investig.* **39** 233–9
[98] Al Badi K and Khan S A 2014 Formulation, evaluation and comparison of the herbal shampoo with the commercial shampoos *Beni-Suef Univ. J. Basic Appl. Sci.* **3** 301–5
[99] Barve K and Dighe A 2016 Hair conditioner *The Chemistry and Applications of Sustainable Natural Hair Products* ed K Barve and A Dighe (Cham: Springer) pp 37–44
[100] D'Souza P and Rathi S K 2015 Shampoo and conditioners: what a dermatologist should know? *Indian J. Dermatol.* **60** 248
[101] Ferreira A, Vecino X, Ferreira D *et al* 2017 Novel cosmetic formulations containing a biosurfactant from lactobacillus paracasei *Colloids Surf.*, B **155** 522–9

[102] Morita T, Fukuoka T, Imura T and Kitamoto D 2013 Production of mannosylerythritol lipids and their application in cosmetics *Appl. Microbiol. Biotechnol.* **97** 4691–700
[103] Men S Y, Huo Q L, Shi L *et al* 2020 Panax notoginseng saponins promote cutaneous wound healing and suppress scar formation in mice *J. Cosmet. Dermatol.* **19** 529–34
[104] Rincon-Fontan M, Rodríguez-López L, Vecino X *et al* 2018 Design and characterization of greener sunscreen formulations based on mica powder and a biosurfactant extract *Powder Technol.* **327** 442–8
[105] Resende A H M, Farias J M, Silva D D B, Rufino R D, Luna J M, Stamford T C M and Sarubbo L A 2019 Application of biosurfactants and chitosan in toothpaste formulation *Colloids Surf.,* B **181** 77–84
[106] Falk N A 2019 Surfactants as antimicrobials: a brief overview of microbial interfacial chemistry and surfactant antimicrobial activity *J. Surfactants Deterg.* **22** 1119–27
[107] Pardhi D S, Panchal R R, Raval V H, Joshi, Poczai P, Almalki W H and Rajput K N 2022 Microbial surfactants: a journey from fundamentals to recent advances *Front. Microbiol.* **13** 982603
[108] Cheng K C, Khoo Z S, Lo N W, Tan W J and Chemmangattuvalappil N G 2020 Design and performance optimisation of detergent product containing binary mixture of anionic-nonionic surfactants *Heliyon* **6** e03861
[109] Yu Y, Zhao J and Bayly A E 2008 Development of surfactants and builders in detergent formulations *Chin. J. Chem. Eng.* **16** 517–27
[110] Adetunji A I and Olaniran A O 2020 Production and potential biotechnological applications of microbial surfactants: an overview *Saudi J. Biol. Sci.* **28** 669–79
[111] Celik P A, Manga E B, Cabuk A and Banat I M 2021 Biosurfactants' potential role in combating COVID-19 and similar future microbial threats *Appl. Sci.* **11** 334
[112] Puchta R 1984 Cationic surfactants in laundry detergents and laundry aftertreatment aids *J. Am. Oil Chem. Soc.* **61** 367–76
[113] Bozetine I, Ahmed Zaïd T, Chitour C E and Canselier J P 2008 Optimization of an alkylpolyglucoside-based dishwashing detergent formulation *J. Surfactants Deterg.* **11** 299–305
[114] Reichel M, Schlicht A, Ostermeyer C *et al* 2014 Efficacy of surface disinfectant cleaners against emerging highly resistant gram-negative bacteria *BMC Infect. Dis.* **14** 292
[115] Boyce J M 2018 Alcohols as surface disinfectants in healthcare settings *Infect. Control Hosp. Epidemiol.* **39** 323–8
[116] Yamamoto N, Miyazaki Y and Sakuda K 1998 Sensory evaluation of carpet cleaner containing essential oil and the effect on mites *J. Wood Sci.* **44** 90–7
[117] Flyvholm M A 1993 Contact allergens in registered cleaning agents for industrial and household use *Occup. Environ. Med.* **50** 1043–50
[118] Banat I M, Makkar R S and Cameotra S S 2000 Potential commercial applications of microbial surfactants *Appl. Microbiol. Biotechnol.* **53** 495–508
[119] Natarajan G, Rajan T P and Das S 2020 Application of sustainable textile finishing using natural biomolecules *J. Nat. Fibers* **19** 4350–67
[120] Bahria H and Erbil Y 2016 UV technology for use in textile dyeing and printing: photocured applications *Dyes. Pigm.* **134** 442–7
[121] Hashem M, Ibrahim N A, El-Shafei A, Refaie R and Hauser P 2009 An eco-friendly—novel approach for attaining wrinkle-free/soft-hand cotton fabric *Carbohydr. Polym.* **78** 690–703

[122] Harsito C, Prabowo A R, Prasetyo S D and Arifin Z 2021 Enhancement stability and color fastness of natural dye: a review *Open Eng.* **11** 548–55

[123] Maina P, Ollengo M A and Nthiga E W 2019 Trends in leather processing: a review *Int. J. Sci. Res. Pub.* **9** 212–23

[124] Ben Rejeb I, Khemir H, Messaoudi Y, Miled N and Gargouri M 2022 Optimization of enzymatic degreasing of sheep leather for an efficient approach and leather quality improvement using fractional experimental design *Appl. Biochem. Biotechnol.* **194** 2251–68

[125] Kanagaraj J, Senthilvelan T, Panda R C and Kavitha S 2015 Eco-friendly waste management strategies for greener environment towards sustainable development in leather industry: a comprehensive review *J. Clean. Prod.* **89** 1–17

www.ingramcontent.com/pod-product-compliance
Ingram Content Group UK Ltd.
Pitfield, Milton Keynes, MK11 3LW, UK
UKHW051900100926
8556IPUK00005B/712

9 780750 35987